双语教材

Cell Biology
细胞生物学

（第二版）

主编 李 瑶 吴超群 沈大棱

编者（以姓氏笔画为序）

曲志才　曲阜师范大学
任　华　华东师范大学
刘治学　上海大学
杨仲南　上海师范大学
李秀兰　曲阜师范大学
李　瑶　复旦大学
吴超群　复旦大学
余庆波　上海师范大学
沈大棱　复旦大学
张　森　上海师范大学
张　儒　同济大学
明　凤　复旦大学
高志芹　潍坊医学院
黄　燕　复旦大学
戴亚蕾　同济大学

復旦大學出版社

内容简介

本教材是在第一版的基础上，由复旦大学生命学院细胞教学组牵头，组织国内多年从事细胞和分子生物学科研及教学的教授和专家改编完成。在系统阐述细胞各部分的结构和功能的基础上，重点介绍了细胞内外的物质运输、信息传递、能量转换、周期调控、分化发育、癌变、免疫、衰老与凋亡等细胞的重大生命活动。正文全文用英文撰写，语言上尽量做到简练与通俗，科学上做到精确与先进，既介绍细胞生物学的基本概念和基本原理，又反映各领域的发展前沿。全书为彩色印刷，各章结尾附有思考题，附录为细胞生物学常用词汇的中文解释，便于读者自学同时掌握中英文的专业知识。本教材可供综合性大学、师范院校、医学院校、农林院校的本科生和研究生使用，也可供教师、科研人员与科教管理人员参考。

为使用教材的教师免费提供PPT课件（详情见书尾附页）。

前　言

　　细胞生物学是一门迅猛发展的生命科学的重要基础学科，有关细胞的基础知识和相关研究是现代生命科学的基石。本教材的目的是使大学本科学生初步掌握本课程基本内容、基本原理、基本知识，为深入学习生命科学各科奠定基础。

　　为了让中国的大学生毕业后尽快完成从学习到研究的转变，为了我国开展细胞生物学双语教学，由本小组主持并邀请国内外综合、师范、农学、医学院校10多名教授专家和美国科学院院士一起研讨分工编写了双语教材，于2006年正式出版双语教材 *Cell Biology*，并配合教材编写了中英文两套ppt课件。本双语教材受到采用学校师生的普遍欢迎，出版6年来，重印4次，发行量达到8,800册以上，并获2011年度上海市高校优秀教材二等奖。

　　由于细胞生物学发展迅速，知识面广，内容更新快，国外原版书每3~4年改版一次。另外，我们的第一版教材在使用过程中也发现了一些不足。为了与国际进展接轨，从去年开始，我们着手修订原教材。为了更好地完成任务，又邀请了几位刚从海外归国的年轻学者加入了改编工作，为本小组增添了活力。在内容上，基本保持了原书的框架，但做了部分调整，删除了和细胞生物学教学要求没有直接关系的原第二章，增加了细胞生物学应该有的重要内容"细胞质与核糖体"独立章节。因此，其他章节根据具体情况作了更新和调整，参考了最新的国外教科书和科学杂志、重要科学论文，适当介绍学科发展的前沿，在学习知识的同时也提高了同学们的专业英语水平，使学生增强专业论文的阅读能力；对专业词汇注明了中文解释，有助于学生的理解。考虑到不同院校的本科教学要求和学时数的限制，在内容上尽可能拓宽知识面，同时避免与其他学科如生化、分子生物学等学科的重复，强调基础知识，尽量控制其深度和难度。为了提高学习效果，本版改为彩图版。因此，本教材普遍适合于农、林、医、师范及综合性大专院校的本科双语教学和中文教学的参考书。

　　由于多方面的原因，参与第一版编写的部分编委未能加入第二版修订工作，但正是有了第一版的基础，才使得第二版的改编工作得以顺利完成，在此表示衷心的感谢！

　　由于本教材由多位作者共同参与编写，虽经轮流传阅，相互修改，最后由主编统稿而成。但由于水平有限，时间仓促，难免有疏漏和错误之处，欢迎热心的同行及读者批评指正，以便我们在今后的修订工作中不断改进。

编　者
2012年10月

Contents

Chapter 1 Introduction to cell biology 1
   1.1　What is cell biology? 1
   1.2　The cell theory 2
   1.3　Cell is the basic unit of life 4
   1.4　The prokaryotic cell 7
   1.5　The eukaryotic cell 8
   1.6　Modern cell biology 11
   1.7　The technology of cell biology 15
   1.8　Training the scientists of tomorrow 19

Chapter 2 Cell membrane and cell surface 22
   2.1　Components and structure of cell membranes 22
   2.2　Transmembrane transport 30
   2.3　Cell adhesion molecules and cell junction 39
   2.4　Extracellular matrix and cell wall 44

Chapter 3 Cytoplasm, ribosomes and RNAs 49
   3.1　Structure and functions of cytoplasm 49
   3.2　Ribosome 52
   3.3　Ribozyme 60
   3.4　Non-coding RNA 62

Chapter 4 Endomembrane system, protein sorting and vesicle transport 67
   4.1　Overview of endomembrane system 68
   4.2　Endoplasmic reticulum 70
   4.3　Golgi apparatus 80
   4.4　Lysosomes and peroxisomes 83
   4.5　Molecular mechanisms of vesicular transport 86
   4.6　Secretory pathways 92
   4.7　Endocytic pathways 95

Chapter 5 Mitochondria and chloroplasts ... 99
  5.1 Mitochondria and oxidative phosphorylation ... 99
  5.2 Chloroplasts and photosynthesis ... 108
  5.3 The origins of chloroplasts and mitochondria ... 119

Chapter 6 Cytoskeleton ... 122
  6.1 Microtubules ... 122
  6.2 Actin filaments ... 129
  6.3 Intermediate filaments ... 137

Chapter 7 Cell communication and signaling ... 142
  7.1 Signaling components ... 142
  7.2 The role of intracellular receptor: signaling of nitric oxide ... 148
  7.3 Signaling through G-protein-coupled cell-surface receptors ... 149
  7.4 Signaling through enzyme-coupled cell-surface receptors ... 154
  7.5 Signal network system ... 159
  7.6 Cell signaling and the cytoskeleton ... 161
  7.7 Cell communication in plants ... 163

Chapter 8 Nucleus and chromosomes ... 166
  8.1 The nucleus of a eukaryotic cell ... 166
  8.2 The nuclear envelope ... 168
  8.3 The nuclear pore complex ... 170
  8.4 Chromatin and chromosomes ... 175
  8.5 Nucleolus and ribosome biogenesis ... 186
  8.6 The nuclear matrix ... 191

Chapter 9 Cell cycle and cell division ... 195
  9.1 An overview of the cell cycle ... 195
  9.2 Regulation of the cell cycle ... 197
  9.3 Cell division ... 206

Chapter 10 Cell differentiation ... 221
  10.1 Introduction ... 221
  10.2 Cells with potency of differentiation ... 223
  10.3 Stem cells ... 225
  10.4 Controls of cell differentiation ... 229

10.5	The major cell differentiation systems	237

Chapter 11 Senescence and apoptosis ····· 244
    11.1 Senescence ····· 244
    11.2 Apoptosis ····· 252
    11.3 Senescence or apoptosis ····· 260

Chapter 12 Cells in immune response ····· 263
    12.1 The immune system ····· 263
    12.2 The organs of the immune system ····· 263
    12.3 Cells in the innate immune system ····· 266
    12.4 Cells in the adaptive immune system ····· 271
    12.5 Innate and adaptive immune responses ····· 273
    12.6 Immunological memory ····· 283

Chapter 13 Cancer cells ····· 284
    13.1 Basic knowledge about cancer cell ····· 284
    13.2 Carcinogenesis ····· 288
    13.3 The genes involved in cancer ····· 292
    13.4 The genetic and epigenetic changes of cancer ····· 295
    13.5 Treatment of cancer ····· 299

References ····· 303

Glossary ····· 307

Chapter 1

Introduction to cell biology

1.1 What is cell biology?

Cell biology is the application of molecular biological approaches to an understanding of life at the cellular level. Knowledge of the molecular basis of cell structure, cell function and cell interactions is fundamental to an understanding of whole organisms, since the properties of organisms are dependent upon the properties of their constituent cells.

1.1.1 Cell biology is the basis of modern biology

Cell biology is modern biology, an academic discipline which studies the structure and physiological properties of cells, as well as their behaviors, interactions, and environment on a microscopic and molecular level. But two main features should be stressed in the modern cell biology.

(1) **Study the molecules within cells** Cell biology is a modern science, which is rooted in an understanding of the molecules within cells, and of the interactions between cells that allow construction of **multicellular organism**s. The more we learn about the structure, function, and development of different organisms, the more we recognize that all life processes exhibit remarkable similarities.

Cell biology concentrates on: ① macromolecules and reactions, investigated by biochemists; ② the processes described by cell biologists; ③ the gene control pathways identified by molecular biologists and geneticists.

(2) **Study the molecular similarities and differences between cell types** Understanding the composition of cells and how cells works is fundamental to all of the biological sciences. Appreciating the similarities and differences between cell types is particularly important to the fields of molecular cell biology. These fundamental similarities and differences provide a unifying theme, allowing the principles learned from studying one cell type to be extrapolated and generalized to other cell types. Research in cell biology is closely related to genetics, biochemistry, molecular biology and developmental biology.

1.1.2 Cell biology is always developing

All the concepts of cell biology continue to be derived from computational experiments and laboratory experiments. The powerful experimental tools that allow the study of living cells and organisms at higher and higher levels are being developed constantly. In this chapter, we address the current state of cell biology and look forward to what further exploration will uncover in the twenty-first century.

In this millennium, two gathering forces will reshape cell biology: ①The **genomics**, the complete DNA sequence of many organisms; ②The **proteomics**, the knowledge of all the possible shapes and functions that proteins employ.

1.2 The cell theory

The cell theory is the basis of molecular cell biology, and this theory is known as one of the three indispensable theories upon which the science of biology is built. These theories are: ① The theory of evolution; ② The cell theory; ③ The theory of equilibrium thermodynamics.

The cell theory, or cell doctrine, states that all organisms are composed of similar units of organization, called cells. The concept was formally articulated in 1839 by Schleiden and Schwann, and has remained as the foundation of modern biology. The idea predates other great paradigms of biology including Darwin's theory of evolution (1859), Mendel's laws of inheritance (1865), and the establishment of comparative biochemistry (1940).

Ultrastructural research and modern molecular biology have added many tenets to the cell theory, but it remains as the preeminent theory of biology. The cell theory is to biology what atomic theory is to physics.

Just as an atom is the smallest particle of a chemical element, which can exist either alone or in combination and still possess the chemical and physical properties of that element, so then, a cell is the smallest entity, which can exhibit the characteristic of life.

1.2.1 Formulation of the cell theory

In 1663, an English scientist, Robert Hooke, discovered cells in a piece of cork, which he examined under his primitive microscope(Figure 1-1). Actually, Hooke only observed cell walls because cork cells are dead and without cytoplasmic contents. Hooke drew the cells he saw and also coined the word "cell". The word cell is derived from the latin word "cellula" which means small compartment. Hooke published his findings in his famous work, "Micrographia: Physiological Descriptions of Minute Bodies made by Magnifying Glasses (1665)."

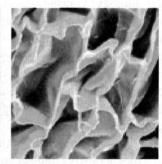

Figure 1-1 Hooke and his microscope. In 1663 Hooke examined under his primitive microscope (left). The small cell structures did not show up well or remained invisible (middle). The electron microscope not only showed more detail of previously known parts of the cell but also revealed new parts. Cells and cell structures can now be examined at magnifications of up to 500,000 times and more (right).

Ten years later, Anton van Leeuwenhoek (1632~1723), a Dutch businessman and a contemporary of Hooke used his own (single lens) monocular microscopes and was the first person to observe bacteria and protozoa. He looked at everything from rain water to tears. He saw moving objects that he termed "animalcules." The tiny creatures appeared to be swimming.

Between 1680 and the early 1800's, it appears that not much was accomplished in the study of cell structure. This may be due to the lack of quality lens for microscopes and the dedication to spend long hours of detailed observation over what microscopes existed at that time. Leeuwenhoek did not record his methodology for grinding quality lenses and thus microscopy suffered for over 100 years.

It is upon the works of Hooke, Leeuwenhoek, Oken, and Brown that Schleiden and Schwann built their cell theory. It was the German professor of botany at the university of Jena, Dr. Schleiden, who brought the nucleus to popular attention, and to asserted its all importance in the function of a cell.

The location of these nuclei at comparatively regular intervals suggested that they are found in definite compartments of the tissue, as Schleiden had shown to be the case with vegetables; indeed, the walls that separated such cell-like compartments one from another were in some cases visible. Soon Schwann was convinced that his original premise was right, and that all animal tissues are composed of cells not unlike the cells of vegetables. Adopting the same designation, Schwann propounded what soon became famous as the cell theory. So expeditious was his observations that he published a book early in 1839, only a few months after the appearance of Schleiden's paper. A most important era in cell biology dates from the publication of his book in 1839.

Schwann summarized his observations into three conclusions about cells: ①The cell is the unit of structure, physiology, and organization in living things; ②The cell retains a dual existence as a distinct entity and a building block in the construction of organisms; ③Cells form by free-cell formation, similar to the formation of crystals (**spontaneous generation**).

For a long time, people believed in spontaneous generation. They believed flies came from rotting meat and frogs from mud. It took a hundred years and many experiments to disprove those ideas and confirm that every cell comes from a pre-existing cell.

1.2.2 Modern tenets of cell theory

For the first 150 years, the cell theory was primarily a structural idea. This structural view, which is found in most textbooks, describes the components of a cell and their fate in cell reproduction. Since the 1950's, however, cell biology has focused on DNA and its informational features. Today we look at the cell as a unit of self-control. The description of a cell must include ideas about how genetic information is converted to structure and function.

The modern tenets of the cell theory include: ①All known living things are made up of cells; ②The cell is structural and functional unit of all living things; ③All cells come from pre-existing cells by division (not by spontaneous generation); ④Cells contain hereditary information which is passed from cell to cell during cell division; ⑤All cells are basically the same in chemical composition; ⑥All energy flow (metabolism & biochemistry) of life occurs within cells.

1.3 Cell is the basic unit of life

According to the cell theory, all living things are composed of one or more cells. Cells fall into **prokaryotic** and **eukaryotic** types. Prokaryotic cells are smaller (as a general rule) and lack much of the internal compartmentalization and complexity than eukaryotic cells are. No matter which type of cells we are considering, all cells have certain features in common: cell membrane, DNA, cytoplasm, and ribosome.

1.3.1 The basic structure of cell

The cell is one of the most basic units of life. There are millions of different types of cells. There are cells that are organisms onto themselves, such as microscopic amoeba and bacteria cells. And there are cells that only function when part of a larger organism, such as the cells that make up your body.

The cell is the smallest unit of life in human bodies. In the body, there are brain cells, skin cells, liver cells, stomach cells, and the list goes on. All of these cells have unique functions and features. All have some recognizable similarities (Figure 1-2).

(1) Plasma membrane All cells have a "skin", called the **plasma membrane**, protecting it from the outside environment. The cell membrane regulates the movement of water, nutrients and wastes into and out of the cell. Inside of the cell membrane are the working parts of the cell.

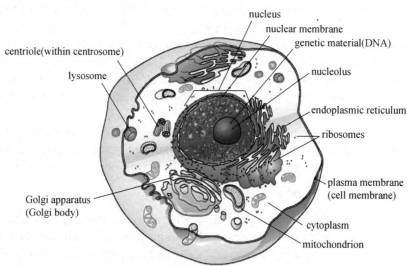

Figure 1-2 **The structure of an animal cell** (From: http://what-when-how.com)

(2) Nucleus At the center of the cell is the cell **nucleus**. The cell nucleus contains the cell's DNA, the genetic code that coordinates protein synthesis.

(3) Organelles There are many organelles inside of the cell — small structures that help carry out the normal operations of the cell. One important cellular organelle is the ribosome. **Ribosomes** participate in protein synthesis. The transcription phase of protein synthesis takes places in the cell nucleus. After this step is complete, the mRNA leaves the nucleus and travels to the cell's ribosomes, where translation occurs. Another important cellular organelle is the **mitochondrion**. Mitochondria (many mitochondrion) are often

referred to as the power plants of the cell because many of the reactions that produce energy take place in mitochondria. Also important in the life of a cell are the **lysosomes**. Lysosomes are organelles that contain enzymes that aid in the digestion of nutrient molecules and other materials.

1.3.2 Differences between plant cell and animal cell

There are many different types of cells. One major difference in cells occurs between plant cells and animal cells. While both plant and animal cells contain the structures discussed above, plant cells have some additional specialized structures.

Plants do not have a skeleton for support and yet plants don't just flop over in a big spongy mess. This is because of a unique cellular structure called the cell wall. The cell wall is a rigid structure outside of the cell membrane composed mainly of the polysaccharide cellulose. The cell wall gives the plant cell a defined shape which helps support individual parts of plants.

Plant cells contain an organelle called the chloroplast. The chloroplast allows plants to harvest energy from sunlight. Specialized pigments in the chloroplast (including the common green pigment chlorophyll) absorb sunlight and use this energy to complete the chemical reaction.

1.3.3 Origin of the cell

In 1950, then-graduate student Stanley Miller designed an experimental test and recovered amino acids from C, H, O and N in abundance. Subsequent modifications of the atmosphere have produced representatives or precursors of all four **organic macromolecular** classes. The interactions of these molecules would have increased as their concentrations increased. Reactions would have led to the building of larger, more complex molecules. A pre-cellular life would have begun with the formation of nucleic acids. Chemicals made by these nucleic acids would have remained in proximity to the nucleic acids. Eventually the pre-cells would have been enclosed in a lipid-protein membrane, which would have resulted in the first cells.

But the question is how did the cell acquire a cell membrane? There are many theories that address this question but they fall into two categories, the thought requires that DNA or RNA be present; the other thought does not require DNA or RNA. There are no clear-cut answers to the nucleic acid question or the origin of a cell membrane, but there are a lot of theories. The most attractive theory is "RNA world theory". The RNA world theory describes that RNA is a close relative of DNA and it has been recently shown that RNA can act in an enzyme-like manner. In the RNA world scenario, RNA came first, playing the role of both DNA and enzyme proteins. This would make the first cell's chemistry very different from today's cells and would require its being superceded by today's cell's chemistry.

1.3.4 Three things make cell different from non-cell system

Life requires a structural compartment separate from the external environment in which macromolecules can perform unique functions in a relatively constant internal environment. These "living compartments" are cells. The cells differ from non-cell systems through three things.
- The capacity for replication from one generation to another. Most organisms today use DNA as the hereditary material, although recent evidence (ribozyme) suggests

that **RNA** may have been the first nucleic acid system to have formed. Nobel laureate Walter Gilbert refers to this as the RNA world.
- The presence of enzymes and other complex molecules essential to the processes needed by living systems. Miller's experiment showed how these could possibly form.
- A membrane that separates the internal chemicals from the external chemical environment. This also delimits the cell from not-cell areas.

1.3.5 Microscope is needed to visualize cells

The small size of cells makes the use of microscopes necessary to view them (Figure 1-3). If two objects are too close together, they start to look like one object. With normal human vision the smallest objects that can be resolved (i.e., distinguished from one another) are about 200 μm (0.2 mm) in size.

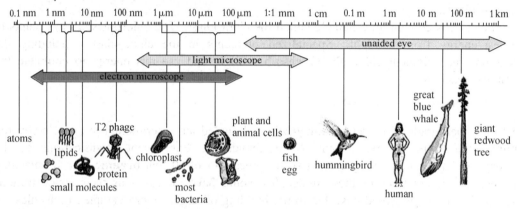

Figure 1-3 The relative sizes of biological objects ranging from atoms to tree (From: Farabee M J. 2002)

Light microscopes use glass lenses and visible light and typically have a resolving power of 0.2 μm (0.2×10^{-6} m). Resolution depends on the wavelength of the illuminating light, but in general, resolution is about 1,000 times better than that of an unaided human eye. Living or killed and fixed cells may be viewed with light microscopes.

Electron microscopes have magnets, rather than glass lenses, to focus an electron beam. The wavelength of the electron beam is far shorter than that of light, and the resulting image resolution is far greater. This image is not visible without the use of either film or a fluorescent screen. Resolution is about 0.5 nm or 400,000 times finer than that of the human eye. Subcellular features can be seen only if the cells are killed and fixed with special fixatives and stains.

1.3.6 Techniques are developed to observe molecules inside cell

In addition the optical and electron microscope, scientists are able to use a number of other techniques to probe the mysteries of the animal cell. Cells can be disassembled by chemical methods and their individual organelles and macromolecules isolated for study. The process of cell fractionation enables the scientist to prepare specific components, the mitochondria for example, in large quantities for investigations of their composition and functions. Using this approach, cell biologists have been able to assign various functions to specific locations within the cell. However, the era of fluorescent proteins has brought microscopy to the forefront of biology by enabling scientists to target living cells with highly localized probes for studies that don't interfere with the delicate balance of life processes.

1.3.7 From prokaryotes to eukaryotes

Every cell has a plasma membrane, a continuous membrane that surrounds the fluids and other structures of a cell. The membrane is composed of a lipid bilayer with proteins floating within it and protruding from it. Living organisms can be classified into one of two major categories based on the location within the cell where the most genetic material is stored: prokaryotes and eukaryotes (Figure 1-4).

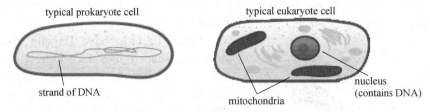

Figure 1-4 Typical structures of prokaryotes and eukaryotes

Prokaryotes, such as archaea and bacteria are small, relatively simple cells surrounded by a membrane and a cell wall, with a circular strand of DNA containing their genes. They have no nucleus or other membrane-bounded compartments. They lack distinct organelles, although some do have invaginated membrane structures.

Eukaryotes have a membrane-bounded nucleus and usually have other membrane-bounded compartments or organelles as well. The complex eukaryotic cell ushered in a whole new era for life on Earth, because these cells evolved into multicellular organisms.

Evidence supports the idea that eukaryotic cells are actually the descendents of separate prokaryotic cells that joined together in a symbiotic union.

1.4 The prokaryotic cell

Prokaryotes are cells without a distinct nucleus. They have genetic material but that material is not enclosed within a membrane. Prokaryotes include bacteria and cyanophytes. The genetic material is a single circular DNA strand and is located within the cytoplasm. Recombination happens through transfers of plasmids (short circles of DNA that pass from one bacterium to another). Prokaryoytes do not engulf solids, nor do they have centrioles or asters. Prokaryotes have a cell wall made up of peptidoglycan.

Common features of prokaryotic cells: ①All have a plasma membrane; ②All have a region called the nucleoid where the DNA is concentrated; ③The cytoplasm (the plasma membrane-enclosed region) consists of the nucleoid, ribosome (the molecular protein synthesis machines), and a liquid portion called the cytosol.

1.4.1 Specialized features of some prokaryotic cells

Most prokaryotic cells have a cell wall just outside the plasma membrane. The cell wall functions to prevent plasma membrane lysis (bursting) when cells are exposed to solutions with lower solute concentrations than the cell interior. It also protects the membrane.

In bacteria (but not in archaea), the cell wall is made of a polymer of amino sugars called peptidoglycan, which is covalently cross-linked to form one giant molecule around the entire cell. Some bacteria have another membrane outside the cell wall, a polysaccharide-rich phospholipid membrane. This membrane has embedded proteins that make it more permeable than the interior membrane. For some bacteria, this capsule provides a means to escape detection by the immune systems of the animals they infect. The capsule can prevent drying out of the cell and help the bacterium attack other cells. The capsule is not essential to cell life; if the cell loses the capsule, it can survive (Figure 1-5).

Some bacteria, including **cyanobacteria**, can carry on photosynthesis, that is, they have the ability to collect solar energy. Cyanobacteria have chlorophyll in the infolded plasma membrane for this purpose. Some bacteria have **mesosomes**, which are involved in cell division or in certain energy-releasing reactions. Like the photosynthetic membrane system, they are formed from plasma membrane infolding.

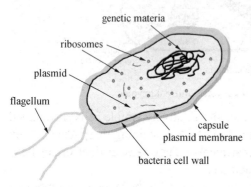

Figure 1-5 A typical bacterial cell. The bacterial cell is an example of a prokaryotic cell.

Some bacteria have **flagella**, locomotory structures shaped like a corkscrew. They spin like a propeller to move the bacteria. The flagella bear no structural similarity to the flagella found in eukaryotic cells, such as sperm cells. Some bacteria have piliform, threadlike structures that help bacteria adhere to one another during mating or to other cells for food and protection. Recent evidence suggests that some prokaryotes have an internal filamentous helical structure just below the plasma membrane. The proteins that make up this structure are similar to actin, a major component of the eukaryotic cytoskeleton.

1.4.2 Bacteria occupy major group of living organisms

Bacteria (singular, bacterium) are a major group of living organisms, most are microscopic and unicellular, with a relatively simple cell structure lacking a cell nucleus, cytoskeleton, and organelles such as mitochondria and chloroplasts. Their cell structure is further described in the article about prokaryotes, because bacteria are prokaryotes, in contrast to organisms with more complex cells, called eukaryotes. The term "bacteria" has variously applied to all prokaryotes or to a major group of them, depending on ideas about their relationships.

1.5 The eukaryotic cell

Animals, plants, fungi, and protists have a membrane-bounded nucleus in each of their cells and are classified as eukaryotes.

The autogenous and endosymbiotic hypotheses relate to possible ways in which eukaryote cells may have evolved from prokaryotic ancestors. The latter hypothesis is favored by most biologists. Briefly, this hypothesis suggests that the first eukaryotic cells resulted from symbiotic consortia between various types of prokaryotic cells. For example, the chloroplast resulted when a photosynthetic prokaryote entered another cell and survived. The mitochondrion is presumed to have evolved in a similar manner, but between non photosynthetic cells.

1.5.1 Eukaryotic cells share common features

Eukaryotic cells vary from animals, plants, to fungi and protists, but they have some common features which make them different from the prokaryotic cells.

(1) Common features   The different types of eukaryotic cells show some similar features, they are: ①Eukaryotic cells tend to be larger than prokaryotic cells; ②Each of eukaryotic cells has a membrane-bounded nucleus; ③Eukaryotic cells have a variety of membrane-bounded compartments called **organelles**; ④Eukaryotes have a protein scaffolding called the cytoskeleton, which provides shape and structure to cells, among other functions.

(2) Compartmentalization is the key to eukaryotic cell function   The subunits, or compartments, within eukaryotic cells are called organelles, which are responsible for specialized functions of the cell. The central organelle nucleus contains most of the cell's genetic material (DNA). The **mitochondrion** is a power plant and industrial park for the storage and conversion of energy. The endoplasmic reticulum and **Golgi apparatus** make up a compartment where proteins are packaged and sent to appropriate locations in the cell. The **lysosome** and vacuole are cellular digestive systems where large molecules are hydrolyzed into usable monomers. The **chloroplast** performs photosynthesis in plant cells. Membranes surrounding these organelles keep away inappropriate molecules that might disturb organelle function. They also act as traffic regulators for raw materials into and out of the organelle.

(3) Organelles can be studied by microscopy or chemical analysis   Cell organelles were first detected by light and electron microscopy.

- The target specific macromolecules can be used to determine the chemical composition of organelles.
- The process of cell fractionation, another means by which cells can be examined.
- Microscopy and cell fractionation can be used as complements to each other, giving a complete picture of the structure and function of each organelle.

1.5.2 Plant cell structure

A plant has two organ systems: ①the shoot system; ②the root system. The shoot system is above ground and includes the organs such as leaves, buds, stems, flowers (if the plant has any), and fruits (if the plant has any). The root system includes those parts of the plant below ground, such as the roots, tubers, and rhizomes.

Plant cells are formed at **meristems**, and then develop into cell types which are grouped into tissues. Plants have only three tissue types: **dermal**, **ground**, and **vascular**. Dermal tissue covers the outer surface of herbaceous plants. Dermal tissue is composed of epidermal cells, closely packed cells that secrete a waxy cuticle that aids in the prevention of water loss. The ground tissue comprises the bulk of the primary plant body. Parenchyma, collenchyma, and sclerenchyma cells are common in the ground tissue. Vascular tissue transports food, water, hormones and minerals within the plant. Vascular tissue includes xylem, phloem, parenchyma, and cambium cells.

Like other eukaryotes, the plant cell is enclosed by a plasma membrane, which forms a selective barrier allowing nutrients to enter and waste products to leave. Unlike other eukaryotes, however, plant cells have retained a significant feature of their prokaryote ancestry, a rigid **cell wall** surrounding the plasma membrane. The cytoplasm contains specialized organelles, each of which is surrounded by a membrane. Plant cells differ from animal cells in that they lack centrioles and organelles for locomotion (cilia and flagella), but they do have additional specialized organelles. Chloroplasts convert light to chemical energy, a single large vacuole acts as a water reservoir, and plasmodesmata allow cytoplasmic substances to pass directly from one cell to another. There is only one nucleus and it contains all the genetic information necessary for cell growth and reproduction. The other organelles occur in multiple copies and carry out the various functions of the cell, allowing it to survive and participate in the functioning of the larger organism (Figure 1-6).

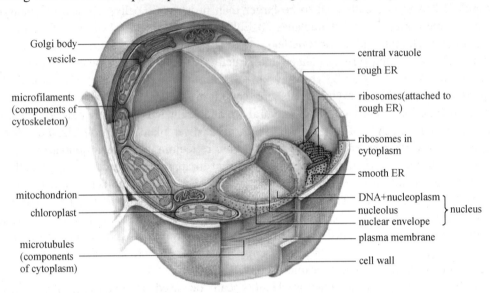

Figure 1-6 **A typical structure of plant cell** (From: Rosemary Richardson)

1.5.3 Animal cell structure

Animals are a large and incredibly diverse group of organisms. Making up about three-quarters of the species on Earth, they run the gamut from corals and jellyfish to ants, whales, elephants, and, of course, humans. Being mobile has given animals, which are capable of sensing and responding to their environment, the flexibility to adopt many different modes of feeding, defense, and reproduction. Unlike plants, however, animals are unable to manufacture their own food, and therefore, are always directly or indirectly dependent on plant life.

Animal cells are typical of the eukaryotic cell, enclosed by a plasma membrane and containing a membrane-bound nucleus and organelles. Unlike the eukaryotic cells of plants and fungi, animal cells do not have a cell wall. This feature was lost in the distant past by the single-celled organisms that gave rise to the kingdom Animalia. Most cells, both animal and plant, range in size between 1 and 100 micrometers and are thus visible only with the aid of a microscope (Figure 1-7).

Most animal cells are **diploid**, meaning that their chromosomes exist in homologous pairs. Different chromosomal ploidies are also, however, known to occasionally occur. The proliferation of animal cells occurs in a variety of ways. In instances of sexual reproduction, the cellular process of meiosis is first necessary so that haploid daughter cells, or gametes, can be produced. Two haploid cells then fuse to form a diploid zygote, which develops into a new organism as its cells divide and multiply.

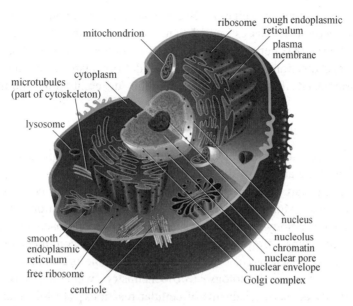

Figure 1-7 The typical structure of an animal cell (From: Audesirk T, Audesirk G, Byers BE. 2002)

1.6 Modern cell biology

Cell biology is an integrated view of cells at work, involving essential cell biology, molecular biology of cell and cell systems biology.

1.6.1 Essential cell biology

Essential cell biology is the scientific discipline that studies the cell as an individual unit and as a contributing part of a larger organism, and a science that studies the properties of cells: the functional structures within them, including organelles as mitochondria, chloroplasts, the Golgi apparatus, microsomes, and many others, the interactions and communication among cells, and the biochemical phenomena, which are shared by all cells.

(1) **Study the cells as a unit of life**   A single cell is often considered a complete organism in itself, such as a bacterium or yeast. An individual cell is capable of digesting its own nutrients and transforming simple nutritional substances into cellular protoplasm, producing some of its own useable energy, and eventually replicating itself. A cell may be viewed as an enclosed vessel composed of smaller subcell parts, organelles, each of which has specific metabolic functions. Within the cell, countless chemical reactions take place simultaneously, all of them are controlled, so they contribute to the processes we call life and the eventual procreation of the cell.

(2) **Study the cells in multicellular organism**   In a multi-cellular organism, individual cells, by differentiating in order to acquire particular functions, can become the building blocks of the larger multi-cellular organisms. In order to do this, each cell keeps in continuous communication with its neighboring cells. Cooperative assemblies of like cells make up tissues, and a collaboration between tissues helps forms organs, the functional units of a whole organism as complex as a human being.

The smallest known cells are a group of tiny bacteria called mycoplasma. Some of these single-celled organisms are spheres about 0.3 μm in diameter with a total mass of 10^{-14}

g (g is a metric system unit of weight). But human cells are about 400 times larger, they are about 20 μm across. It was estimated that a human being may be composed of more than 75,000,000,000,000 cells (7.5×10^{13} cells), divided to more than 200 types with special structure and function. Cells in a multi-cellular organism are organized in strict roles to be fully harmonized the whole body, through differentiation, communication and interaction.

1.6.2 Molecular biology of cell

Although biochemistry and molecular biology, as disciplines, might have made substantial development without the advent of the cell theory, each science has strongly influenced the other almost from the start. Contemporary cell biology, often referred to as Molecular Cell Biology, or Cell and Molecular Biology, is then a compilation of four separation disciplines: ①Cell physiology, which takes a comparative approach looking at how cells answer universal problems, from water conservation to cell communication; ② Systemic physiology, which is the science of organ systems and whole organism physiology, such as insect, plant, fish or human physiology; ③Biochemistry, which looks at the chemical and physical commonality of the mechanisms of cellular reactions; ④Molecular biology, which studies the properties of organisms through the constituent molecules.

(1) **Genetic control in cells** The molecular basis of genetic control in cells, particularly in eukaryotic cells is one of the most basic active areas of molecular cell biology. Of particular interest is the understanding of the regulation mechanisms involved in the development of multicellular organisms. In most well-known case studies, such as in the fruit fly *Drosophila melanogaster*, it has been shown that regulation among the genes that control early development and cell differentiation.

Genes are the blueprint of our bodies, a blueprint that creates the variety of proteins essential to any organism's survival. These proteins, which are used in countless ways by our bodies, are produced by genetic sequences, i.e. our genes, as described in the cell biology section, protein synthesis pages. Some other genes that will be functional during specialization determine the physical characteristics of the cell, i.e. long and smooth for a muscle cell or indented like a goblet cell.

(2) **Three flows within cells** Cells can be described by the processes that take place within them. The functions of the cell can be broken down into three basic flows: ①flow of information; ②flow of mass; ③ flow of energy.

None of these three flows operates in isolation from the other two. The flow of mass and flow of energy are extremely closely linked, as biological molecules are broken apart to make other biological molecules or to harvest energy.

1) Flow of information: The flow of information in the cell is described simplistically by the Central Dogma of Molecular Biology — "DNA makes DNA makes RNA makes protein" (Figure 1-8).

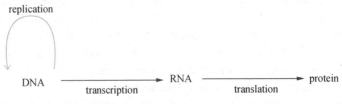

Figure 1-8 Central dogma of molecular biology

The Central Dogma of Molecular Biology contains:
- Chromosomal DNA is used as a template to make more chromosomal DNA.
- Chromosomal DNA is used as a template to make RNA.
- mRNA is used as a template from which proteins are made.
- Other relationships will become apparent as we cover the course material.

2) Flow of mass: Simple molecules absorbed by cells are incorporated into larger molecules and/or used to make other molecules. The characteristic particular to material transformation is the necessity for the living cell to maintain a non-equilibrium state while flow of matter occurs in directions towards equilibrium driven by thermodynamic potentials. In contrast, in passive, irreversible flow of matter thermodynamic driving force results in flow until equilibrium is reached.

The mammalian cell has solved this problem by controlled dissipation of thermodynamic driving force via enzyme networks and replenishment of driving force by oxidation of foodstuffs.

3) Flow of energy: Some processes (anabolic) are energy consuming. Other processes are energy producing (catabolic). Energy intermediates are the molecules, adenosine triphosphate (ATP) and adenosine diphosphate (ADP), which are used to allow endergonic reactions to occur in cells.

1.6.3 Systems biology of the cell

Currently, biologists tend to study the cell structure and behaviors as the ultimate objective to systemically understand the effect of the molecules on the whole cell, and the effect of the individual cell on the whole organism as well.

Molecular alterations in cells as elementary units of cell systems, tissues and organs may manifest as disease at the organism level. They are detectable as altered molecular cell phenotypes by single cell flow, image or chemical cytometry. Molecular cell phenotypes result from genotype and exposure influences within the heterogeneity of various cell types, but also within given cell types according to cell cycle or functional status.

(1) Molecular cytomics Molecular **cytomics** is an entry to cell systems biology, and a cell systems research, which aims at the understanding of the molecular architecture and functionality of cell systems (**cytomes**) by single-cell analysis in combination with exhaustive bioinformatic knowledge extraction.

The cytomics concept has been significantly advanced by a multitude of current developments. Amongst them are confocal and laser scanning microscopy, multiphoton fluorescence excitation, spectral imaging, fluorescence resonance energy transfer (FRET), fast imaging in flow, optical stretching in flow, and miniaturised flow and image cytometry within laboratories on a chip or laser microdissection, as well as the use of bead arrays.

The elucidation of the molecular pathways from the 20~40 thousands genes of the sequenced human genome via investigation of genetic networks and molecular pathways up to the cellular and organismal phenotypes is highly complex and time consuming.

The concept of a **human cytome project** systematizes this approach by analyzing in a first step the multitude of multiparametric cytometric data from large patient groups available at many clinical sites and establishing a structure for public databases for the standardized relational knowledge. In a second step molecular disease pathways are investigated and mathematically modeled by biomedical cell systems biology. Leukemia/lymphomas, cancers, rheumatoid diseases, allergies and infections are of high medical interest while stem cell differentiation, cell cycle regulation, cell proteomics and cell organelle functionality are attractive for basic research.

Human cytome project will immediate medical use, facilitate access to the detection of new drug targets, increase research speed and the stimulation for advanced technological developments.

(2) **Cellular genomics**　Cell **genomics** is the measurement of gene expression in single cells, including what is the transcriptional status of single cells and how their transcription patterns might change in response to external stimuli.

Gene expression profile. Gene expression is a complex process involving a cascade of events that starts with interactions between a multitude of transcription factors and gene promoter sequences, RNA polymerase binding and elongation, and RNA processing and maturation. The approach for gene-expression profiling in single cells is conceptually simple, but in practice it is technically challenging.

The focuses of cell genomics. The research of cellular genomics emphases following aspects: ①the spatial analysis of linkage between genetic loci as a means of detecting translocations; ②the study of nuclear organization, namely whether active genes occupy different positions relative to inactive genes; ③the organization of active genes in relation to chromosome territories; ④ the study of gene expression during different stages of differentiation and the cell cycle.

(3) **Cellular proteomics**　The genome sequences of important model systems are available and the focus is now shifting to large-scale experiments enabled by this data. Following in the footsteps of genomics, we have functional genomics, proteomics, and even metabolomics, roughly paralleling the biological hierarchy of the transcription, translation, and production of small molecules.

Cellular proteomics is a main part of proteomics. Cellular proteomics is a main part of proteomics, and initially concerned with determining the structure, expression, localization, biochemical activity, interactions, and cellular roles of as many proteins as possible. There has been great progress owing to novel instrumentation, experimental strategies, and bioinformatics methods. The area of protein-protein interactions has been especially fruitful.

Protein expression profiling in single cell. An established and widely accessible strategy for protein profiling is two-dimensional gel electrophoresis(2DE), which displays changes in protein expression and post-translational modifications(PTMs) on the basis of protein staining intensities and electrophoretic mobility. An alternative strategy for protein profiling, variously referred to as multidimensional protein identification technology (MudPIT), multidimensional LC-MS/MS (multidimensional liquid chromatography coupled with tandem mass spectrometry), or "bottom up" shotgun proteomics. Several studies have successfully identified novel signal transduction targets by selectively activating or inhibiting pathways and screening molecular responses by the methods mentioned above, such as MAP kinase signaling in human erythroleukemia cells, and Rho GTPase signaling in a melanoma cell.

Cellular proteomics assays. Cellular proteomic systems are increasingly being exploited to assess the activity of proteins in a single cell, and this promise to provide unparalleled advances towards annotation of gene function and the identification of novel drug targets. Genes can be systematically evaluated for their participation in signaling pathways as well as cellular processes and physiological phenotypes. Then, integration of extracellular signals often involves the crosstalk between signal cascades that has been suggested to share some common traits with neural networks.

(4) **Cellular epigenomics**　**Epigenetics** study the heritable changes in gene function that

occur without a change in the DNA sequence. The **epigenome** is the sum of both the chromatin structure and the pattern of DNA methylation, which is the result of an interaction between the genome and the environment.

Epigenomics is a whole genome approach to epigenetics. Takes a whole-genome approach to study environmental or developmental epigenetic effects, primarily DNA methylation, on gene function. Thus, epigenomics focuses on those genes whose functions are determined by external factors.

An approach that views the imprinting, metabolic networks, genetic hierarchies in embryonic development, and epigenetic mechanisms of gene activation in a cell and other complex phenotypes from the genomic level down, rather than from the genetic level up, can provide powerful insights into the functional interrelationships of genes in health and disease.

(5) In silico cell   "In silico cell" is also named as "e-cell", which is produced in or by means of a computer simulation. This must originally have been a computer scientist's joke, but it now appears in print often enough that it has to be added to our list of modern Latinisms. The E-cell system permits the construction of model structure, such as gene regulatory and/or biochemical networks equivalent to a cell system or a part of the cell and perform both quantitative and qualitative analysis.

"In silico cell" is the computational model of living cell. Systems biology involves the use of computer simulations of cellular subsystems (such as the networks of metabolites and enzymes which comprise metabolism, signal transduction pathways and gene regulatory networks) to both analyze and visualize the complex connections of these cellular processes. Artificial life or virtual evolution attempts to understand evolutionary processes via the computer simulation of simple (artificial) life forms.

"In silico cell" reflects the molecular networks within a cell. Cellular components interact with each other to form networks that process information and evoke biological responses. A deep understanding of the behavior of these networks requires the development and analysis of mathematical models, which has become an attractive focal point for engineers, computer scientists, mathematicians, and systems biologists.

1.7   The technology of cell biology

Despite the considerable success in sequencing the human genome and other genomes, the function of many thousands of genes in humans remains unknown. Developments of techniques can screen thousands of genes simultaneously for their effect on the biological processes in living cells — cell division, signaling pathways and gene expression, for example and would result in major benefits to both the basic biomedical sciences and the pharmaceutical industry.

The details of how cells are able to grow, reproduce, differentiate and communicate with each other are highly important in modern biology. The investigation of these processes in living cells at the sub-cellular and molecular level requires a number of new technological developments.

1.7.1   Cell culture

Cell culture is one of the most widely used techniques in life sciences. This Cell Culture workshop will cover the basic techniques used for maintaining animal cells in culture: aseptic technique, counting cells, subculturing, cryopreservation (freezing) and thawing.

Participants will obtain hands-on training in all techniques listed above. Lecture and discussion sessions will include the techniques mentioned above as well as the following topics: cell culture equipment, contamination and optimization of growth conditions.

1.7.2 Flow cytometry

Flow cytometry is a means of measuring certain physical and chemical characteristics of cells or particles as they travel in suspension one by one past a sensing point. In one way flow cytometers can be considered to be specialized fluorescence microscopes. The modern flow cytometer consists of a light source, collection optics, electronics and a computer to translate signals to data. In most modern cytometers the light source of choice is a laser which emits coherent light at a specified wavelength. We can measure physical characteristics such as cell size, shape and internal complexity and, of course, any cell component or function that can be detected by a fluorescent compound can be examined. So the applications of flow cytometry are numerous, and this has led to the widespread use of these instruments in the cellular biological and medical fields.

Flow cytometric applications include: phenotypic analysis、Sterile sorting of transfectants、DNA analysis、Assessment of apoptosis、Functional studies.

1.7.3 Functional bio-imaging of cell

In the longer-term, additional multi-parameter imaging-measurement techniques could also be developed and applied to single living cells inside animals.

(1) **Modern light microscopy** Modern light microscopy has become a most powerful analytical tool for studying molecular processes in cells. Recent advances combining sample preparation, microscope design and image processing allow the generation of "multidimensional" image data, simultaneously reporting the three-dimensional distribution and concentrations of several different molecules within cells and tissues at multiple time points with sub-micron spatial resolution and sub-second temporal resolution. Molecular interactions and processes that were approached by biochemical analyses *in vitro* can now be monitored in live cells.

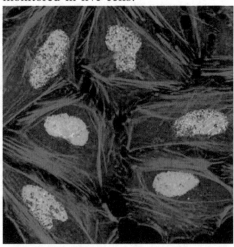

Figure 1-9 A confocal image of cancer cells.
Confocal immunofluorescent analysis of HeLa cells (From: http://www.cellsignal.com/products/3440.html)

(2) **Confocal microscopy** Biological specimens are three-dimensional objects. In confocal microscopy, a pinhole aperture is inserted at the optical focal plane in the microscope. This passes only light going through the structure of interest and blocks all reflected light from the rest of the specimen, resulting in a clear image. Movable mirrors scan the area of good focus across the specimen in a raster pattern, creating a large, in-focus image of a thin slice of the specimen. The raster scan then moves vertically, and another slice is scanned. After multiple scans, computer processing of the images produces a three-dimensional image of the specimen (Figure 1-9).

(3) **Two-photon microscopy** In two-photon microscopy fluorophores are excited simultaneously by two photons of twice the wavelength of the

single, smaller-wavelength photons used in traditional confocal microscopy. Longer wavelength photons have lower energies and penetrate further into tissue than short wavelength photons, making the technique ideal for imaging deep into live tissue samples.

With two-photon imaging, lower energy photons illuminate the specimen. A dye molecule that absorbs two of the lower energy photons within a short period of time will receive energy above the threshold for fluorescence. To increase the chances that a dye molecule will absorb two photons within the short time period, the laser is pulsed causing all photons to arrive within a few seconds. The laser beam is focused on the area of interest, causing all of the photons to illuminate just the area of interest. These two techniques put all of the photons on target at the same time.

The technique could allow researchers in many fields of biology to track migrating cells in a form of real-time movie, which biologists have discovered are common in many types of tissue, ranging from nerves to lymph nodes. To date, such long-range migrations have been inferred from observations of chemically fixed tissue at different stages of development.

(4) **Live cell Image analysis** Live-cell image analysis started with the earliest microscopists. Although most of these measurements were based on manual inspection and intervention, with the advent of fluorescence microscopy, many studies also involved quantitative imaging of living cells either using video or CCD cameras. CCD camera uses a small, rectangular piece of silicon rather than a piece of film to receive incoming light, this silicon piece is called a charge-coupled device (CCD). In the early years of live-cell microscopy, methods for segmentation and tracking of cells were rapidly developed and adapted from other areas. Nowadays, techniques for fully automated analysis and time—space visualization of time series from living cells involve either segmentation or tracking of individual structures, or continuous motion estimation.

A major requirement of an automated, real-time, computer vision-based cell tracking system is an efficient method for segmenting cell images. The usual segmentation algorithms proposed in the literature exhibit weak performance on live unstained cell images, which can be characterized as being of low contrast, intensity-variant, and unevenly illuminated.

By modern microscopy system, life scientists are able to conduct three-dimensional time-lapse studies with shorter exposure times and more frequent exposure opportunities, allowing them to capture a more complete picture of intracellular activity. This system now offers live-cell researchers the shortest exposure time of any microscopy system with the greatest molecular-level resolution available. Improved and more frequent image sampling over time dramatically increases the chance that a researcher might discover a new key structure or function occurring in the cell they are viewing.

1.7.4 Bioinformatics and cell biology

Bioinformatics is set to play a significant role in cell biology immediately and in the foreseeable future through the development of software to organize the multitude of data emerging from simultaneous time — course measurements and allow interpretation of that data in the context of cell function. An example is in the construction of 3-D images. At present this software has been written with the medical imaging in mind and will need to be modified to suit the particular application.

1.7.5 RNA interference technology

RNA interference (RNAi) is a molecular biological technique that has been extensively used to deplete specific components in cells and has been invaluable in determining the requirements of a given protein in a particular cellular process. In the response, small segments of double stranded RNA direct specific message RNA cleavage resulting in little or no protein expression. The initial use of RNAi technology was published in 1998 where Mello and colleagues showed double-stranded RNA-mediated genetic interference in Caenorhabditis elegans. Researchers are still exploring the pathway mechanisms and components, but it appears to be conserved in other organisms including plants and humans.

1.7.6 Antisense technique

The **antisense** technique is tools we are using in an attempt to selectively block the gene expression. The primary idea of antisense technique is to interfere with the information flow from gene to protein, and to do so in a very specific manner. Antisense oligodeoxynucleotides (ODN) have been used selectively to arrest translation of target mRNA into functional proteins in a wide variety of systems from cell-culture to animals. The action of antisense technique depends on the ability of short synthetic ODNs, which has been taken up by the living cells, to hybridize selectively to the target mRNA, and thereby interfere with ribosomal activity, activate ribonuclease H-mediated degradation of mRNA, or both.

1.7.7 Gene microarray

Gene microarray, or DNA micro chips, is a parallel quantification of large numbers of messenger RNA transcripts using microarray technology, which promises to provide detailed insight into cellular processes involved in the regulation of gene expression. This should allow new understanding of signaling networks that operate in the cell and of the molecular basis and classification of disease.

Within the department of molecular genetics, DNA chip technology is a core activity. This technology, which is also synonymous with DNA micro-array analysis, aims at taking a global view of gene activities, thus enabling the researcher to simultaneously follow the expression levels of thousands of genes in a single experiment. The purpose of this is to gain molecular understanding of the biological action of trans cription network.

Gene microarray are devices not much larger than postage stamps. They are based on a glass substrate wafer and contain many tiny spots—400,000 is common. Each holds DNA from a different gene. The array of cells makes it possible to carry out a very large number of genetic tests on a sample at one time. The process of testing the gene expression patterns of an individual is sometimes called microarray profiling.

The DNA microchip is also a revolutionary new tool used to identify mutations in genes in the genomics level and to gain molecular understanding of the biological action of drug candidate molecules to help select and develop optimal drugs.

1.7.8 Coming technological developments

Understanding the subcellular localisation of activity and the dynamics of the activity as well as being able to quantify these measurements represent significant technical challenges. In addition, it is often necessary to study a wide range of parameters simultaneously in living cells under normal physiological conditions. To assist the greater understanding of cell function, the following technological developments will need to occur over the next five years.

(1) **New chemically based technologies include**   ①The development of a range of new non-invasive biological assays for specific biochemical processes e. g., new enzyme assays specific for gene expression to allow the study of dynamic processes in living cells; ②The development of peptide tags for the localization of proteins in cells; ③The synthesis of new chemical compounds for monitoring specific cellular processes in living cells.

(2) **New types of equipment**   ①More sensitive and versatile confocal microscopes; ②Cameras which are specific for particular assays combining for example, higher quantum efficiency, operation at faster speeds with broader wavelength sensitivity and low background noise; ③Cameras with greater versatility, which can be used for a variety of assays combining for example, higher quantum efficiency, operation at faster speeds with broader wavelength sensitivity and low background noise; ④Digital cameras able to record numerous simultaneous events; ⑤Detectors which are specific for the type of reporter or probe being used; ⑥Detectors need to work at the appropriate speed with the minimum of background noise; ⑦Specific new software to control the microscope and camera in order to collect and subsequently to analyze multi-parameter time course data.

1.8   Training the scientists of tomorrow

1.8.1   Cell biology is an important basic course

Since cells are ubiquitous in life, cell biology must be of fundamental importance to biologists. The birth of the cloned sheep Dolly and the culmination of the Human Genome Project are indicative of the entry into a new century, which belongs to the life sciences. The cell biology is the study of the fundamental life activities and regulation mechanisms of the cell.

1.8.2   The principles in the course of cell biology

The intent of modern molecular cell biology is to interpret the properties of the whole organism through the structure of its own cell's component molecules. The cell biology also is the study of the life processes, which occur within the makeup of a cell and its molecules.

(1) **Understanding the cell from molecular view**   The goals of the molecular cell biology lie in understanding the cell theory from molecular view, which by itself cannot explain the development and unity of the multi-cellular organism. A cell is not necessarily an independently functioning unit, and a plant or an animal is not merely an assembly of individual cells.

(2) **Understand the cell from systemic view**   Molecular cell biology is not so self-inclusive as to eliminate the concept of the organism as more than the sum of its parts. But the study of a particular organism does require the investigation of cells, as both individuals and groups. A plant or an animal governs the division of its own cells; the correct cells must divide, become differentiated, and then be integrated into the appropriate organ system at the right time and place. Breakdown of this process results in a variety of abnormalities, one of which may be cancer. When a cell biologist studies the problem of the regulation of cell division, the ultimate objective is to understand the effect of the individual cell on the whole organism.

1.8.3 What dose cell biology show students

Cell biology will not only show students what we know about the cell, but also show that how we know and why we do about the cell.

(1) **What we know about the cell** The main content of the molecular cell biology course includes: ①The structure and function of the cell; ②The organelles of the cell; ③The mechanisms of important activities of the cell; ④The regulation of the activities; ⑤The advance in cellular research.

This course covers some modern topics of cell biology, such as the proliferation and differentiation of cells; single transduction in the cell; gene expression and regulation in the eukaryote cell; and the origin and evolution of cells. Thus cell biology is an important basic course for all students of life science.

(2) **How we know what we know about the cell** It is critical to present to students the experimental basis of our understanding — to show them how we know what we know. Hopefully this will demonstrate the dynamic nature of science and prepare them not only to engage actively in scientific research and teaching, but also to become educated members of a public that increasingly is asked to deal with complex issues such as environmental toxins, genetically modified foods, and human gene technology.

(3) **Why we do what we do about the cell** Of course, we want students to learn not only how we know what we know, but why we do what we do. As in other cell biology books, our coverage of basic cell biology, medical topics, biotechnology, human biology is integrated throughout. We know that these topics may be of particular interest to students.

The cell biologists focus recently on systems cell biology, epigenetics of the cell and functional cellular imaging. These researches lead to full understanding the cell activity and the cellular roles in organ as well as organism. With the development of research, more interests will produce further research projects to further our knowledge about the cell biology. The awareness of what we do and why we do is essential for a cell biologist of tomorrow.

1.8.4 who is a cell biologist

Cell biologists are scientists who study cells, which are generally considered to be the building blocks of life because all living things are composed of one or more cells.

(1) **What do they work on** Cell biologists works in the field of cell biology, which is also known as cytology. Cell biology studies all aspects of cells, from their interactions with each other and their environments to their cellular and atomic composition. A cell biologist will generally work to conduct experiments to better understand the precise nature of cells.

(2) **What are they interested in** Cell biologists are generally very concerned with investigating both the processes and structures that occur and exist within cells. There are almost limitless combinations of structures and processes that any given cell can contain; they vary based on the purpose of the cell.

Cell biologists are tends to be interested in the relationship between structures and functions. Cell biologists also tend to seek similarities between different kinds of cells or between cells belonging to different species. A cell biologist is also typically very interested in proteins; there are many different kinds of proteins within cells, and each has a purpose within the cell.

Chapter 1 Introduction to cell biology

Summary

Cell biology is the modern biology of the cell. It concerns on the cell which is the units of life and a complex system. Current genomic and proteomic knowledge shapes the cell biology, leading to the expanse of the contents of cell biology from essential cell biology, to molecular cell biology and systemic cell biology. The development of modern cell biology depends on the progress of contemporary biotechnology. Cell biology is also an integrated view of cells at work, involving essential cell biology, molecular biology of cell and cell systems biology.

Development of techniques which can screen thousands of genes simultaneously for their effect on the biological processes in living cells — cell division, signaling pathways and gene expression, for example, would result in major benefits to both the basic biomedical sciences and the pharmaceutical industry.

Questions

1. What does cell biology concern?
2. Why we say the cell is a complex system?
3. What is cell theory?
4. What are the contemporary views on cell biology?
5. What cause cell diversity?
6. How many modern techniques are employed in cell biology research?

(吴超群)

Chapter 2

Cell membrane and cell surface

All cells are surrounded by a membrane layer called **cell membrane or plasma membrane**. Cell membranes consist of membrane lipid and proteins and are crucial for the life of the cell by defining the cell boundaries, therefore, maintaining the differences between the intra-and extracellular environment. On the surface of cell plasma membrane, structures enriched with sugar groups form cell coat, also termed **glycocalyx** or **extracellular matrix**, which is functionally important for the cell-cell communication and cell-environment communication.

Within a eukaryotic cell, membranes compartmentalize the cell into smaller subcompartments with specialized functions termed organelles, such as endoplasmic reticulum, vesicle, Golgi body, mitochondria, and chloroplasts. All these membranes form closed structures and a related net called cytoplasmic membrane system. The plasma membrane and cytoplasmic membrane system belong to biomembrane system. The most important character of biomembrane is selectively permeable (semipermeability). It forms an adjustable barrier between the cell and the extracellular environment or between inside and outside of organelles in eukaryotic cell. Prokaryotic cells have very simple internal structures. The cells contain no internal membrane limited subcompartments. Most enzymes and metabolites are thought to diffuse freely within the single internal aqueous compartment. The replication of DNA and the production of ATP, take place at the plasma membrane.

In this chapter, the components, structures and functions of biomembrane will be described and the plasma membrane will be mainly focused on. Cell surface, extracellular matrix, cell wall, trans-membrane transport, and cell junction will be discussed in detail. Other structures with biomembrane, such as nuclear envelope, mitochondrion, chloroplast, endoplasmic reticulum, Golgi body and vacuole, will be discussed in pertinent chapters.

2.1 Components and structure of cell membranes

All cell membranes have a common structure: a thin film of lipids and protein molecules held together mainly by non-covalent interactions (Figure 2-1). Besides, a small quantity of saccharide chains covalently binds to the membrane proteins (glycoproteins) and lipids (glycolipids). All kinds of phospholipid molecules in biomembrane contain a hydrophobic core called "nonpolar tail" and a hydrophilic core called "polar head". This molecular character enables phospholipids to have the ability to form a bilayer framework structure of biomembrane. In this framework, phospholipids spontaneously aggregate to bury their hydrophobic tails in the interior and expose their hydrophilic heads to water. The bilayer structure is maintained by hydrophobic and van der Waals interactions between the lipid chains.

Chapter 2   Cell membrane and cell surface

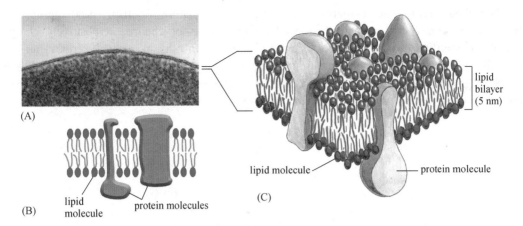

Figure 2-1   The structure of plasma membrane. (A) Structure of human red blood cell plasma membrane under electron microscope. (B) and (C) Schematic view of a plasma membrane (B: two; C: three dimension) (From: Alberts B, et al. Molecular Biology of the Cell. 2008)

2.1.1   Lipids

Lipids contained in a typical biomembrane are phospholipids (include phosphoglycerides, sphingolipids) and steroids. All three classes of lipids are **amphipathic** molecules having a polar (hydrophilic) head and a hydrophobic tail.

(1) Phosphoglycerides   **Phosphoglycerides**, derivatives of glycerol 3-phosphate, are the most abundant class of lipids in most membranes. A typical phosphoglyceride molecule consists of a polar head attached to the phosphate group and a hydrophobic tail composed of two fatty acyl chains esterified to the two hydroxyl groups in glycerol phosphate (Figure 2-2). The two fatty acyl chains may differ in the number of carbons (commonly 16 or 18) and their degree of saturation (0, 1, or 2 double bonds). The length and the saturation degree of the fatty acid tails are important in regulating the fluidity of the membrane.

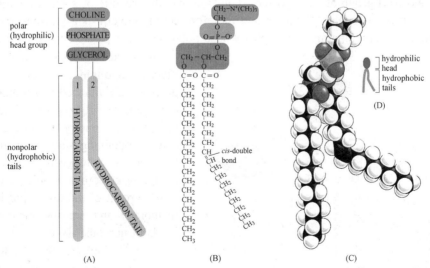

Figure 2-2   Structure of phosphatidylcholine, a typical phosphoglyceride molecule. (A) Scheme of molecular structure. (B) Formula. (C) Space-filling model. (D) Symbol. (From: Alberts B, et al. Molecular Biology of the Cell. 2008)

Phosphoglycerides are classified according to the nature of their head groups. In phosphatidylcholines, for example, the head group consists of choline, a positively charged alcohol, is esterified to the negatively charged phosphate. In other phosphoglycerides, an OH-containing molecule such as ethanolamine, serine, or the sugar derivative inositol is linked to the phosphate group (Figure 2-3 A~C). The negatively charged phosphate group and the positively charged or the hydroxyl groups on the head group interact strongly with water.

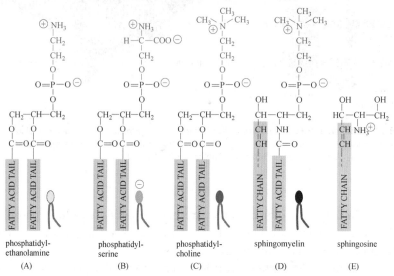

Figure 2-3   Structure of major phospholipid molecules in mammalian plasma membrane (From: Alberts B, et al. Molecular Biology of the Cell. 2008)

(2) **Sphingolipids**   The second class of membrane lipid is **Sphingolipid**. Sphingolipid is derived from sphingosine and contains a long-chain fatty acid attached to the sphingosine amino group (Figure 2-3 D~E). Sphingosine is an amino alcohol with a long hydrocarbon chain. In sphingomyelin, the most abundant sphingolipid, phosphocholine is attached to the terminal hydroxyl group of sphingosine. Thus the overall structure of sphingomyelin is quite similar to that of phosphatidylcholine and therefore sphingomyelin is a phospholipid. Other sphingolipids are amphipathic glycolipids whose polar head groups are sugars (Figure 2-4).

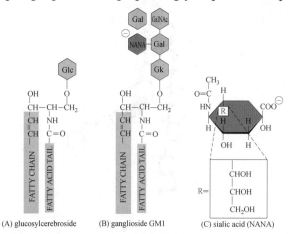

Figure 2-4   Structure of glycolipid molecules in plasma membrane

Glucosylcerebroside, the simplest glycosphingolipid, contains a single glucose unit attached to sphingosine. In the complex glycosphingolipids called gangliosides, one or two branched sugar chains containing sialic acid groups are attached to sphingosine. Glycolipids, including glyceroglycolipid, constitute 2~10 percent of the total lipid in plasma membrane and are the most abundant lipids in nervous tissue.

(3) **Cholesterol**   **Cholesterol** and its derivatives constitute the third important class of membrane lipids, the steroids. The basic structure of steroids

is a four-ring hydrocarbon and cholesterol has a hydroxyl substituent on one ring (Figure 2-5). Although the composition of cholesterol is almost entirely hydrocarbon, it is amphipathic since its hydroxyl group can interact with water. The cholesterol molecules orient themselves in the bilayer with their hydroxyl group close to the polar head groups of adjacent phospholipid molecules and their rigid, platelike steroid rings interacting with-and partly immobilizing-the first few CH_2 groups of the hydrocarbon chains of the phospholipid molecules, and therefore makes the lipid bilayer less deformable in this region and less permeable for small water-soluble molecules. Cholesterol is especially abundant in the plasma membranes of mammalian cells but is absent from most prokaryotic cells. At the high concentrations found in eukaryotic plasma membranes, cholesterol prevents the hydrocarbon chains from coming together and crystallizing and therefore keeps the bilayer fluidity.

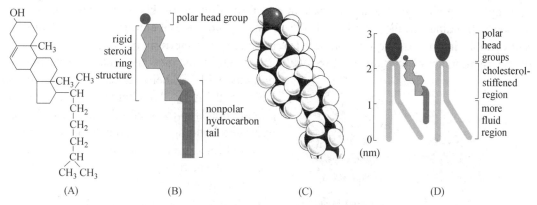

Figure 2-5 **Structure and function of cholesterol molecules in mammalian plasma membrane.** (A) Formula. (B) Symbol. (C) Space-filling model. (D) Interaction between phospholipids and cholesterols in plasma membrane (From: Alberts B, et al. Molecular Biology of the Cell. 2008)

2.1.2 Proteins

The hydrophobic core is an impermeable barrier that prevents the diffusion of water-soluble solutes across the membrane. This simple barrier function is modulated by the presence of membrane proteins that mediate the transport of specific molecules, ions and water across this otherwise impermeable bilayer. The proteins associated with a particular membrane are responsible for its distinctive activities.

The lipid bilayer presents a unique two-dimensional hydrophobic environment for membrane proteins either inserted in or linked to the surface of the bilayer. Membrane proteins can be classified into three categories (Figure 2-6): integral proteins (buried within the bilayer), lipid-anchored proteins and peripheral proteins (associated with the exoplasmic or cytosolic leaflet of the bilayer).

Integral proteins (transmembrane proteins) contain three different segments

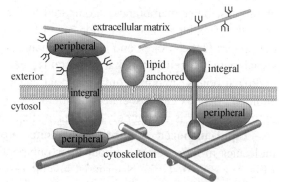

Figure 2-6 **Diagram of proteins associated with the lipid bilayer** (From: Lodish H, et al. Molecular Cell Biology. 2008)

called domains: cytosolic domain, exoplasmic domain and transmembrane domain. The cytosolic and exoplasmic domains have hydrophilic exterior surfaces that interact with the aqueous solutions on the two faces of the membrane. Most transmembrane proteins are glycosylated with a complex branched sugar group attached to amino acid side chains which invariably are localized to the exoplasmic domains and generally bind to external signaling proteins, ions, and small metabolites (e. g., glucose, fatty acids), and to adhesion molecules on other cells or in the external environment. The cytosolic domain lying along the cytosolic face of the plasma membrane anchors cytoskeleton proteins and triggers intracellular signaling pathways. In most transmembrane proteins, their transmembrane regions contain many hydrophobic amino acids and form one or more α helices which are embedded in membranes. The hydrophobic side chains protrude outward from the helix and form van der Waals interactions with the fatty acyl chains in the bilayer. In contrast, the carbonyl and amino groups taking part in the formation of backbone peptide bonds form hydrogen bonding in the interior of the α helix, and are therefore shielded from the hydrophobic interior of the membrane. The structure of bacteriorhodopsin, a protein found in the membrane of certain photosynthetic bacteria, illustrates the general structure of proteins with α helix transmembrane region (Figure 2-7A). An alternative way for the peptide bonds in the lipid bilayer to satisfy their hydrogen-bonding requirements is to arrange the multiple transmembrane domains as β sheets that are rolled up into a closed barrel (a so-called β barrel). This form of multipass transmembrane structure is seen in OmpX, a porin protein (Figure 2-7B). Segments containing about 20 ~ 30 amino acids with a high degree of hydrophobicity are long enough to span a lipid bilayer as α helices and they can often be identified in hydropathy plots.

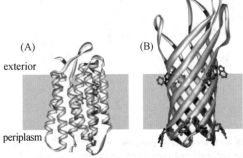

Figure 2-7 **Structural basis of transmembrane proteins.** (A) α helix model of bacteriorhodopsin (From: Luecke H, et al. J Mol Biol, 1999, 291:899). (B) β barrel model of one subunit of OmpX (From: Schulz GE. Curr Opin Struc Biol, 2000, 10:443)

However, hydropathy plots cannot identify the membrane-spanning segments of a β barrel, as 10 amino acids or fewer are sufficient to traverse a lipid bilayer as an extended β strand. Most multipass transmembrane proteins in eukaryotic cells and in the bacterial plasma membrane are constructed as transmembrane α helices whereas β barrel proteins are largely restricted to bacterial, mitochondrial, and chloroplast outer membranes. Compared to the β barrel, the helix transmembrane domains are rather flexible which can slide against each other, allowing conformational changes in the protein that can open and close ion channels, transport solutes, or transduce extracellular signals into the cell. In β barrel proteins, each β strand binds rigidly to its neighbors via hydrogen bonds, making conformational changes within the wall of the barrel unlikely. In addition, α helix can associate with the helix in another integral protein to form a coiled-coil dimmer, which is a common mechanism for creating dimeric membrane proteins.

Lipid-anchored membrane proteins are bound covalently to one or more lipid molecules in three forms without their polypeptide chains entering the phospholipid bilayer (Figure 2-8): first, the N-terminal glycine residue of a membrane protein is anchored to the cytosolic face of a membrane via a fatty acyl group (acyl anchor). Second, a cysteine residue at or near the C-terminus is anchored to membranes by an unsaturated fatty acyl group (prenyl anchor). Third, some cell-surface proteins and heavily glycosylated proteoglycans of the extracellular matrix are bound to the exoplasmic face of the plasma membrane by glycosylphosphatidylinositol (GPI).

Peripheral membrane proteins do not interact with the hydrophobic core of the phospholipid bilayer, but are usually bound indirectly to the membrane by interactions with integral proteins or directly by interactions with lipid head groups.

2.1.3 Membrane carbohydrate

About 2~10 percent of membrane content is made up of carbohydrates. These carbohydrates occur as oligosaccharide chains covalently bound to membrane proteins (glycoproteins) and lipids (glycolipids). They also occur as the

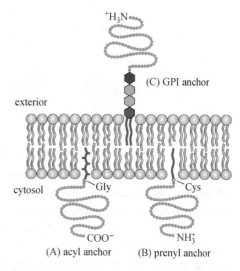

Figure 2-8 **Covalent anchoring of plasma membrane proteins to the bilayer** (From: Lodish H, et al. Molecular Cell Biology. 2008)

polysaccharide chains of integral membrane proteoglycan molecules found mainly outside the cell, as part of the extracellular matrix. The carbohydrate layer protects cells against mechanical and chemical damage and also helps to keep cells at a distance, preventing unwanted protein-protein interactions. The membrane carbohydrates help membrane proteins to form and maintain the correct packing and three-dimensional configures and to transfer to a correct position. The oligosaccharide side chains of glycoprotein and glycolipids usually contain fewer than 15 sugars, but are enormously diverse in their arrangement and are often branched and bonded together by various covalent linkages. Both the diversity and the exposed position of the oligosaccharides on the cell surface make them especially well suited to function in specific cell-recognition processes. For example, some glycolipids are important blood group determinants, and plasma membrane-bound lectins recognize specific oligosaccharides on cell-surface and mediate a variety of transient cell-cell adhesion processes in sperm-egg interaction, blood clotting, lymphocyte recirculation, and inflammatory response.

2.1.4 Structure characters of plasma membrane

The phospholipid molecules diffuse freely in biomembrane and form a bilayer framework structure with their hydrophobic nonpolar tails buried in the interior and their hydrophilic heads exposed to water. Depending on their shape, the cone-shaped lipid molecules can form spherical micelles with the tails inward whereas the cylinder-shaped lipids form double layer sheets with the hydrophobic tails sandwiched between the hydrophilic head groups. A mixture of two shapes can form **liposomes**, bilayer spherical vesicles with diameter spanning 25 nm to 1 μm (Figure 2-9).

(1) **Fluid mosaic model of plasma membrane structure**   As to the structural organization of plasma membrane, several models have been proposed. Singer and Nicolson established the well-accepted **fluid mosaic model** in 1972. Fluid-mosaic model presents membranes as dynamic structures in which both lipids and associated proteins are mobile and capable of moving within the membrane to engage in interactions with other membrane

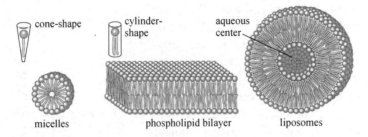

Figure 2-9 Arrangement of lipid molecules in an aqueous environment

molecules. The lipid bilayers form the basis of the membranes; a large quantity of proteins either span the bilayer or are attached to either side of the lipid membrane; the membranes are asymmetrical, in which lipid and protein molecules diffuse more or less easily.

(2) **Lipid raft model** **Lipid rafts** model has been recently established as a complementation for the former fluid mosaic model. In this model, some regions of the plasma membrane are especially rich in cholesterol, glycosphingolipid, and GPI-anchored membrane proteins, which are organized in small (10 ~ 200 nm) glycolipoprotein microdomains, termed lipid rafts (Figure 2-10). These specialized membrane microdomains compartmentalize cellular processes by serving as organizing centers for the assembly of signaling molecules, influencing membrane fluidity and membrane protein trafficking, and regulating neurotransmission and receptor trafficking.

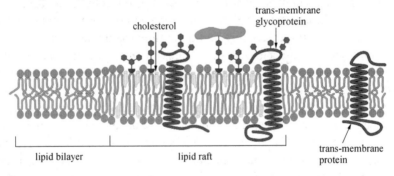

Figure 2-10 Structure of lipid raft in plasma membrane

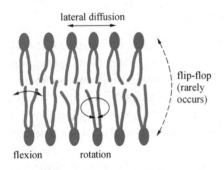

Figure 2-11 The possible movements of phospholipids in a membrane (From: Alberts B, et al. Molecular Biology of the Cell. 2008)

(3) **Membrane fluidity** **Membrane fluidity**, referring the viscosity of the lipid bilayer, is the physical state of the lipid molecules within the cell membrane and is essential for the membrane function. It allows molecules to diffuse rapidly in the plane of the bilayer and interact with each other, thus many basic cellular processes such as cell movement, cell growth, cell division, secretion etc. can take place. Certain types of movement (e.g., diffusion and rotation) of molecules within the membrane are more frequent and rapid than the others (e.g., "flip-flop") (Figure 2-11). This is because in order for "flip-flop" to occur, the hydrophilic head group of the lipid must overcome the internal hydrophobic sheet of the

membrane, which is thermodynamically unfavorable. Lipids provide the matrix in which integral proteins of a membrane are embedded, thus the physical state of the lipid is important in determining the mobility of integral proteins.

Temperature and membrane composition are the main determinants for membrane fluidity. At relatively warm temperature (e.g., 37℃), the lipid bilayer is in a relatively fluid state. In this case, individual phospholipids keep their long axes in a parallel arrangement and rotate around their axes or move laterally within the plane of the bilayer. When lowering the temperature slowly, there comes a point (transition temperature) when the lipid is converted to a frozen crystalline gel in which the movement of the phospholipid fatty acid chains is greatly limited. The length and the degree of unsaturation of hydrocarbon tails determine how closer and regular the phospholipids are packed and therefore the viscosity and fluidity of the membrane. The longer the hydrocarbon chain is, the more likely the tails interact with one another and the less fluid the membrane is. As for unsaturation, the double bond in the chain creates a small kink in the hydrocarbon tail, which makes it more difficult for the tail to pack against one another. Therefore, the larger the proportion of unsaturated hydrocarbon tails is, the more fluid the membrane is. In most eukaryotic cells, the higher concentrations of cholesterol in the plasma membrane help to prevent the hydrocarbon chains from coming together and crystallizing and therefore keep the bilayer fluidity. Membrane fluidity needs to be maintained within certain limits for cell to function normally. For example, when temperature drops, the cells respond metabolically: enzymes remodel membranes to make the cell more cold-resistant by desaturating single bonds in fatty acyl chains to form double bonds. In addition, the cell changes the types of phospholipids being synthesized in favor of ones containing more unsaturated fatty acids.

Membrane protein mobility was first demonstrated using cell fusion technique (Figure 2-12). In such studies, mouse and human cells were fused, and the locations of specific proteins of the plasma membrane were tracked using various fluorescent dyes labeled antibodies against certain mouse or human proteins. At the onset of fusion, the plasma membrane appeared half human and half mouse. As the time of fusion progresses, the membrane proteins from two species were seen to gradually move within the membrane into the opposite hemisphere and finally distribute uniformly in the entire hybrid cell membrane.

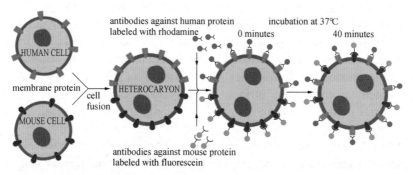

Figure 2-12 Membrane protein mobility revealed by cell fusion technique (From: Alberts B, et al. Molecular Biology of the Cell. 2008)

These early studies seem to show that integral membrane proteins were capable of moving freely within the membrane. However, other techniques such as fluorescence recovery after

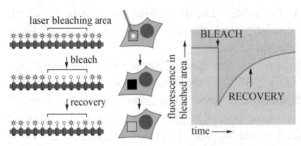

Figure 2-13 Membrane protein mobility revealed by FRAP technique

photobleaching (FRAP) show that there are restrictions on protein mobility (Figure 2-13). In FRAP assays, cells with a particular membrane protein labeled with a fluorescent probe are placed under the microscope, and irradiated by a laser beam that bleaches the fluorescent molecules in its path, leaving a circular spot on the cell surface lacking of fluorescence. If the labeled proteins are mobile, the random movements of these molecules should produce a gradual reappearance of fluorescence in the irradiated blank circle. The rate of the fluorescence recovery indicates the diffusion rate of the mobile molecules. The extent of fluorescence recovery represents the percentage of the labeled molecules that are free to diffuse. The result of FRAP studies indicated that a significant portion of membrane proteins (30% ~ 70%) were not free to diffuse back into the irradiated circle, indicating that the movement of certain proteins on cell membrane are restricted. Some membrane proteins assemble into large complexes, in which individual protein molecules are relatively fixed and diffuse very slowly. In epithelial tissue, such as those that line the gut or the tubules of the kidney, certain plasma membrane enzymes and transport proteins are confined to the apical surface of the cells, whereas others are confined to the basal and lateral surfaces. Such restricted movement of proteins is controlled by a specific type of intercellular junction such as tight junction. Clearly, the membrane proteins that form these intercellular junctions cannot be allowed to diffuse laterally in the interacting membranes.

(4) **Membrane asymmetry** All cellular membranes have an internal face designated as the cytosolic face (the surface oriented toward the interior of the compartment) and an external face designated as the exoplasmic face (the surface presented to the environment). Therefore cell membranes are asymmetrical: the two halves of the bilayer often contain different types of phospholipids and glycolipids. In addition, the proteins embedded in the bilayer have a specific orientation.

The asymmetry of the lipid bilayer was found by taking advantage of lipid-digesting enzymes that cannot penetrate the plasma membrane and subsequently are only able to digest lipids that reside in the external monolayer of the bilayer. When intact human red blood cells are treated with these enzymes, about 80% of the membrane's phosphatidylcholine (PC) is hydrolyzed, whereas only approximately 20% of the phosphatidylethanolamine (PE) and less than 10% of its phosphatidylserine (PS) are attacked. These data suggest that the external monolayer has relatively higher concentration of PC and sphingomyelin (SM) and inner monolayer has higher concentration of PE and PS. Glycolipids are exclusively located in the outer surface monolayer. Lipid asymmetry is functionally important, especially in converting extra cellular signals into intracellular ones. The phospholipid asymmetry of plasma membrane can be used to distinguish between live and dead cells. When animal cells undergo apoptosis, PS normally confined to the cytosolic monolayer rapidly translocates to the extracellular monolayer and signals neighboring cells such as macrophages to clean up the dead cells.

2.2 Transmembrane transport

The plasma membrane is a selectively permeable barrier between the cell and the extracellular environment to maintain a constant internal environment. The selective permeability ensures that essential molecules such as ions, glucoses, amino acids, and lipids readily enter the cell, metabolic intermediates remain in the cell, and waste components leave the cell. Few molecules can diffuse across a membrane, however, most molecules and all ions require a process, termed transmembrane transport, to across cellular membranes. This process is mediated by selective **membrane transport proteins** embedded in the phospholipid bilayer.

2.2.1 Overview of trans-membrane transport

The phospholipid bilayer of biomembranes is essentially impermeable to most water-soluble molecules, ions, and waters itself. There are different mechanisms of transmembrane transport for various molecules and ions. In order to define a type of transmembrane transport, some characters below should be considered: ①Whether need the aid of specific transport proteins? ②If the direction of movement is against concentration gradient of molecules or ions? ③Whether require the energy supplied by ATP hydrolysis or other types of energy, such as light? ④If it is driven by movement of a cotransported ion down its gradient?

According to all of above, the transmembrane transports are divided into two major types: **passive transport** and **active transport** (Figure 2-14). More detailed classifications and comparisons are listed in table 2-1. **Cotransport** is a special type of active transport, within which the movement of one type of ion or molecule against its concentration gradient is coupled with the movement of one or more different ions down its concentration gradient.

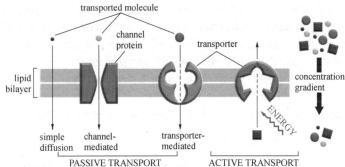

Figure 2-14 Comparison between passive and active transport
(From: Alberts B, et al. Molecular Biology of the Cell. 2008)

Table 2-1 Mechanisms for transporting ions and molecules across cell membranes

property	passive diffusion	facilitated diffusion	active transport	cotransport
requiring specific transport protein	no	yes	yes	yes
solute transported against its gradient	no	no	yes	yes
coupled to ATP hydrolysis	no	no	yes	no
driven by movement of an ion down its gradient	no	no	no	yes
examples	O_2, CO_2, steroid hormones, many drugs	glucose and amino acids (uniporters); ions and water (channels)	ions, small hydrophilic molecules, lipids (ATP-powered pumps)	glucose and amino acids (symporters); various ions and sucrose (antiporters)

All transport proteins are transmembrane proteins containing multiple membrane-spanning segments that generally are α helices. There are two main classes of membrane transport proteins, transporters and channels (Figure 2-14).

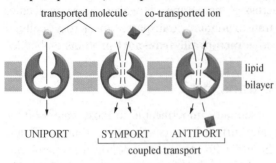

Figure 2-15　Three types of transporters (From: Alberts B, et al. Molecular Biology of the Cell. 2008)

Transporters (also called carriers or permeases) bind the specific solute and undergo a series of conformational changes to transfer the bound solute across the membrane. Three types of transporters have been identified (Figure 2-15). **Uniporters** transport a single type of molecule down its concentration gradient via facilitated diffusion. Glucose and amino acids cross the plasma membrane into most mammalian cells with the aid of uniporters. In contrast, **antiporters** and **symporters** couple the movement of one type of ion or molecule against its concentration gradient with the movement of one or more different ions down its concentration gradient. These proteins are often called cotransporters and play roles in active transport.

In contrast, **Channels** interact much more weakly with the solute and form aqueous pores that extend across the lipid bilayer; when pores open, these pores allow specific solutes (usually inorganic ions of appropriate size and charge) to pass through. Transport through channels occurs at a much faster rate than that mediated by transporters. Although water can diffuse across lipid bilayer, all cells contain specific channel proteins (called water channels or aquaporins) that greatly increase membrane permeability to water.

2.2.2　Passive transport

(1) passive diffusion and facilitated diffusion　**Passive transport** includes both passive diffusion and facilitated diffusion (Figure 2-14). In this process, molecules move down their electrochemical concentration gradient and no metabolic energy is required. In the case of transport of a single uncharged molecule, the difference in the concentration on the two sides of the membrane drives passive transport and determines its direction. If the solute carries a net charge, both its concentration gradient and the electrical potential difference across the membrane influence its transport. The concentration gradient and the electrical gradient combine to form a net driving force, the **electrochemical gradient**, for each charged solute. A few molecules, such as O_2, CO_2, and small, uncharged polar molecules, such as urea and ethanol, can diffuse across an artificial membranes composed of pure phospholipid and cholesterol, suggesting that such molecules also can diffuse across cellular membranes without the aid of transport proteins, a process called **passive diffusion or simple diffusion**. In contrast, many polar molecules, such as, glucose, amino acids, ions and water, are transported across membrane by a protein-mediated movement, namely **facilitated diffusion**. The rate of facilitated diffusion within which the transported molecules never enter the hydrophobic core of the phospholipid bilayer is far higher than passive diffusion.

Transport action of facilitated diffusion can occur via a limited number of uniporter molecules. The best-understood uniporter, GLUT1 (glucose transporter) alternates between two conformational states: state A with the glucose-binding site facing the outside of the membrane and state B with the glucose-binding site facing inside (Figure 2-16). Through sequential conformational changes, the uniporter GLUT1 facilitates the unidirectional transport of glucose down its concentration gradient. Such transport is specific and there is a maximum transport rate v_{max} that is achieved when the concentration gradient across the membrane is very large and each uniporter is working at its maximal rate.

(2) Ion channels and the electrical properties of membranes
Channel protein-assisted transport, also called **assist diffusion**, is another major type facilitated diffusion and transports water or ions and hydrophilic small molecules. Channel proteins form a hydrophilic passageway across the membrane through which multiple water

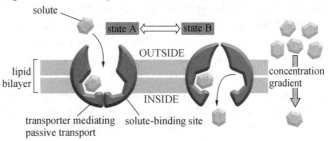

Figure 2-16　Uniporter mediated passive movement of a glucose solute (From: Alberts B, et al. Molecular Biology of the Cell. 2008)

molecules or inorganic ions, such as K^+, Na^+, Ca^{2+} and Cl^-, diffuse rapidly down their electrochemical gradients across the membrane at a very rapid rate (up to 106 ions/s). Above 100 kinds of ion channels have be discovered on various cell plasma membrane and are essential for many cell functions. For example, the ion channels propagate the leaf-closing response of mimosa plant; the single-celled *Paramecium* needs ion channels to reverse direction after collision. In particular neurons and muscles have used many different ion channels to receive, conduct and transmit signals.

Ion channels have two distinct characters compared to simple aqueous pores. First they show specific ion selectivity. The pores of ion channels are very narrow to force permeating ions into intimate contact with the walls of the channels so that only ions of appropriate size and charge can pass. K^+ channel is probably the most common channel found in the plasma membrane of animal cells. They conduct K^+ 10,000 fold better than Na^+, yet the two ions are both featureless spheres and have similar diameters (0.133 nm and 0.095 nm, respectively). A single amino acid substitution in the pore of K^+ channel can result in a loss of ion selectivity and cell death. Since Na^+ is smaller than K^+, the size of the pore cannot be the correct explanation for the channel selectivity. The resolve of a bacterial K^+ channel structure reveals the reason (Figure 2-17).

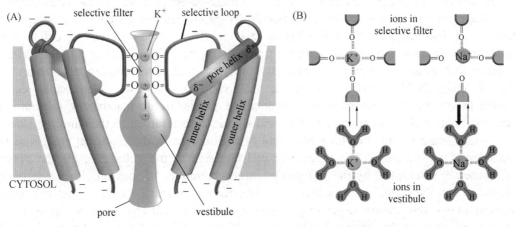

Figure 2-17　**The structure and ion selectivity of a bacteria K^+ channel** (From: Alberts B, et al., Molecular Biology of the Cell. 2008)

The bacterial K^+ channel consists of four identical transmembrane subunits forming a central pore across the membrane. Negatively charged amino acids concentrated at the cytosolic entrance to the pore are thought to attract cations and repel anions, making the channel cation-selective. Each subunit contributes two transmembrane α helices, which are tilted outward in the membrane and together form a cone, with its wide end facing the outside of the cell where K^+ ions exit from the channel. The polypeptide chain that connects the two transmembrane helices forms a short α helix (the pore helix) and a crucial loop that protrudes into the wide section of the cone to form the selectivity filter. The selectivity loops from the four subunits form a short, rigid, narrow pore (0.3 nm), which is lined by the carbonyl oxygen atoms of their polypeptide backbones. The crystal structure shows two K^+ ions line in single file within the selectivity filter, separated by about 0.8 nm. A K^+ ion losing all of its bound water molecules has a diameter of 0.27 nm. It enters the filter and interacts with the carbonyl oxygens which are spaced at the exact distance to accommodate a K^+ ion. In contrast, a dehydrated Na^+ ion which has a much smaller diameter (0.19 nm), cannot enter the filter because the carbonyl oxygens are too far away from the smaller Na^+ ion to stabilize it in the pore. As a result, the smaller Na^+ ions cannot overcome the higher energy barrier required to penetrate the pore.

Prokaryotic and eukaryotic cells have water channels, or aquaporins, embedded in their plasma membrane to allow water to move readily across this membrane. Aquaporins are especially abundant in cells that must transport water at particularly high rates, such as the epithelial cells of the kidney. To avoid disrupting ion gradients across membranes, they have to allow the rapid passage of water molecules while completely blocking the passage of ions. The crystal structure of an aquaporin reveals the water channels having a narrow pore that allows water molecules to traverse the membrane in single file, following the path of carbonyl oxygens that line one side of the pore. Hydrophobic amino acids line the other side of the pore. The pore is too narrow for any hydrated ion to enter, and the energy cost of dehydrating an ion would be enormous because the hydrophobic wall of the pore cannot interact with a dehydrated ion to compensate for the loss of water.

The second distinct character of ion channels is that they are so gated that allows them to open briefly and then close again. Some ion channels are open much of the time and are referred to as nongated channels. Most ion channels open only in response to specific chemical or electrical signals which lead to the conformation changes of the channel protein and are referred to as gated channels, include **voltage-gated** channel (signal is a change in the voltage across the membrane), **mechanically-gated** channel (signal is a mechanical stress) and **ligand-gated** channel (Figure 2-18). The ligand can be an extracellular mediator-specifically a neurotransmitter (transmitter-gated channels) or an intracellular mediator such as an ion (ion-gated channels) or a nucleotide (nucleotide-gated channels). The ion channels also have an automatic inactivating mechanism. When cells receive continuous stimuli, the channel recloses rapidly and remains in this inactivated state until the membrane potential or chemical and electrical signals have returned to their initial value.

A membrane potential arises when there is a difference in the electrical charge on the two sides of a membrane. The magnitude and direction of the voltage across the plasma membrane are determined by the differences in concentrations of ions on either side of the membrane and their relative permeabilities. The resting potential of an animal cell varies between 20 mV and 120 mV depending on the organism and cell type. The K^+ gradient has a major influence on this potential. Due to the effects of Na^+-K^+ pump introduced later in this chapter, Na^+ is pumped out of the cell and K^+ is pumped into the cell, thereby

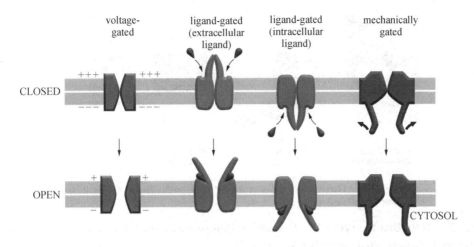

Figure 2-18 The gating of ion channels (From: Alberts B, et al. Molecular Biology of the Cell. 2008)

establishing steep gradients of these two ions across the plasma membrane. Because of these gradients, one would expect that K^+ would leak out of the cell and Na^+ would leak inward through their respective ion channels. However, the vast majority of the ion channels that are open in the plasma membrane of a resting cell are selective for K^+ (often referred to as K^+ leak channels). Because of the present of K^+ leak channels, outflow of K^+ through the membrane leaves an excess of negative charges on the cytoplasmic side of the membrane. K^+ comes almost to equilibrium when an electrical force exerted by an excess of negative charge attracting K^+ into the cell balances the tendency of K^+ to leak out down concentration gradient. The membrane potential is the manifestation of this electrical force and can be calculated from the steepness of the K^+ concentration gradient. Besides, the gradients of other ions (and the disequilibrating effects of ion pumps) also have a significant effect: the more permeable the membrane for a given ion, the more strongly the membrane potential tends to be driven toward the equilibrium value for that ion. Consequently, changes in a membrane's permeability to ions can cause significant changes in the membrane potential. This is one of the key principles relating the electrical excitability of cells to the activities of ion channels.

2.2.3 Active transport

Like the membrane proteins mediatedly facilitated diffusion, active transports are mediated by specific membrane proteins. The difference is that active transports move ions or molecules against their concentration gradient and need the energy supplied by ATP hydrolysis, or by light energy, or by tranporter coupled the uphill transport of one solute across the membrane to the downhill transport of another (Figure 2-19).

(1) **ATP-driven pumps** **ATP-driven pumps** are **ATPases** that use the energy of ATP hydrolysis to move ions or small molecules across a membrane against a chemical concentration gradient or electric potential gradient or both. All ATP-powered pumps are transmembrane proteins with one or more binding sites for ATP located on the cytosolic face of the membrane. Although these proteins commonly are called **ATPases**, they normally do not hydrolyze ATP into ADP and Pi unless ions or other molecules are simultaneously transported; therefore, the energy stored in the phosphoanhydride bond is rather used to move ions or other molecules uphill against an electrochemical gradient.

The general structures of the three classes of ATP-powered pumps are depicted in

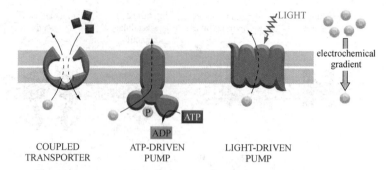

Figure 2-19 Three ways driving active transport (From: Alberts B, et al. Molecular Biology of the Cell. 2008)

figure 2-20. Note that the types P-, F-, and V-pumps transport ions only, whereas the ABC transporters transport small molecules primarily.

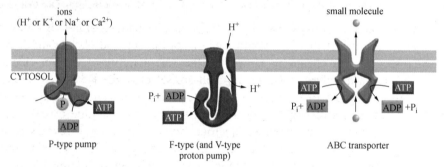

Figure 2-20 Three types of ATP-driven Pumps (From: Alberts B, et al. Molecular Biology of the Cell. 2008)

1) P-type ion pump: possesses two identical catalytic α subunits that contain an ATP-binding site. Most also have two smaller β subunits that usually have regulatory functions. During the transport process, at least one of the α subunits is phosphorylated (hence the name "P" type), and the transported ions are thought to move through the phosphorylated subunit.

This class includes the Na^+-K^+ ATPase in the plasma membrane. The pump operates as an ATP-driven antiporter, actively pumping each time 3 Na^+ out of the cell against its steep electrochemical gradient and pumping 2 K^+ in. Since three positively charged ions are pumped out of the cell and only two are pumped in, the Na^+-K^+ ATPase is electrogenic and maintains the low cytosolic Na^+ and high cytosolic K^+ concentrations in animal cells. It drives a net current across the membrane tending to create an electrical potential with the inside of the cell negative. The Na^+ gradient across the membrane established via the Na^+-K^+ ATPase drives the transport of most nutrients into cells and is also crucial in regulating cytosolic pH and osmolarity. Cells expend considerable energy to establish the ion gradients. For example, to transport ions, up to 25 percent of the ATP produced by nerve and kidney cells is consumed, and up to 50 percent of that in human erythrocytes; in both cases, most of the ATP is used to power the Na^+-K^+ pump.

Certain Ca^{2+} ATPases are P-type ion pumps and pump Ca^{2+} ions out of the cytosol into the external medium; others pump Ca^{2+} from the cytosol into the endoplasmic reticulum or into the sarcoplasmic reticulum, which is found in muscle cells. Another member of the P class, found in acid-secreting cells of the mammalian stomach, transports protons (H^+

ions) out of the cell and K$^+$ ions into the cell. The H$^+$ pump that maintains the membrane electric potential in plant, fungal, and bacterial cells also belongs to this class.

2) V-type and F-type ion pumps: are similar in structure which is more complicated than that of P-type pumps. V-and F-type pumps contain several different transmembrane and cytosolic subunits. All known V-and F-type pumps transport only protons, in a process that does not involve a phosphoprotein intermediate. V-type pumps generally function to maintain the low pH of plant vacuoles and of lysosomes and other acidic vesicles in animal cells by pumping protons from the cytosolic to the exoplasmic face of the membrane against a proton electrochemical gradient. F-type pumps are found in bacterial plasma membranes and in the inner membrane of mitochondria and chloroplasts. In contrast to V-type pumps, F-type pumps generally function to power the synthesis of ATP from ADP and Pi by movement of protons from the exoplasmic to the cytosolic face of the membrane down the proton electrochemical gradient.

3) ABC (ATP-binding cassette) transporters: referred to as the **ABC superfamily**, include several hundred different transport proteins found in organisms ranging from bacteria to humans. Each ABC protein is specific for a single substrate or a group of related substrates, which may be ions, sugars, amino acids, phospholipids, peptides, polysaccharides, or even proteins. All ABC transport proteins share a structural organization consisting of four "core" domains: two transmembrane domains, forming the passageway through which transported molecules cross the membrane, and two cytosolic ATP-binding domains. ATP binding leads to dimerization of the two ATP-binding domains, and ATP hydrolysis leads to their dissociation. Such structural changes in the cytosolic domains are thought to be transmitted to the transmembrane segments, driving conformational changes that alternately expose substrate-binding sites on one side of the membrane and then on the other. As a consequence, small molecules are transported across the bilayer.

ABC transporters constitute the largest family of membrane transport proteins. In *E. coli*, 78 genes (nearly 5% of the bacterium genes) encode ABC transporters and animal genomes encode more. The bacterial ABC transporters are used for both import and export, whereas those identified in eukaryotes seem mostly specialized for export. ABC transporters are of great clinical importance. The first eukaryotic ABC transporters were discovered because of their ability to pump hydrophobic drugs out of the cytosol. One of these transporters is the multidrugresistance (MDR) protein, the overexpression of which in human cancer cells can make the cells simultaneously resistant to a variety of chemically unrelated cytotoxic drugs that are widely used in cancer chemotherapy. Similar phenomenon occurs with the antimalarial drug *chloroquine*. The protist *Plasmodium falciparum* develops a drug resistance to *chloroquine* by amplifying a gene encoding an ABC transporter that pumps out the *chloroquine*.

(2) Cotransporters   Cotransporters mediate coupled reactions in which an energetically unfavorable reaction (i. e., uphill movement of molecules) is coupled to an energetically favorable reaction (see "symporter" and "antiporter" in figure 2-15). Cotransporters use the energy stored in an electrochemical gradient, which differs from ATP pumps that use energy directly from hydrolysis of ATP. Establishment of electrochemical gradients by Na$^+$-K$^+$ pumps or H$^+$ pumps consumes ATP, so, the energy supplied by ATP is indirectly utilized by cotransporter to operate the transport. The tight coupling between the transfer of two solutes allows these coupled transporters to harvest the energy stored in the electrochemical gradient of one solute to transport the other. This process sometimes is referred to as secondary active transport.

The cotransporters fall in two types, antiporters and symporters. Symporters, driven by ions, transfer molecules in or out of the cell or organelle across the membrane in the same direction. There are concurrent two binding sites on the surface of symporters, one for the ions and another for the molecules. For example, there are two types of glucose and amino acid carriers that enable gut epithelial cells to transfer glucose and amino acid across the gut lining: one is a Na^+-driven symporters that mediate the glucose and amino acid to enter the cells against their concentration gradient; another is a passive transport carrier protein mediating the glucose and amino acid out of the cells down the concentration gradient (Figure 2-21). In epithelial cells which absorb nutrients from the gut, transporters are nonuniformly distributed in the plasma membrane and thereby facilitate the transcellular transport of the absorbed nutrients across the epithelial cell layer into blood. Na^+-linked symporters located in the absorptive domain of the plasma membrane actively transport nutrients into the cell, building up concentration gradients for these solutes across the cell membrane. Na^+ independent transport proteins in the basal and lateral domain allow the nutrients to leave the cell passively down their concentration gradients.

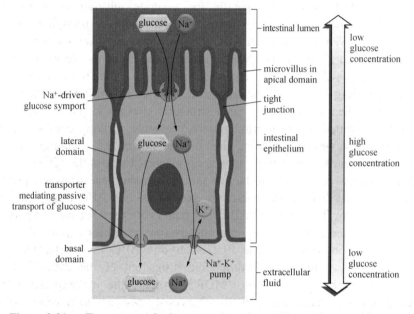

Figure 2-21　Two types of glucose and amino acid carriers enable gut epithelial cells to transfer glucose and amino acid across the gut lining (From: Alberts B, et al. Molecular Biology of the Cell. 2008)

Proteins within different intracelluar compartments require an optimal pH in order to function correctly. Lysosomal enzymes in lysosomes function best at the low pH (5), whereas cytosolic enzymes in the cytosol function best at neutral pH (7.2). It is therefore crucial that cells control the pH of their intracellular compartments. Most cells have one or more types of Na^+-driven antiporters in their plasma membrane that help to maintain the cytosolic pH at about 7.2. These transporters use the energy stored in the Na^+ gradient to pump out excess H^+, which either leaks in or is produced in the cell by acid-forming reactions. One of the antiporters that uses a Na^+-H^+ exchanger, which couples an influx of Na^+ to an efflux of H^+. Another is a Na^+-driven Cl^--HCO_3^- exchanger that couples an influx of Na^+ and HCO_3^- to an efflux of Cl^- and H^+.

There are similarities and differences in carrier-mediated solute symport in animal and plant cells. An electrochemical gradient of Na^+, generated by Na^+-K^+ pump, is often used to drive the active transport of solutes across the animal cell plasma membrane, whereas in plant, bacteria and fungi cells, an electrochemical gradient of H^+, usually set up by H^+ ATPase, is often used for this purpose. The electrochemical H^+ gradient drives the active transport of many sugars and amino acids across the plasma membrane and into bacterial cells. One well-studied H^+-driven symporter is lactose permease, which transports lactose across the plasma membrane of *E. coli*. The permease consists of 12 loosely packed transmembrane α helices. During the transport cycle, some of the helices undergo sliding motions that cause them to tilt. These motions alternately open and close a crevice between the helices, exposing the binding sites for lactose and H^+, first on one side of the membrane and then on the other.

2.3 Cell adhesion molecules and cell junction

In multicellular organisms, it is often necessary for adjacent cells within a tissue to be hold together. Cells in tissues can adhere directly to one another (cell-cell adhesion) through specialized integral membrane proteins called **cell-adhesion molecules** (**CAMs**) that often cluster into specialized cell junctions. Cells can also adhere indirectly (cell-matrix adhesion) through the binding of adhesion receptors in the plasma membrane to components of the surrounding **extracellular matrix** (**ECM**) that cells secreted into the spaces between them. Such connections create pathways for cell communication, allowing cells to exchange the signals that coordinate their functions. Attachments to other cells and to extracellular matrix control the orientation of the internal structure of each cell.

2.3.1 Cell adhesion molecules

CAMs can be classified into 4 main groups: **the cadherins, the integrins, the selectins** and **the immunoglobulin (Ig) superfamily** (Figure 2-22 and Table 2-2). A CAM on one cell can directly bind to the same kind of CAM on an adjacent cell (homophilic binding) or to a different class of CAM (heterophilic binding). CAMs can be broadly distributed along the plasma membranes that contact other cells or clustered in discrete patches or spots called cell junctions. Cell-cell adhesions can be tight and long lasting or relatively weak and transient. The associations between nerve cells in the spinal cord or the metabolic cells in the liver exhibit tight adhesions. In contrast, immune cells in the blood show only weak, short-lasting interactions, allowing them to roll along and pass through a blood vessel wall to fight an infection within a tissue.

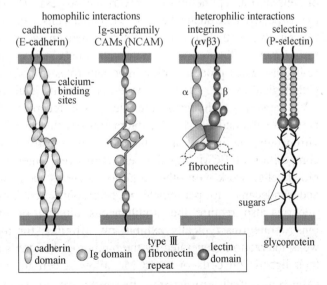

Figure 2-22 Major families of cell-adhesion molecules (CAMs) (From: Lodish H, et al. Molecular Cell Biology. 2008)

Table 2-2 Cell adhesion molecules

family	ligands recognized	stable cell junction
cadherins	homophilic interactions	adheren junctions and desmosomes
integrins	extracellular matrix	focal adhesions and hemidesmosomes
	members of Ig superfamily	no
selectins	carbohydrates	no
Ig superfamily	integrins	no
	homophilic interactions	no

(1) **Cadherins** Cadherins are key molecules in cell-cell adhesion and cell signaling, and play a critical role during tissue differentiation. Cadherins take their name from their dependence on Ca^{2+} ions (calcium-dependent adhesion). In structure, they share cadherin repeats in N-terminus, which are the extracellular Ca^{2+}-binding domains. The C-terminal cytosolic domain of cadherins is linked to the actin cytoskeleton by a number of cytosolic adapter proteins. These linkages are essential for strong adhesion. For example, disruption of the interactions between cadherins and two common adapter proteins (α- or β-catenin) that link these cadherins to actin filaments dramatically reduces cadherin mediated cell-cell adhesion. Three classical cadherin family members are closely related in sequence and are named according to the main tissues in which they were found: E-cadherin is present on many types of epithelial cells; N-cadherin on nerve, muscle, and lens cells; and P-cadherin on cells in the placenta and epidermis. There are also a large number of nonclassical cadherins with distantly related sequence.

Cadherins generally bind to one another homophilically: the head of one cadherin molecule binds to the head of a similar cadherin on an opposite cell. This selection enables mixed populations of different types of cells to sort out from one another according to the specific cadherins they express, and it helps to control cell rearrangement during development. Epithelial-mesenchymal transition (EMT) and mesenchymal-epithelial transition (MET) are typical examples concerning the function of E-cadherin. The assembly of cells into an epithelium is a reversible process. By switching on expression of cadherins, the dispersed and unattached mesenchymal cells can come together to form an epithelium. Conversely, epithelial cells can change their character, disassemble, and migrate away from their parent epithelium as separate individuals at least in part by inhibiting expression of the cadherins. Such processes play important roles in normal embryonic development, as well as pathological events, such carcinogenesis, during adult life.

(2) **Integrins** Integrins are membrane receptors that mediate mainly the attachment between cells and ECM. Via interactions with the ECM components, such as fibronectin, vitronectin, collagen, and laminin, inte grins pass information about the chemical composition of ECM into the cell. However, integrins, particularly expressed by certain blood cells, work with other adhesion proteins such as cadherins, Ig superfamily, selectins and syndecans, to participate in heterophilic cell-cell interactions. Ligands binding to integrins also require the simultaneous binding of divalent cations (positively charged ions). Integrins typically exhibit low affinities for their ligands. However, the multiple weak interactions generated by the binding of hundreds or thousands of integrin molecules to their ligands on cells or in ECM, allow a cell to remain firmly anchored to its ligand-expressing target. Moreover, the weakness of individual integrin-mediated interactions facilitates cell migration.

Integrins are obligate heterodimers containing two distinct chains, α and β subunits.

The α and β subunits each penetrate the plasma membrane and possess small cytoplasmic domains. Like other cell-surface adhesive molecules, the cytosolic region of integrins interacts with adapter proteins that in turn bind to the cytoskeleton and intracellular signaling molecules. In addition to their adhesion function, integrins can mediate signaling, involved in processes as diverse as cell survival, cell proliferation, and programmed cell death. The engagement of integrins with their extracellular ligands can influence the cytoskeleton and intracellular signaling pathways through adapter proteins bound to the integrin cytosolic region. Conversely, intracellular signaling pathways can alter, from the cytoplasm, the structure of integrins and consequently their abilities to adhere to their extracellular ligands and regulate cell-cell and cell-matrix interactions.

(3) Selectins Selectins are expressed on white blood cells, blood platelets, and endothelial cells. They bind heterophilically to carbohydrate groups on cell surfaces, mediate transient cell-cell adhesions in the bloodstream and control the binding of white blood cells to the endothelial cells lining blood vessels, thereby enabling the blood cells to migrate out of the bloodstream into sites of inflammation.

Selectins contain a conserved Ca^{2+}-dependent lectin domain, which is located at the distal end of the extracellular region of the molecule and recognizes oligosaccharides in glycoproteins or glycolipids. There are at least three types: L-selectin on leukocytes (white blood cells), P-selectin on platelets and endothelial cells that can be locally activated by inflammatory responses, and E-selectin on activated endothelial cells. Selectins do not act alone, but collaborate with integrins, which strengthen the binding of the blood cells to the endothelium. The selectins mediate a weak adhesion allowing the white blood cell rolling along the surface of the blood vessel until the blood cell activates its integrins.

(4) Ig superfamily (IgSF) Numerous transmembrane proteins characterized by the presence of multiple immunoglobulin domains in their extracellular regions constitute the IgSF of CAMs. Such proteins mediate Ca^{2+}-independent cell-cell adhesion both homophilically and heterophilically. ICAMs (intercellular cell adhesion molecules) or VCAMs (vascular cell adhesion molecules) on endothelial cells both mediate heterophilic binding to integrins, another IgSF member NCAM (neural cell adhesion molecule) expressed on neurons mediates homophilic binding. Most IgSF mediate the specific interactions of lymphocytes with cells required for an immune response (e.g., macrophages, other lymphocytes, and target cells). However, some IgSF members, such as VCAM and NCAM mediate adhesion between nonimmune cells. Molecules such as NCAM seem to contribute more to the fine tuning of these adhesive interactions during development and regeneration, playing a role in various specialized adhesive phenomena, such as for blood and endothelial cells.

2.3.2 Cell junctions

The physical attachment between cell and cell or between cell and matrix are diverse in structure. There are mainly three types of junctions: **tight junctions**, **anchoring junctions** (including **adherens junction**, **focal adhesion**, **desmosome**, **hemidesmosome**) and **gap junctions** (**plasmodesmata**, a specified gap junction in plant cells). These cell junctions not only transmit physical forces, but also define the diverse functions of different tissues (Figure 2-23).

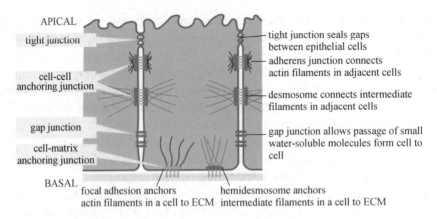

Figure 2-23 Typical cell junctions found in vertebrate epithelial tissues
(From: Alberts B, et al. Molecular Biology of the Cell. 2008)

(1) **Tight junction** Tight junctions are the closest known contacts between adjacent cells which are found wherever and are particularly common in epithelial cells such as those lining the small intestine. They are formed by interactions between strands of transmembrane proteins (occludin and claudins) on adjacent cells that continue around the entire circumference of the cell. The adjacent cells are pressed together so tight that no intercellular space left (Figure 2-24).

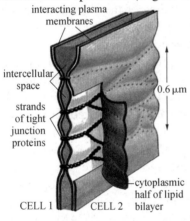

Tight junctions play at least three main roles: ①Form seals that prevent the free passage of molecules between the cells of epithelia; ② Prevent leakage of molecules across the epithelium though the gaps between cells; ③Separate the apical and basolateral domains of plasma membrane by preventing the free diffusion of lipids and proteins between them, and help establishing and maintaining cell polarity.

(2) **Anchoring junction** Anchoring junction anchors cells together allowing the tissue to be stretched without tearing. There are four main types: adherens junction, desmosome junctions binding one epithelial cell to another, while focal adhesions and hemidesmosomes binding cells to the ECM.

Figure 2-24 A structure model of tight junction (From: Alberts B, et al. Molecular Biology of the Cell. 2008)

1) **Cell to cell anchoring junction**: The adherens junctions and desmosome are cell-cell junctions in which cadherins are linked to actin bundles and intermediate filaments, respectively (Figure 2-25). At an adherens junction, cadherin molecules bind to actin filaments via several linker proteins (catenin α, β) and form a continuous adhesion belt between each of the interacting cells. This belt is close beneath the apical face of the epithelium. The potentially contractile nature of actin network provides the motile force for folding of epithelial cell sheets into tubes, vesicles and other related structures. Such epithelial movement is very important in embryonic development, such as the formation of neural tube structures which later develop into the central nervous system.

Desmosome junctions confer great tensile strength to the epithelia. Desmosomes contain two specialized cadherin proteins, desmoglein and desmocollin, whose cytosolic domains are distinct from those in the classical cadherins. The cytosolic domains of desmosomal cadherins interact with plakoglobin (similar in structure to β-catenin) and the plakophilins. These adapter proteins, whi ch form the thick cytoplasmic plaques characteristic of desmosomes, in turn interact with intermediate filaments.

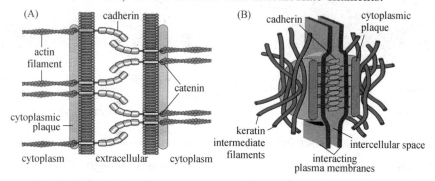

Figure 2-25　Structure components of cell to cell anchoring junctions. (A) Adherens junctions. (B) Desmosomes

2) Cell-matrix anchoring junction: There are two types of cell-matrix anchoring junctions, focal adhesion and hemidesmosomes, which are established by interactions between integrins and the cytoskeleton. Hemidesmosome junctions anchor the keratin intermediate filaments in an epithelial cell to the extracellular basal lamina through intergrin proteins in the basal plasma membrane of the epithelial cells. Such structures, just like half a desmosome, can attach and anchor the epithelial cells to the underlying tissue. In focal adhesions, integrins connect the extracellular matrix and cytoskeletal actin filaments mainly in nonepithelial cells and transduce adhesion-dependent signals for cell growth and cell motility.

(3) Gap junction　Gap junctions bridge between adjacent cells so as to created direct connection between the cytoplasm of one with that of the other.

1) Gap junction in animal tissues: In animal tissues, gap junctions are constructed by four-pass transmembrane proteins called connexins. Six connexins assemble to form a connexon with an open hydrophilic pore in its center. A connexon in one cell then aligns with the connexon in an adjacent cell forming a channel between the two cell interiors that connects the two cytoplasms (Figure 2-26A). Most cells in animal tissues-including epithelial cells, endothelial cells, smooth muscle and heart muscle cells communicate by gap junctions. Gap junctions allow ions and small molecules (molecular mass less than 1,000 daltons) including signal molecules cAMP and Ca^{2+} to pass freely between adjacent cells but prevent the passage of macromolecules, such as proteins, nucleic acids and polysaccharides. Gap junctions create metabolic and electrical coupling between the cells. For example, between heart muscle cells, gap junctions provide coupling that allows electrical signal of excitation to spread through the tissue, triggering coordinated contraction of the heart cells.

2) Plant cell adhesion and plasmodesmata: Plant cells are imprisoned within rigid cell walls composed of an extracellular matrix rich in cellulose and other polysaccharides. Adhesion between plant cells is mediated by their cell walls. A specialized pectin-rich region of the cell wall called the **middle lamella** acts as mucilage to hold adjacent cells together. Plant cells have only one class of intercellular junctions, plasmodesmata (Figure 2-26B), which function analogously to animal cell gap junctions.

At each plasmodesma, the plasma membrane of one cell is continuous with that of the adjacent cell creating an open channel between the two cytoplasms. An extension of the smooth endoplasmic reticulum passed through the center of channel is a narrower cylindrical structure,

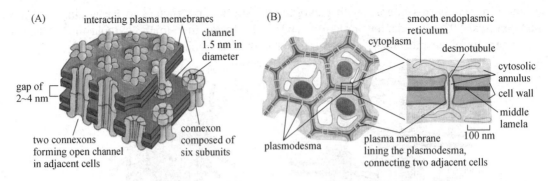

Figure 2-26 Structure models of gap junctions. (A) Gap junctions in animal cells. (B) Plasmodesma in plant cells.

the desmotubule, leaving a ring of surrounding cytoplasm through which ions and small molecules (molecular mass less than 800 daltons) can pass from cell to cell. In addition, plasmodesma can expand in response to appropriate signals, permitting the regulated passage of macromolecules, such as mRNA and virus RNA, between neighboring cells.

2.4 Extracellular matrix and cell wall

The extracellular matrix (ECM) is a complex meshwork of proteins and polysaccharides secreted by cells into the spaces between them. The ECM plays important roles in cell-cell signaling, wound repair, cell adhesion and tissue function.

2.4.1 Composition of the ECM

In plants, the ECM (cell wall) is primarily composed of cellulose surround each cell and is more important in cell organization which gives strength to the plants. In arthropods and fungi, the ECM is largely composed of chitin. In animal cells, the relative volumes of cells versus matrix vary greatly among different animal tissues and organs. In connective tissues, such as bone, cartilage or tendon, the extracellular matrix is plentiful and cells are bound to the matrix and sparsely distributed within it. However, many organs are composed of very densely packed cells with relatively little matrix. All epithelial cells as well as some other types (e.g., smooth muscle cells) are attached to a thin basal lamina (also known as the basement membrane, a specialized ECM structure).

In vertebrates, the ECM is made of three abundant components: **collagens**, structure proteins that often form fibers and provide mechanical strength and resilience; **proteoglycans**, a unique type of glycoproteins that cushion cells and bind a wide variety of extracellular molecules; and soluble **multiadhesive matrix proteins** (e.g., fibronectin and laminin), which bind to and cross-link cell-surface adhesion receptors and other ECM components.

(1) Structure proteins   Collagens and elastins are the main structure proteins in ECM and serve as a structural scaffold. Collagens are the most abundant proteins in mammals and make up about 25% to 35% of the whole-body protein content. They have wide variety of types. Bone utilizes type Ⅰ collagen (also the major collagen in skin), cartilage utilizes type Ⅱ, while basement membrane utilizes type Ⅳ. Collagens Ⅰ and Ⅱ form long fibrillar arrays, while type Ⅳ forms a mesh-type structure, reminiscent of a chain-link fence.

Collagens are trimeric proteins made from three polypeptides called collagen α chains which are wrapped around one another to form a triple-stranded helical rod at approximately 300 nm long and 1.5 nm in diameter (Figure 2-27). The collagen triple helix can form because of an unusual abundance of three amino acids: glycine, proline, and a modified form of proline called hydroxyproline. They make up the characteristic repeating motif Gly-X-Y. X and Y can be any amino acid but are often proline and hydroxyproline and less often lysine and hydroxylysine. Collagen molecules assemble into higher-order polymers called collagen fibrils, which are thin structures (10 ~ 300 nm in diameter) with several hundreds of micrometers long. Collagen fibrils have characteristic cross-striations every 67 nm, reflecting the regularly staggered packing of the individual collagen molecules in the fibril. After the fibrils have formed in the extracellular

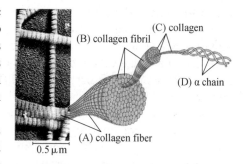

Figure 2-27 Structure of collagens

space, they are greatly strengthened by the formation of covalent cross-links between lysine residues of the constituent collagen molecules. Collagen fibrils often aggregated into collagen fibers several micrometers in diameter which are visible as cablelike bundles under the light microscope.

Elastins are ECM structure proteins and present in connective tissues. They are rich in hydrophobic amino acids such as glycine and proline, which form mobile hydrophobic regions bounded by crosslinks between lysine residues. These proteins are elastic and provide flexibility to many tissues such as skin, arteries, and lungs.

(2) **Proteoglycans**   Proteoglycans are also glycoproteins but consist of much more carbohydrate than protein. They are huge clusters of carbohydrate chains, glycosaminoglycans (GAGs), often attached covalently to a protein backbone. One GAG is composed of a repeating disaccharide (chondroitin sulfate, keratan sulfate, heparan sulfate, hyaluronic acid) with a -A-B-A-B-A-structure (A and B represent two different sugars) and is highly acidic due to the presence of both sulfate and carboxyl groups attached to the sugar rings. Proteoglycans can be assembled into huge complexes by linkage of the core proteins to a nonsulfated GAG, hyaluronic acid. The sulfated GAGs are negatively charged and bind huge numbers of cations, which in turn bind large numbers of water molecules. Therefore, proteoglycans form a porous, hydrated gel like packing material filling the extracellular space constructed by the adjacent collagen molecules and buffer compression forces. Together, collagens and proteoglycans give ECM strength and resistance to deformation.

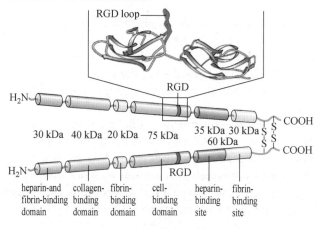

Figure 2-28 Structure of a human fibronectin

(3) **Multiadhesive matrix proteins**   Fibronectins are abundant multiadhesive matrix proteins that play a key role in migration and cellular differentiation. A human fibronectin molecule consists of two similar, but nonidentical, polypeptides joined by a pair of disulfide bonds located near the C-termini (Figure 2-28). Each polypeptide is composed of a linear series of distinct modules, which are organized into several larger functional units. Each unit possesses binding sites for numerous components of the ECM such as collagens, proteoglycans, and other fibronectin. Through these binding sites fi-

bronectins interact and link these diverse molecules into a stable, interconnected network. Fibronectins also possess binding sites, the tripeptide RGD sequence (Arg-Gly-Asp), for integrins which are ECM receptors located on the cell plasma membrane.

Laminins are a family of multiadhesive proteins in the basal lamina, a specialized ECM structure. Laminins consist of three different polypeptide chains linked by disulfide bonds and organized into a molecule resembling a cross with three short arms and one long arm (Figure 2-29). Like fibronectin, extracellular laminins can bind to integrins and greatly influence a cell's potential for migration, growth, and differentiation. Laminins also bind to type IV collagen, proteoglycans, other laminin molecules, and other components of basement membranes to form a fibrous two-dimensional network. These networks give basement membranes both strength and flexibility.

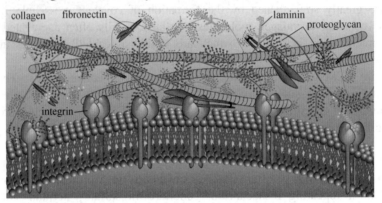

Figure 2-29 Connection between cells and the ECM

2.4.2 Connecting cells to the ECM

Most normal vertebrate cells cannot survive unless they are anchored to the extracellular matrix. This anchorage dependence is often lost when a cell turns malignant. The components of the ECM, such as fibronectin, laminin, proteoglycans, and collagen are capable of binding to receptors situated on the cell surface (Figure 2-29). The integrin is the most important family of receptors that attach cells to their extracellular microenvironment.

Integrins are part of a large family of cell adhesion receptors which are involved in cell-ECM and cell-cell interactions. Integrins play a key structural role in cells, binding to various types of ECM proteins via their extracellular domain and to the actin filaments of the cytoskeleton via their intracellular portion. In addition, integrins are the main way that cells both bind to and respond to the extracellular matrix and therefore are important initiators and modulators of signal-transduction processes.

The basement membrane is a specialized ECM structure as a thin sheet of fibers that underlies the epithelium, which lines the cavities and surfaces of organs including skin, or the endothelium, which lines the interior surface of blood vessels. Most of the ECM components in the basal lamina are synthesized by the cells that rest on it. Four ubiquitous protein components are found in basal lamina: laminins, type IV collagen, entactin (also called nidogen) and perlecan. Type IV collagens are trimeric molecules with both rod like and globular domains that form a two-dimensional network. Entactin is a rod like molecule that cross-links type IV collagen and laminin and helps incorporate other components into the ECM. Perlecan is a large multidomain proteoglycan that binds to and cross-links many

ECM components and cell-surface molecules. The primary function of the basement membrane is to anchor the epithelium to its loose connective tissue underneath. As a mechanical barrier, the basement membrane prevents malignant cells from invading the deeper tissues. Early stages of malignancy that are thus limited to the epithelial layer by the basement membrane are called carcinoma in situ.

2.4.3 Cell wall

Prokaryotic cells and plant cells both have a rigid cell wall made up of polysaccharides. The cell wall provides and maintains the shape of these cells, and serves as a protective barrier.

(1) **The peptidoglycan cell wall** All bacteria except mycoplasmas have a cell wall containing peptidoglycan, also called murein. Peptidoglycan is a vast polymer consisting of sugars and amino acids that forms a mesh-like layer outside the plasma membrane of bacteria (but not archaea). A peptidoglycan monomer consists of two joined amino sugars, N-acetylglucosamine (NAG) and N-acetylmuramic acid (NAM). Peptidoglycan serves a structural role in the bacterial cell wall, giving structural strength, as well as counteracting the osmotic pressure of the cytoplasm. Peptidoglycan is also involved in binary fission during bacterial cell reproduction. The peptidoglycan layer is substantially thicker in Gram-positive bacteria (20 to 80 nm) than in Gram-negative bacteria (7 to 8 nm) and the presence of high levels of peptidoglycan is the primary determinant of the characterization of bacteria as gram-positive. Some anti-bacterial drugs such as penicillin interfere with the production of peptidoglycan by binding to bacterial enzymes known as penicillin-binding proteins or transpeptidases.

(2) **The plant cell wall** The plant cell walls were the first cellular structures to be observed with a light microscope that performing numerous vital functions. The cell wall is located outside the plasma membrane and gives the enclosed cell its characteristic polyhedral shape. Cell walls protect the cell against damage from mechanical abrasion and pathogens, and serve collectively as a type of "skeleton" for the entire plant. Like the ECM at the surface of an animal cell, a plant cell wall mediates cell-cell interactions and is a source of signals that alter the activities of the cells that it contacts.

Cell wall is a secretion of the cell with sugar as 90% of its composition. Cellulose is the most abundant organic molecule providing the fibrous component of the cell wall, and proteins and pectin provide the matrix. Like starch, cellulose is a polysaccharide with glucose as its monomer. The matrix of the cell wall is composed of three types of macromolecules: ①Hemicelluloses are branched polysaccharides whose backbone consists of one sugar, such as glucose, and side chains of other sugars, such as xylose. Hemicellulose bind to the surfaces of cellulose microfibrils, cross-linking them into a resilient structural network. ②Pectins are a heterogeneous class of negatively charged polysaccharides containing galacturonic acid. Like the GAGs in ECM, pectins hold water and thus form an extensive hydrated gel that fills in the spaces between the fibrous elements.

Cell walls arise as a thin cell plate that forms between the plasma membranes of newly formed daughter cells following cell division. In addition to providing mechanical support and protection from foreign agents, the cell wall of a young, undifferentiated plant cell must be able to grow in conjunction with the enormous growth of the cell it surrounds. The walls of growing cells are called primary walls, and they possess an extensibility that is lacking in the thicker secondary walls present around many mature plant cells. The transformation from primary to secondary cell wall occurs as the wall increases in cellulose content and, in most cases, incorporates a phenol-containing polymer called lignin. Lignin provides structural support and is the major component of wood. The lignin in the walls of water-conducting cells of the xylem provides the support required to move water through the plant.

Summary

All cells are surrounded by cell membrane. Prokaryotes cells contain no internal membrane limited subcompartments, while in the eukaryotes a cell was partitioned into smaller subcompartments termed organelles, such as endoplasmic reticulum, vesicle, Golgi body, mitochondria, and chloroplasts. Each organelle is surrounded by one or more membrane. All this membranes are formed by biomembrane. The basic compositions of all biomembranes are lipids, proteins and a small quantity of saccharide. The most important character of biomembrane is selective permeable (semipermeability), that makes the biomembrane to form a barrier between the cell and the extracellular environment, or between inside and outside of organelles. This barrier controls the movement of molecules or ions between the two sides of the membrane.

The cell-cell interactions are mediated by four types of cell adhesion proteins, cadherins, integrins, selectins and Ig like CAMs. The cadherins link the cytoskeletons of adjacent cells at stable cell-cell junctions, and integrins link cell-matrix junctions. There are three classes of junctions in animal cells: tight junction, gap junction and anchoring junction. Plant cells have only one class of intercellular junctions — plasmodesmata. The extracellular matrix (ECM) is the material found around cells. Biochemically the ECM of vertebrates is composed of complex mixtures of proteins, proteoglycans and in the case of bone, mineral deposits. The interaction between a cell and ECM maintains the normal growth and response of the cell.

Questions

1. When viewed by electron microscopy, the lipid bilayer is often described as looking like a railroad track. Explain how the structure of the bilayer creates this image.
2. Biomembranes contain many different types of lipid molecules. What are the three main types of lipid molecules found in biomembranes? How are the three types similar, and how are they different?
3. Lipid bilayer is considered to be two-dimensional fluids. What does this mean? What drives the movement of lipid molecules and proteins within the bilayer? How can such movement be measured? What factors affect the degree of membrane fluidity?
4. Explain the following statement: The structure of all biomembranes depends on the chemical properties of phospholipids, whereas the function of each specific biomembrane depends on the specific proteins associated with that membrane.
5. Name the three groups into which membrane-associated proteins may be classified. Explain the mechanism by which each group associates with a biomembrane.
6. Compare the features distinguishing active transport from passive transport, uniport transport from passive diffusion.
7. Compare the arrangement of integral membrane proteins in a tight junction versus a gap junction.
8. Which types of cell junctions contain actin filaments? Which contain intermediate filaments? Which contain integrins? Which contain cadherins?
9. How are integrins able to link the cell surface with materials that make up the ECM? What is the significance of the presence of an RGD motif in an integrin ligand?
10. What are the roles of each component that makes up a plant cell wall?

(张 儒 刘治学)

Chapter 3

Cytoplasm, ribosomes and RNAs

All cells have four things in common. They all have a cell membrane, DNA and RNA, cytoplasm and ribosomes. They are all very important in the function of the cell. Ribosomes are scattered in the cytoplasm or attached to the endoplasmic reticulum in cytoplasm.

3.1 Structure and functions of cytoplasm

Cytoplasm is the thick, gel-like semitransparent fluid that is found in both plant and animal cell. It surrounds the nucleus of a cell and is bounded by the cell membrane. Cytoplasm was discovered in 1835. Discovery of different organelles found in cytoplasm can be attributed to different scientists, but no single scientist can be credited for discovering cytoplasm.

3.1.1 Structure of cytoplasm

Cytoplasm contains about 80% water and is usually clear in color. It is more like a viscous gel than a watery substance, but it liquefies when shaken or stirred. Cytoplasm is the substance of life that serves as a molecular soup in which all of the cell's organelles are suspended and held together by a fatty membrane.

Cytoplasm is formerly referred to as protoplasm. It contains three parts: the cytosol, the organelles and the cytoplasmic inclusions. The most basic part of cytoplasm is the the cytosol, and it takes up most of the space of the cell.

(1) Cytosol   The cytosol is the portion not within membrane-bound organelles, and is a gel, with a network of fibers dispersed through water. It is a complex mixture of cytoskeleton filaments, ribosome, dissolved molecules, and water that fills much of the volume of a cell. The inner, granular mass is referred to as the endoplasm and the outer, clear and glassy layer is called the cell cortex or the ectoplasm. Cytosol accounts for almost 70% of the total cell volume.

The dissolved molecules in cytosol include dissolved nutrients and waste products, including small organic molecules, such as sugars, amino acids, fatty acid. It also contains dissolved macromolecules, such as RNAs and proteins that the cell uses for growth and reproduction. The cytosol also contains many salts and is an excellent conductor of electricity, which therefore creates a medium for the vesicles, or mechanics of the cell.

The cytosol functions to suspend and to hold into place the organelles within the cell. The cytosol helps materials move around the cell by moving and churning through a process called cytoplasmic streaming.

(2) Organelles   The organelles are microscopic semi-organs that facilitate a number of im-

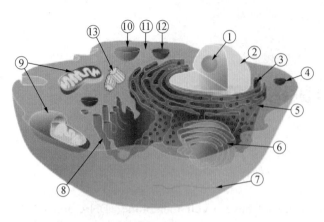

Figure 3-1 Schematic showing the cytoplasm, with major organelles of a typical animal cell. ①nucleolus; ②nucleus; ③ribosomes; ④vesicle; ⑤rough endoplasmic reticulum (ER); ⑥Golgi apparatus; ⑦cytoskeleton; ⑧smooth ER; ⑨mitochondria; ⑩vacuole; ⑪cytosol; ⑫lysosome; ⑬centrioles within centrosome.

portant metabolic reactions, such as synthesizing or breaking down macromolecules, producing energy and cell mitosis or meiosis. The major membrane-bounded organelles that are suspended in the cytosol are Golgi bodies, mitochondria, endoplasmic reticulum, vacuoles, lysosomes, and chloroplasts in plant cells (Figure 3-1), each organelle is bounded by a fatty membrane, and has some specific functions. There are also non-membrane structures including cytoskeleton and ribosome suspended in the cytosol. Without the function of these organelles, cells would wither and die, and life would not be possible.

Cytoplasm is the home of the cytoskeleton, a network of cytoplasmic filaments that are responsible for the movement of the cell and give the cell its shape. The cytoplasm, as seen through an electron microscope, appears as a three-dimensional lattice of thin protein-rich strands. These lattices serve to interconnect and support the other organelles or macromoleculars in the cytoplasm. All the contents of the cells of prokaryotes are contained within the cytoplasm. Within the cells of eukaryotes the contents of the cell nucleus are separated from the cytoplasm, which is called the nucleoplasm.

(3) **Cytoplasmic inclusions**　The inclusions are small particles of insoluble substances suspended in the cytosol. For example, proteasomes exists inside all eukaryotes and archaea, and in some bacteria.

A huge range of inclusions exist in different cell types, and range from crystals of calcium oxalate or silicon dioxide in plants, to granules of starch, glycogen, or polyhydroxybutyrate, which can store energy. Lipid droplets are spherical droplets composed of lipids and proteins that are used in both prokaryotes and eukaryotes as a way of storing fatty acids and sterols. Lipid droplets make up much of the volume of adipocytes, which are specialized lipid-storage cells, but they are also found in a range of other cell types.

3.1.2 Functions of cytoplasm

In general, the function of cytoplasm is mostly mechanical in nature. It provides support to the internal structures by being a medium for their suspension. It maintains the shape and consistency of the cell. It stores many chemicals that are inevitable for life. It is in this cytoplasm, that vital metabolic reactions, like, anaerobic glycolysis and protein synthesis, take place.

(1) **Metabolic pathways and biochemical reactions**　Cytoplasm is the site of many vital biochemical reactions crucial for maintaining life. It is within the cytoplasm that many of the most basic, and most important, facets of biology take place, such as many metabolic

pathways including glycolysis, and processes such as cell division, expansion and growth.

Cytoplasm provides a medium in which the organelles can remain suspended. Besides, cytoskeleton found in cytoplasm gives the shape to the cell, and facilitates its movement. It also assists the movement of different elements found within the cell. Movement of calcium ions in and out of the cytoplasm is thought to be a signaling activity for metabolic processes.

The enzymes found in the cytoplasm breaks down the macromolecules into small parts so that it can be easily used by the other organelles like mitochondria. For example, mitochondria cannot use glucose present in the cell, unless it is broken down by the enzymes into pyruvate. They act as catalysts in glycolysis, as well as in the synthesis of fatty acid, sugar and amino acid. Cell reproduction, protein synthesis, anaerobic glycolysis, cytokinesis are some other vital functions that are carried out in cytoplasm.

(2) **Ubiquitin-proteasome system**   The proteasomes included in cytoplasm, is a cylindrical complex containing a "core" of four stacked rings around a central pore. Each ring is composed of seven individual proteins. The inner two rings are made of seven β subunits that contain three to seven protease active sites. The outer two rings each contain seven α subunits whose function is to maintain a "gate" through which proteins enter the barrel. Proteins are tagged for degradation with a small protein called ubiquitin. The tagging reaction is catalyzed by enzymes called ubiquitin ligases. Once a protein is tagged with a single ubiquitin molecule, this is a signal to other ligases to attach additional ubiquitin molecules. The α subunits of proteasome are controlled by binding to "cap" structures or regulatory particles that recognize polyubiquitin tags attached to protein substrates. The overall system of ubiquitination and proteasomal degradation is known as the ubiquitin-proteasome system.

The main function of the proteasome is to degrade unneeded or damaged proteins by proteolysis, a chemical reaction that breaks peptide bonds. Enzymes that carry out such reactions are called proteases. Proteasomes are part of a major mechanism by which cells regulate the concentration of particular proteins and degrade misfolded proteins. The degradation process yields peptides of about seven to eight amino acids long, which can then be further degraded into amino acids and used in synthesizing new proteins.

The proteasomal degradation pathway is essential for many cellular processes, including the cell cycle, the regulation of gene expression, and responses to oxidative stress. The importance of proteolytic degradation inside cells and the role of ubiquitin in proteolytic pathways was acknowledged in the award of the 2004 Nobel Prize in Chemistry to Aaron Ciechanover, Avram Hershko and Irwin Rose.

(3) **The functions of organelles**   Organelles present in cytoplasm have their specific vital functions; for example, mitochondria produce and store energy, while endoplasmic reticulum facilitates synthesis and transport of protein, production of steroids, production as well as storage of glycogen, etc. The functions of Golgi apparatus include modification, packaging, transportation and processing of macromolecules, such as proteins, and lipids. Lysosomes contain digestive enzymes and hence, they digest food particles, damaged or worn-out organelles and also virus and bacteria. However, the smooth operation of all these functions depend on the existence of cytoplasm, as it provides the medium for carrying out these vital processes.

(4) **Maintenance and movement**   One of the most important functions of cell cytoplasm is to maintain the shape of the cell and suspension of organelles. Another function that cytoplasm sometimes involves is allowing cell movement. By squeezing organelles to a particular part of the cell, cytoplasm can cause the cell to move within the blood flow. In

human, this allows white blood cells to get the parts of the body where they need to be in order to operate. In basic organisms, such as the amoeba, this provides their only means of transportation. In plants, this process, called cytoplasmic streaming, allows them to optimize cell organelles for the collection of sunlight necessary for photosynthesis.

(5) Storage   Cytoplasm acts as a storage space for the chemical building blocks of the body, storing protein, lipid, oxygen and other substances until they can be used by the organelles, and storing the waste byproducts of metabolic reactions, such as carbon, until they can be disposed of. These stores are the cytoplasmic inclusions.

3.2 Ribosome

The ribosome is a ribonucleoprotein particle composed of RNA and protein, and serves as the site of protein synthesis. Specifically, the ribosome carries out the process of translation, decoding the genetic information encoded in messenger RNA, one amino acid at a time, into newly synthesized polypeptide chains.

Ribosomes were found in the cytoplasm of prokaryotic and eukaryotic cells. Some ribosomes occur freely in the cytosol whereas others are attached to the nuclear membrane or to the endoplasmic reticulum(ER) giving the latter a rough appearance, hence, the name rough ER(rER).

Ribosomes are sometimes referred to as organelles, but the use of the term organelle is often restricted to describing sub-cellular components that include a phospholipid membrane, which ribosomes, being entirely particulate. For this reason, ribosomes may sometimes be described as "non-membranous organelles".

The structure and function of the ribosomes and associated molecules, known as the translational apparatus, has been of research interest since the mid-twentieth century and is still a very active research field today. In mid-1950s, ribosomes were first observed as dense particles or granules by George Palade using electron microscope. The term "ribosome" was proposed by scientist Richard B. Roberts in 1958 to designate ribonucleoprotein particles in sizes ranging from 35 to 100 S. The word ribosome comes from ribo nucleic acid and the Greek: soma (meaning body). For his discovery, George won a Nobel Prize with Albert Claude and Christian de Duve in 1974. Venkatraman Ramakrishnan, Thomas A. Steitz and Ada E. Yonath have been awarded the the 2009 Nobel Prize in Chemistry for their landmark work revealing the detailed structure and mechanism of the ribosome.

3.2.1 Ribosome components

Ribosomes are roughly spherical. With a diameter of ~20 nm, they can be seen only with the electron microscope. They can make up 25% of the dry weight of cells (e.g., pancreas cells) that specialize in protein synthesis(A single pancreas cell can synthesize 5 million molecules of protein per minute).

Each ribosome consists of two subunits of many proteins and RNAs of substantial length. The smaller subunit binds to the mRNA pattern, while the formation of peptide bonds occurs in the large subunit where the acceptor-stems of the tRNAs are docked. When a ribosome finishes reading an mRNA molecule, these two subunits split apart.

Ribosomes from bacteria, archaea and eukaryotes differ in their size, composition and the ratio of protein to RNA. Prokaryotic ribosomes are around 20 nm in diameter and are composed of 65% ribosomal RNA and 35% ribosomal proteins. Eukaryotic ribosomes are between 25 and 30 nm in diameter and the ratio of rRNA to protein is about 1. The unit of measurement is the Svedberg unit, a measure of the rate at which the particles are spun down in the ultracentrifuge, rather than size and accounts for why fragment names do not add up (70 S is made of 50 S and 30 S).

(1) **Ribosomes in prokaryotes**   Prokaryotes have 70 S ribosomes, each consisting of a small and a large subunit. The small subunit of the prokaryotic ribosome sediments at 30 S. It is composed of has a 16 S rRNA subunit (consisting of ~1,540 nucleotides, ~500 kDa) bound to 21 proteins(S1 ~ S21). The large subunit of the prokaryotic ribosome sediments at 50 S. It is composed of of a 5 S rRNA subunit (~120 nucleotides, 39 kDa), a 23 S rRNA subunit (~2,900 nucleotides, 946 kDa) and 34 proteins(L1 ~ L34)(Figure 3-2).

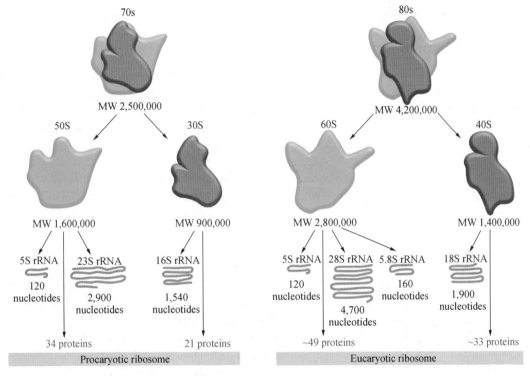

Figure 3-2 **Composition of ribosomes of prokaryote and eukaryote** (From: Alberts B, et al. 2008)

Other macromolecules in a functioning ribosome include three transfer RNA molecules, messenger RNA, and the nascent protein chain. The total molecular mass is several million Daltons.

(2) **Ribosomes in eukaryotes**   The ribosome of eukaryote sediments at 80 S with two subunits sediment at 40 S and 60 S, it is generally about another million Daltons larger than the prokaryotic one(Figure 3-2). The 40 S subunit has an 18 S rRNA (1,900 nucleotides) and 33 proteins. The large subunit is composed of a 5 S rRNA (120 nucleotides), 28 S rRNA (4,700 nucleotides), a 5.8 S rRNA (160 nucleotides) subunits and about 49 proteins.

(3) **Comparison of ribosomes in prokaryotes, eukaryotes, and mitochondria**   The ribosomes found in chloroplasts and mitochondria of eukaryotes also consist of large and small subunits bound together with proteins into one 70 S particle, which are similar to those of bacteria(Table 3-1).

Table 3-1 Comparison of ribosome structure in bacteria, eukaryotes, and mitochondria

	Bacterial (70 S)	Eukaryotic (80 S)	Mitochondrial (55 S)
Large subunit	50 S	60 S	39 S
rRNAs(1 of each)	23 S (2,900 nts)	25 S (4,700 nts)	16 S (1,560 nts)
	5 S (120 nts)	5 S (120 nts)	
		5.8 S (160 nts)	
Proteins	34	49	48
Small subunit	30 S	40 S	28 S
rRNA	16 S (1,542 nts)	18 S (1,900 nts)	12 S (950 nts)
Proteins	21	33	29

3.2.2 Ribosome structure

Despite differences in components, the basic operations of bacterial, eukaryotic, and mitochondrial ribosomes are very similar. Much of the RNA is highly organized into various tertiary structural motifs, for example pseudoknots that exhibit coaxial stacking (Figure 3-3). The extra RNA in the larger ribosomes is in several long continuous insertions, such that they form loops out of the core structure without disrupting or changing it.

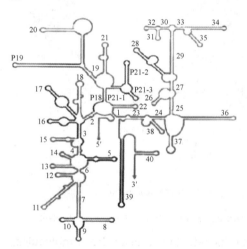

Figure 3-3 The second structure of E. coli 16 S rRNA (From: Cited Kleinsmith, et al. 1995)

The small subunit of the ribosome is the main site of decoding, directing the interaction of the mRNA codon with the anticodon stem-loops of the proper tRNA. The formation of peptide bonds occurs in the large subunit where the acceptor-stems of the tRNAs are docked. However, in the active ribosome the two subunits are in contact via bridges, and the actions in one subunit affect the other as the process of translation advances through the stages of initiation, elongation, and termination.

Affinity label for the tRNA binding sites on the E. coli ribosome allowed the identification of A and P site proteins most likely associated with the peptidyl transferase activity; labeled proteins are L27, L14, L15, L16, L2; at least L27 is located at the donor site. Additional research has demonstrated that the S1 and S21 proteins, in association with the 3'-end of 16 S ribosomal RNA, are involved in the initiation of translation.

Ribosomes have been classified as ribozymes, because the rRNA contributes directly to forming the site of both decoding and catalysis of peptide bond synthesis. Crystallographic work has shown that no ribosomal protein was observed close enough to the site of peptide bond synthesis. This suggests that the proteins in the subunit probably provide the scaffolding needed to maintain the three-dimensional structure of the RNA.

Though similarity, the ribosomes of prokaryotes, eukaryotes, and mitochondria differ in many details of their structure. The differences in structure between bacterial and eukaryotic allow some antibiotics to kill the infected bacteria by inhibiting their ribosomes without any detriment to a eukaryotic host's. Due to the differences in their structures, the bacterial 70S ribosomes are vulnerable to these antibiotics while the eukaryotic 80S ribosomes are not. Even though mitochondria possess ribosomes similar to the bacterial ones, mitochondria are not affected by these antibiotics because they are surrounded by a double membrane that does not easily admit these antibiotics into the organelle.

A number of antibiotics act by inhibiting translation, these include anisomycin, cycloheximide, chloramphenicol, tetracycline, streptomycin, erythromycin, and puromycin, among others.

3.2.3 Function of ribosome

Ribosome is the translation factory. Translation is the process in which mRNA, produced by transcription, is decoded by the ribosome to produce a specific polypeptide in the cytoplasm, that will later fold into an active protein. The basic process of protein production is addition of one amino acid at a time to the end of a protein.

The choice of amino acid type to add is determined by an mRNA molecule. The mRNA comprises a series of codons (three nucleotide subsequence) that dictate to the ribosome the sequence of the amino acids needed to make the protein. Each of those triplets codes for a specific amino acid. The successive amino acids added to the chain are matched to successive nucleotide triplets in the mRNA. In this way the sequence of nucleotides in the template mRNA chain determines the sequence of amino acids in the generated amino acid chain. Addition of an amino acid occurs at the C-terminus of the peptide and thus translation is said to be amino-to-carboxyl directed.

Transfer RNA (tRNAs) are small noncoding RNA chains (74 ~ 93 nucleotides) that transport amino acids to the ribosome. tRNA is shaped like a clover leaf with three loops. It contains an amino acid attachment site on one end and a special section in the middle loop called the anticodon site. The anticodon is an RNA triplet complementary to the mRNA codon that codes for their cargo amino acid (Figure 3-4). Aminoacyl tRNA synthetase catalyzes the bonding between specific tRNAs and the amino acids that their anticodon sequences call for. The amino acid is joined by its carboxyl group to the 3' OH of the tRNA

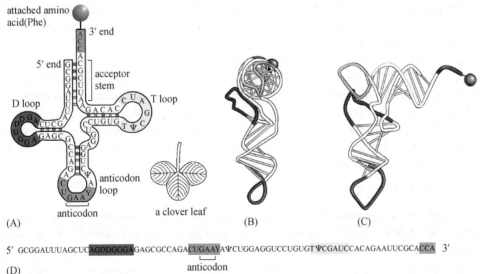

Figure 3-4 Tertiary structure of tRNA (From: Alberts B, et al. 2008)

with an ester bond. The product of this reaction is an aminoacyl-tRNA molecule, which is termed "charged".

This aminoacyl-tRNA travels inside the ribosome, where mRNA codons are matched through complementary base pairing to specific tRNA anticodons. The ribosome facilitates decoding by inducing the binding of tRNAs with complementary anticodon sequences to that of the mRNA. Two subunits of ribosome can bind together to an mRNA chain and use it as a template for determining the correct sequence of amino acids in a particular protein. Amino acids are selected, collected and carried to the ribosome by tRNA containing a complementary anticodon on one end and the appropriate amino acid on the other end, which enter one part of the ribosome and bind to the mRNA chain. The attached amino acids are then linked together by another part of the ribosome as the mRNA passes through.

More than one ribosome may move along a single mRNA chain simultaneously, forming what is called a **polyribosome** or **polysome**, each "reading" its sequence and producing a corresponding protein molecule. Because of the relatively large size of ribosomes, they can only attach to sites on mRNA 35 nucleotides apart.

In Bacteria, translation occurs in the cell's cytoplasm, where the large and small subunits of the ribosome are located, and bind to the mRNA. In Eukaryotes, Ribosomes are classified as being either "free" or "membrane-bound". Free and membrane-bound ribosomes differ only in their spatial distribution, they are identical in structure. Free ribosomes can move about anywhere in the cytosol, but are excluded from the cell nucleus and other organelles. Proteins that are formed from free ribosomes are released into the cytosol and used within the cell. Since the cytosol contains high concentrations of glutathione and is, therefore, a reducing environment, proteins containing disulfide bonds, which are formed from oxidized cysteine residues, cannot be produced in this compartment.

In many cases, the entire ribosome/mRNA complex causing it to bind to the outer membrane of the rough endoplasmic reticulum (ER) and release the nascent protein polypeptide inside for later vesicle transport and secretion outside of the cell. The ribosome making this protein can become "membrane-bound".

Whether the ribosome exists in a free or membrane-bound state depends on the presence of an ER-targeting signal sequence on the protein being synthesized, so an individual ribosome might be membrane-bound when it is making one protein, but free in the cytosol when it makes another protein.

The ribosome contains three RNA binding sites, they are the aminoacyl site (abbreviated A), the peptidyl site (abbreviated P) and the exit site (abbreviated E). With respect to the mRNA, the three sites are oriented 5' to 3' E-P-A, because ribosomes moves toward the 3' end of mRNA. The A site binds the incoming tRNA with the complementary codon on the mRNA (except for the first aminoacyl tRNA, fMet-tRNA^{fMet}, which enters at the P site). The P site is where the peptidyl tRNA is formed in the ribosome. The E site binds the uncharged tRNA before it exits the ribosome.

3.2.4 Translation on ribosomes

Protein synthesis is accomplished through a process called translation, the process of converting the mRNA codon sequences into an amino acid polypeptide by ribosomes. Translation proceeds in four phases: **initiation**, **elongation**, **termination** and **recycling**.

(1) Initiation In prokaryote   The initiation of translation involves the assembly of the initiation complex of the translation system which are: the small ribosomal subunits, the mRNA, the first (formyl) aminoacyl tRNA, three initiation factors (IF1, IF2, and IF3)

and GTP, which help the assembly of the initiation complex.

During translation, a small ribosomal subunit attaches to a mRNA molecule. At the same time an initiator tRNA (aminoacyl-tRNA containing the amino acid N-formylmethionine, fMet-tRNA^{fMet}) molecule recognizes and binds to an AUG codon on the same mRNA molecule. A large ribosomal subunit then joins the newly formed initiation complex. The initiator tRNA resides in one binding P site of the ribosome, leaving the A site open. Protein synthesis begins at a start codon AUG near the 5' end of the mRNA.

The ribosome is able to identify the start codon by use of the Shine-Dalgarno sequence of the mRNA (Kozak box in eukaryotes). The 30 S subunit binds to the mRNA template at a purine-rich region (the Shine Dalgarno sequence) upstream of the AUG initiation codon. The Shine Dalgarno sequence is complementary to a pyrimidine rich region on the 16 S rRNA component on the 30 S subunit. During the formation of the initiation complex, these complementary nucleotide sequences pair to form a double stranded RNA structure that binds the mRNA to the ribosome in such a way that the initiation codon is placed at the P site.

(2) Initiation of eukaryotic translation

1) Cap-dependent initiation: Initiation of eukaryotic translation usually involves the interaction of certain key proteins with a special tag bound to the 5'-end of an mRNA molecule, the 5' cap, named cap-dependent initiation. The protein factors bind the 40 S subunit of ribosome, and these initiation factors hold the mRNA in place. The eukaryotic initiation factor 3 (eIF3) is associated with the small ribosomal subunit, keeping the large ribosomal subunit from prematurely binding. eIF3 also interacts with the eIF4F complex which consists of three other initiation factors: eIF4A, eIF4E and eIF4G. eIF4G is a scaffolding protein which directly associates with both eIF3 and the other two components. eIF4E is the cap-binding protein. It is the rate-limiting step of cap-dependent initiation, and is often cleaved from the complex by some viral proteases to limit the cell's ability to translate its own transcripts. This is a method of hijacking the host machinery in favor of the viral (cap-independent) messages. eIF4A is an ATP-dependent RNA helicase, which aids the ribosome in resolving certain secondary structures formed by the mRNA. There is another protein associated with the eIF4F complex called the Poly (A)-binding protein (PABP), which binds the poly-A tail of most eukaryotic mRNA molecules. This protein plays a role in circularization of the mRNA during translation. This pre-initiation complex (the 40 S subunit and tRNA) accompanied by the protein factors move along the mRNA chain towards its 3'-end, scanning for the "start" codon (typically AUG) on the mRNA. The initiator tRNA charged with Met (Met-tRNA^{Met}) forms part of the ribosomal complex and thus all proteins start with this amino acid (unless it is cleaved away by a protease in subsequent modifications). The Met-tRNA^{Met} is brought to the P site of the small ribosomal subunit by eukaryotic initiation factor 2 (eIF2). It hydrolyzes GTP, and signals for the dissociation of several factors from the small ribosomal subunit which results in the association of the 60 S subunit. Then the complete ribosome (80 S) starts translation elongation.

Regulation of protein synthesis is dependent on phosphorylation of initiation factor eIF2. When large numbers of eIF2 are phosphorylated, protein synthesis is inhibited. This would occur if there is amino acid starvation or there has been a virus infection. However, naturally a small percentage of this initiation factor is phosphorylated. Another regulator is 4EBP which binds to the initiation factor eIF4E on the 5' cap of mRNA stopping protein synthesis. To oppose the effects of the 4EBP, growth factors phosphorylate 4EBP reducing its affinity for eIF4E and permitting protein synthesis.

2) **cap-independent initiation**: There exits cap-independent initiation in eukaryotic translation. The best studied example of the cap-independent mode of translation initiation in eukaryotes is the internal ribosome entry site (IRES) approach. IRES are a RNA structure that allow for translation initiation in the middle of a mRNA sequence as part of the process of protein synthesis.

What differentiates cap-independent translation from cap-dependent translation is that cap-independent translation does not require the ribosome to start scanning from the 5' end of the mRNA cap until the start codon. The ribosome can be trafficked to the start site by ITAFs (IRES trans-acting factors) bypassing the need to scan from the 5' end of the untranslated region of the mRNA. This method of translation has been recently discovered, and has found to be important in conditions that require the translation of specific mRNAs, despite cellular stress or the inability to translate most mRNAs. Examples include factors responding to apoptosis, stress-induced responses.

(3) Elongation Elongation of the polypeptide involves addition of amino acids to the carboxyl end of the growing chain, which is catalyzed by peptidyl transferase (ribozyme). The growing protein exits the ribosome through the polypeptide exit tunnel in the large subunit. Elongation is dependent on elongation factors.

During elongation, each additional amino acid is added to the nascent polypeptide chain in a four-step microcycle. In step1, a correct aminoacyl-tRNA is positioned in the A site of the ribosome, and a spent tRNA dissociates from the E site. In step 2, an new peptide bond is formed. In step 3, the large subunit translocates relative to the mRNA held by the small subunit, thereby shifting the two tRNAs to the E and P sites of large subunit. In step 4, the small subunit translocates carrying its mRNA a distance of three nucectides through the ribosome. (Figure 3-5). This four-step cycle is repeated and each time an amino acid is added the peptide.

The initiator fmet-tRNA occupies the P site in the ribosome, causing a conformational change which opens the A site for the new aminoacyl-tRNA to bind. This binding is facilitated by elongation factor-Tu (EF-Tu), a small GTPase.

When an aminoacyl-tRNA recognizes the next codon sequence on the mRNA, it attaches to the open A site. A peptide bond forms connecting the amino acid of the tRNA in the A site and the amino acid of the charged tRNA in the P site. This process, known as peptide bond formation, is catalyzed by a ribozyme (the 23 S ribosomal RNA in the 50 S ribosomal subunit). Now, the A site has the newly formed peptide, while the P site has an uncharged tRNA.

In translocation stage, the uncharged tRNA in the E site leaves the ribosome via E site, the peptidyl-tRNA in the A site along with its corresponding codon moves to the P site and the codon after the A site moves into the A site. The A binding site becomes vacant again until another aminoacyl-tRNA that recognizes the new mRNA codon takes the open position.

After the new amino acid is added to the chain, the energy provided by the hydrolysis of a GTP bound to the translocase EF-G (in prokaryotes) and eEF-2 (in eukaryotes), moves the ribosome down one codon towards the 3' end. The energy required for translation of proteins is important. For a protein containing n amino acids, the number of high-energy phosphate bonds required to translate it is $4n-1$. The rate of translation varies; it is significantly higher in prokaryotic cells (up to 17~21 amino acid residues per second) than in eukaryotic cells (up to 6~9 amino acid residues per second).

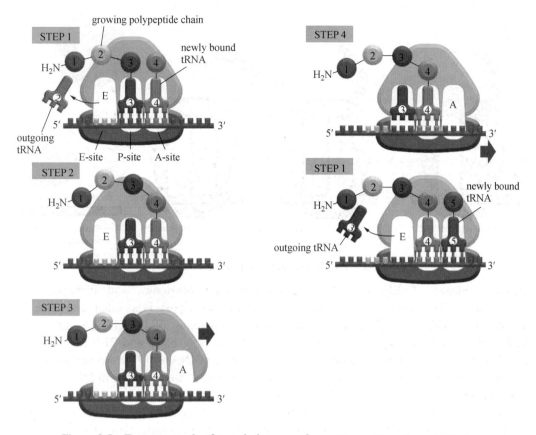

Figure 3-5 Four-step cycle of translating procedures (From: Alberts B, et al. 2008)

The translation machinery works relatively slowly compared to the enzyme systems that catalyze DNA replication. The ribosome will translate the mRNA molecule as more aminoacyl-tRNA bind to the A site, until it reaches a termination codon on the mRNA (UAA, UGA, or UAG).

(4) Termination Termination of the polypeptide occurs when one of the three termination codons moves into the A site. No tRNA can recognize or bind to this codon. Instead, they are recognized by proteins called release factors, namely RF1 (recognizing the UAA and UAG stop codons) or RF2 (recognizing the UAA and UGA stop codons). These factors trigger the hydrolysis of the ester bond in peptidyl-tRNA and the release of the newly synthesized protein from the ribosome. A third release factor RF-3 catalyzes the release of RF-1 and RF-2 at the end of the termination process (Figure 3-6).

(5) Recycling After translational termination, mRNA and P site deacylated tRNA remain associated with ribosomes in post-termination complexes (post-TCs), which must therefore be recycled by releasing mRNA and deacylated tRNA and by dissociating ribosomes into subunits. Hence, ribosome recycling step is responsible for the disassembly of the post-termination ribosomal complex. Once the nascent protein is released in termination, ribosome recycling factor and elongation factor G (EF-G) function to release mRNA and tRNAs from ribosomes and dissociate the ribosome into two subunits. IF3 then replaces the deacylated tRNA releasing the mRNA. All translational components are free for additional rounds of translation.

During prokaryotic termination, RF1 and RF2 promote peptide release, whereas RF3

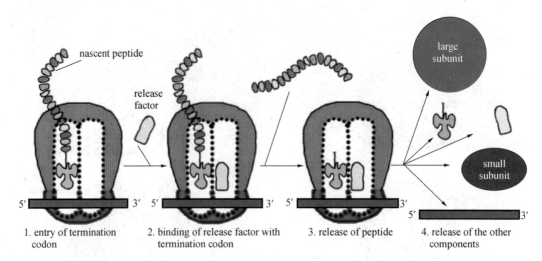

Figure 3-6 Termination of peptide synthesis

mediates release of RF1/RF2 from post-termination ribosomes and dissociates after hydrolyzing GTP, yielding post-termination complexes (post-TCs) that comprise 70 S ribosomes, mRNA, and P site deacylated tRNA. Recycling of post-TCs requires EF-G, RRF (ribosome recycling factor), and initiation factor IF3. EF-G and RRF dissociate post-TCs into free 50 S subunits and 30 S subunits bound to mRNA and P site deacylated tRNA, and IF3 induces release of tRNA from 30 S subunits, after which mRNA dissociates spontaneously.

In eukaryotic recycling, initiation factors eIF3, eIF1, eIF1A, and eIF3j, a loosely associated subunit of eIF3, can promote recycling of eukaryotic post-TCs. eIF3 is the principal factor that promotes splitting of post-termination ribosomes into 60S subunits and tRNA-and mRNA bound 40S subunits. Its activity is enhanced by eIF 3j, 1, and 1A. eIF1 also mediates release of P site tRNA, whereas eIF3j ensures subsequent dissociation of mRNA.

Proteins have a variety of functions. Some will be used in the membrane of the cell, while others will remain in the cytoplasm or be transported out of the cell. Many copies of a protein can be made from one mRNA molecule. This is because several ribosomes can translate the same mRNA molecule at the same time, forming polysomes.

(6) Post-translational modification   Protein undergoes several modifications before becoming a fully functioning protein. These processes are called post-translational modification (PTM). PTM increases the functional diversity of the proteome by the covalent addition of functional groups or proteins, proteolytic cleavage of regulatory subunits or degradation of entire proteins. These modifications include phosphorylation, glycosylation, ubiquitination, nitrosylation, methylation, acetylation, lipidation and proteolysis.

3.3 Ribozyme

A ribozyme is defined as an RNA molecule capable of catalyzing a chemical reaction in the absence of proteins.

3.3.1 Discovery and activity of ribozyme

Until about 30 years ago, all known enzymes were proteins. But then it was discovered that some RNA molecules can act as enzymes; that is, catalyze covalent changes in the structure of substrates (most of which are also RNA molecules). Catalytic RNA molecules are called **ribozyme(RZ)**. Ribozyme means ribonucleic acid enzyme. It may also be called an RNA enzyme or catalytic RNA.

The first ribozyme was discovered in the 1980s by Thomas R. Cech, who was studying RNA splicing in the ciliated protozoan. The term ribozyme was first introduced by Kelly Kruger et al. in 1982. In 1989, Thomas R. Cech and Sidney Altman won the Nobel Prize in chemistry for their "discovery of catalytic properties of RNA".

Ribozyme catalyzes a chemical reaction with its tertiary structure. It contains an active site that consists entirely of RNA. Many natural ribozymes cleave one of their own phosphodiester bonds (self-cleaving ribozymes), or bonds in other RNAs. Some have been found to catalyze the aminotransferase activity of the ribosome. Although most ribozymes are quite rare in the cell, their roles are sometimes essential to life. For example, the functional part of the ribosome, the molecular machine that translates RNA into proteins, is fundamentally a ribozyme. A recent test-tube study of prion folding suggests that an RNA may catalyze the pathological protein conformation in the manner of a chaperone enzyme. Ribozymes have been shown to be involved in the viral concatemer cleavage that precedes the packing of viral genetic material into virions.

3.3.2 Examples of naturally occurring ribozymes

(1) **Peptidyl transferase 23S rRNA** The three-dimensional structure of the large (50 S) subunit of a bacterial ribosome clearly shows that formation of the peptide bond that links each amino acid to the growing polypeptide chain is catalyzed by the 23S RNA molecule in the large subunit.

(2) **RNase P** Almost all living things synthesize an enzyme — called ribonuclease P (RNase P) — that cleaves the head (5') end of the precursors of tRNA. In bacteria, ribonuclease P is a heterodimer containing a molecule of RNA and one of protein. Separated from each other, the RNA retains its catalysis ability with less efficiency than the intact dimmer. Recent findings reveal that the human nucleus RNase P is also required for the normal and efficient transcription of various ncRNAs transcribed by RNA polymerase III, such as tRNA, 5S rRNA, SRP RNA and U6 snRNA genes.

(3) **Spliceosomes** Spliceosomes is a complex of snRNA and protein subunits participating in converting pre-mRNA into mRNA by excising the introns and splicing the exons of most nuclear genes. They are composed of 5 kinds of small nuclear RNA (snRNA) molecules and a large number of protein molecules. It is the snRNA — not the protein — that catalyzes the splicing reactions. There are two main groups of self-splicing RNAs: the group I catalytic intron and group II catalytic intron.

(4) **Leadzyme** Leadzyme is a small, artificially-made RZ which can cleave RNA in the presence of lead. Leadzyme is able to cleave RNA in the presence of lead. The structure of leadzyme has been determined by X-ray crystallography. Although initially created in vitro, it has been proposed that a naturally occurring leadzyme occurs in the 5 S rRNA and further that this may be an important mechanism in lead toxicity.

(5) **Hammerhead ribozyme** Hammerhead ribozyme is an RNA molecule that self-cleaves via a small conserved secondary structural motif termed a hammerhead because of its shape.

Most hammerhead RNAs are subsets of two classes of plant pathogenic RNAs: the satellite RNAs of RNA viruses and the viroids. Viroids are RNA molecules that infect plant cells as conventional viruses do, but are far smaller (246 nucleotides), and are naked; that is, they are not encased in a capsid. The self-cleavage reactions are part of a rolling circle replication mechanism. The hammerhead sequence is sufficient for self-cleavage and acts by forming a conserved three-dimensional tertiary structure.

There are more ribozymes, including HDV ribozyme, Hairpin ribozyme, Mammalian CPEB3 ribozyme, VS ribozyme and glmS ribozyme.

3.3.3 Artificial ribozymes and deoxyribozymes

Since the discovery of ribozymes that exist in living organisms, there has been interest in the study of new synthetic ribozymes made in the laboratory. Scientists have produced ribozymes in the laboratory that are capable of catalyzing their own synthesis under very specific conditions, such as an RNA polymerase ribozyme and self-cleaving RNAs. Some of the artificial ribozymes had novel structures, while some were similar to the naturally occurring hammerhead ribozyme.

Deoxyribozymes or DNA enzymes or catalytic DNA, rDNAzymes are artificially DNA molecules that have the ability to perform a chemical reaction, such as catalytic action. In contrast to the RNA ribozymes, which have many catalytic capabilities, in nature DNA is only associated with gene replication and nothing else. The reasons are that DNA lacks the 2'-hydroxyl group of RNA, which diminishes its chemical reactivity and its ability to form complex tertiary structures, and that nearly all biological DNA exists in the double helix conformation in which potential catalytic sites are shielded. In comparison to proteins built up from 20 different monomers, both RNA and DNA have a much more restricted set of monomers to choose from which limits the construction of interesting catalytic sites. For these reasons DNAzymes exist only in the laboratory. The ability of ribozymes to recognize and to cut specific RNA molecules makes them exciting candidates for human therapy.

3.4 Non-coding RNA

A non-coding RNA (ncRNA) is a functional RNA molecule that is not translated into a protein. The DNA sequence from which a ncRNA is transcribed is often called an RNA gene. The term small RNA (sRNA) is often used for short ncRNA.

3.4.1 Classification

Non-coding RNA belong to several groups and include highly abundant and functionally important RNAs such as tRNA and rRNA, as well as RNA such as snoRNA, microRNA (miRNA), siRNA, piRNAs and the long ncRNA (such as Xist and HOTAIR). Recent transcriptomic and bioinformatic studies suggest the existence of thousands of ncRNA. Since many of the newly identified ncRNA have not been validated for their function, it is possible that many are non-functional.

3.4.2 Small RNA

Among nc RNA, sRNAs have burst on the scene as ubiquitous, versatile repressors of gene expression in plants, animals and many fungi. sRNA are tiny RNA (21 ~ 26 nt), which induce silencing through homologous sequence interactions. There are different types: short interfering (si) RNA, small temporal (st) RNA, heterochromatic siRNA, tiny ncRNA and microRNA. They can control mRNA stability or translation, or target epigenetic modifications to specific regions of the genome. sRNA and RNA-mediated silencing pathways have established a new paradigm for understanding eukaryotic gene regulation and revealed novel host defenses to viruses and transposons(Figure 3-7).

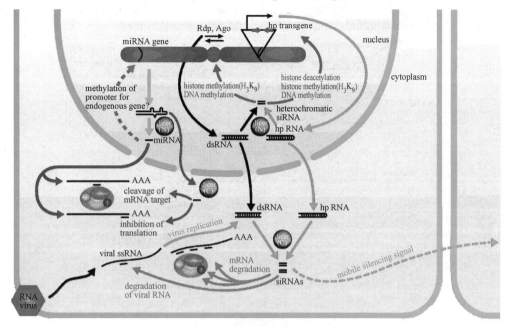

Figure 3-7　The small RNA world (From: Cell Science,2003)

　　Small regulatory RNA are usually generated via processing of longer double-stranded RNA(dsRNA) precursors by an RNase Ⅲ-like enzyme termed Dicer. Dicer acts in complexes with other proteins, including members of the Argonaute family and possibly HEN1, to produce small RNA. Both nuclear(N) and cytoplasmic(C) Dicer activities are found, but both forms may not be present in all organisms.

　　In general, siRNA can be derived from all regions of perfect duplex RNA and, at least in plants, they accumulate in both sense and antisense polarities. Perfect duplex RNA can be produced by transcription of a "hairpin" (hp) transgene, which produces a corresponding hpRNA. Plants make two functionally distinct size classes of small RNA. A shorter class, 21~22 nt, has been implicated in mRNA degradation, and a longer size class, 24~26 nt, in directing DNA methylation and in systemic silencing.

　　MicoRNA are small RNA that downregulate endogenous genes important for implementing developmental programs in animals and plants. The classic miRNA, lin-4 and let-7(originally termed stRNA), were discovered by way of their heterochronic mutant phenotypes in C elegans.

　　Pri-miRNA are usually transcribed either from their own genes or from introns by RNA polymerase Ⅱ. A single pri-miRNA may contain from one to six miRNA precursors. These hairpin loop structures are composed of about 70 nucleotides each. The double-stranded RNA structure of the hairpins in a pri-miRNA is recognized cleaved by the enzyme Drosha to form pre-miRNA(precursor-miRNA) with imperfect duplex RNA, 70~200 nt in length. pre-miRNA hairpins are exported from the nucleus and is cleaved by the RNase Ⅲ enzyme Dicer in cytoplasm. This endoribonuclease interacts with the 3'-end of the hairpin and cuts away the loop joining the 3' and 5' arms, yielding an imperfect miRNA/miRNA duplex about 22 nucleotides in length. Although either strand of the duplex may potentially act as a functional miRNA, only one strand is usually incorporated into the RNA-induced silencing complex (RISC) where the miRNA and its mRNA target interact (Figure 3-8).

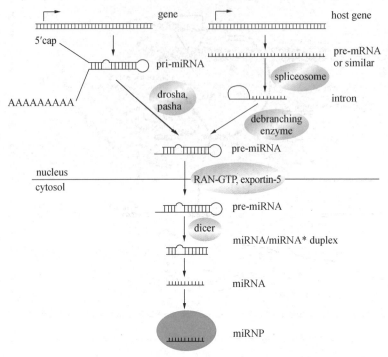

Figure 3-8　The procedure of microRNA synthesis and maturation

3.4.3　Biological roles of ncRNA

ncRNAs are involved in many cellular processes. These range from ncRNA of central importance that are conserved across all or most cellular life through to more transient ncRNAs specific to one or a few closely related species.

(1) ncRNAs in translation　Many of the conserved, essential and abundant ncRNA are involved in translation (Figure 3-9). Ribosomes are the "factories" of translation (see 2.2). tRNA form an "adaptor molecule" between mRNA and protein. The H/ACA box and C/D box snoRNAs are ncRNAs found in archaea and eukaryotes, RNase MRP is restricted to

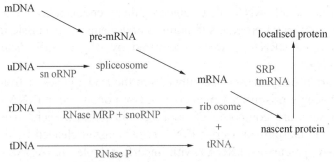

Figure 3-9　An illustration of the central dogma annotated with the processes ncRNA are involved in. RNP are shown in red, ncRNA are shown in blue

eukaryotes, both groups of ncRNA are involved in the maturation of rRNA. The snoRNAs guide covalent modifications of rRNA, tRNA and snRNAs, RNase MRP cleaves the internal transcribed spacer 1 between 18S and 5.8S rRNAs. RNase P matures tRNA sequences. Another ubiquitous RNP called SRP recognizes and transports specific nascent proteins to the endoplasmic reticulum in eukaryotes and the plasma membrane in prokaryotes.

(2) **ncRNAs in RNA splicing**　　In eukaryotes the spliceosome performs the splicing reactions essential for removing intron sequences, this process is required for the formation of mature mRNA. The spliceosome is a RNP often also known as the snoRNP or tri-snoRNP. The ncRNA components of the major spliceosome are U1, U2, U4, U5, and U6. The ncRNA components of the minor spliceosome are U11, U12, U5, U4atac and U6atac.

Another group of introns can catalyse their own removal from host transcripts, these are called self-splicing RNA(see 3.3.2). These ncRNAs catalyze their own excision from mRNA, tRNA and rRNA precursors in a wide range of organisms. snoRNA can also regulate the alternative splicing of mRNA in mammal cells, for example snoRNA HBII-52 regulates the splicing of serotonin receptor 2C.

(3) **ncRNAs in gene regulation**　　The expression of genes are regulated by ncRNA in wide range. In higher eukaryotes miRNA and siRNA reduce gene expression. miRNA and siRNA silence at the post-transcriptional level by virtue of their sequence complementarity to target mRNA. siRNAs associate with an endonuclease-containing complex, RISC (RNA induced silencing complex), and cause degradation of cognate mRNAs in plants and viral RNA. This process is termed RNAi in animals.

miRNAs, which also appear to associate with a RISC-like complex. A single miRNA can reduce the expression levels of hundreds of genes. It can be either partial complementary to the 3′ UTR(untranslated regions) of mRNA and block translation or act in the manner of siRNA and guide mRNA degradation. The choice between these two pathways is probably determined by the degree of complementarily between a given miRNA and its target mRNA. Since most animal miRNA base pair imperfectly with their targets, the predominant mode of silencing in these organisms is translational repression. By contrast, plant miRNAs frequently show perfect complementarity to their target and hence trigger mRNA degradation. However, plant miRNA complementary to 3′ UTRs can probably guide translational repression. Proteins that might determine whether siRNA or miRNA are used as substrates by RISC are various members of the Argonaute (Ago) family.

The ncRNA RNase P has also been shown to influence gene expression. RNase P exerts its role in transcription through association with Pol III and chromatin of active tRNA and 5S rRNA genes.

The B2 RNA is a small non-coding RNA polymerase III transcript that represses mRNA transcription in response to heat shock in mouse cells. B2 RNA inhibits transcription by binding to core Pol II and assembling into preinitiation complexes at the promoter and blocks RNA synthesis.

(4) **ncRNA and genome defense**　　siRNAs are associated with silencing triggered by transgenes, microinjected RNA, viruses, and transposons, and hence can be considered intermediaries in host defense pathways against foreign nucleic acids.

Piwi-interacting RNA(piRNA) expressed in mammalian testes and somatic cells, they form RNA-protein complexes with Piwi proteins. These piRNA complexes (piRCs) have been linked to transcriptional gene silencing of retrotransposons and other genetic elements in germ line cells, particularly those in spermatogenesis.

(5) **ncRNAs and chromosome structure**　　Recent work has identified links between RNAi

and epigenetic alterations of the genome, such as RNA-directed DNA methylation (RdDM), and histone modifications.

Telomerase is an RNP enzyme that adds specific DNA sequence repeats to telomeric regions, which are found at the ends of eukaryotic chromosomes. The enzyme is a reverse transcriptase that carries Telomerase RNA, which is used as a template when it elongates telomeres.

Xist (X-inactive-specific transcript) is a long ncRNA gene on the X chromosome of the placental mammals that acts as major effector of the X chromosome inactivation process forming Barr bodies. An antisense RNA, Tsix, is a negative regulator of Xist. X chromosomes lacking Tsix expression are inactivated more frequently than normal chromosomes. Both Xist and roX operate by epigenetic regulation of transcription through the recruitment of histone-modifying enzymes.

Summary

Cytoplasm is the thick, gel-like semitransparent fluid that is found in both plant and animal cell. It contains three parts, the cytosol, the organelles and cytoplasmic inclusions. Cytoplasm is the site of many vital biochemical reactions crucial for maintaining life. Cytoplasm can maintain the shape of the cell and suspension of organelles. Cytoplasm also acts as a storage space for the chemical building blocks of the body.

The ribosome is a ribonucleoprotein particle composed of RNA and protein, and serves as the site of protein synthesis. Each ribosome consists of two subunits. The ribosome contains three RNA binding sites, the site A, P and E. Translation is the process in which mRNA is decoded by the ribosome to produce a specific polypeptide in the cytoplasm, it proceeds in four phases: initiation, elongation, translocation and termination.

Ribozyme means ribonucleic acid enzyme. Ribozyme catalyzes a chemical reaction with its tertiary structure. It contains an active site that consists entirely of RNA. Many natural ribozymes cleave one of their own phosphodiester bonds, or bonds in other RNA. Some have been found to catalyze the aminotransferase activity of the ribosome. There are synthetic ribozymes made in the laboratory.

A non-coding RNA is a functional RNA molecule that is not translated into a protein. The term small RNA is often used for short ncRNA. ncRNA belong to several groups and include highly abundant and functionally important RNA such as tRNA and rRNA, as well as RNA such as small RNA and the long ncRNA. ncRNA are involved in many cellular processes, such as translation, RNA splicing, gene regulation, genome defense and chromosome structure.

Questions

1. Give short definitions of the following terms: Cytoplasm, Proteasome, Ribosome, Ribozyme, Deoxyribozyme, small RNA, microRNA.
2. Please compare the components and the structure of ribosome between prokaryotes and eukaryotes.
3. Please expatiate the translation procedure.
4. Give 5 examples of naturally occurring ribozymes.
5. What's the function of non-coding RNA?
6. Explain the structures and functions of cytoplasm.

(李 瑶)

Chapter 4

Endomembrane system, protein sorting and vesicle transport

A typical eukaryotic cell carries out thousands of different chemical reactions, many of which are mutually incompatible. As a result cells have developed several strategies for isolating and organizing their chemical reactions. Both prokaryotic and eukaryotic cells organize the different enzymes required to catalyze a series of particular sequence of reactions into a single large protein complex, such as complexes for DNA, RNA, or protein synthesis. A second strategy, which is highly developed in eukaryotic cells, is to confine different metabolic processes, and the proteins required to perform them, within different membrane-enclosed compartments. The major membrane enclosed compartments of a typical eukaryotic cell include nuclei, mitochondria, chloroplasts, endoplasmic reticulum (ER), Golgi apparatus, lysosomes, endosomes and peroxisomes. These organelles are surrounded by the cytosol, which is enclosed by the plasma membrane (Figure 4-1). The main functions of the membrane enclosed compartments of a eukaryotic cell are listed in table 4-1.

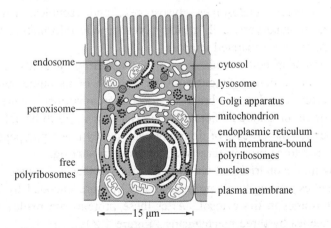

Figure 4-1　A cell from the lining of the intestine contains the basic set of organelles found in most animal cells (From: Alberts B, et al. Molecular Biology of the Cell. 2008)

Table 4-1 The main functions of the membrane-enclosed compartments of a eukaryotic cell

compartment	main function
cytosol	containing many metabolic pathways, protein synthesis
nuclei	contain main genome, DNA and RNA synthesis
endoplasmic reticulum	synthesis of most lipids, synthesis of proteins for distribution to many organelles and to the plasma membrane
Golgi apparatus	modification, sorting, and packaging of proteins and lipids for either secretion or delivery to another organelle
lysosomes	intracellular degradation
endosomes	sorting of endocytosed material
mitochondria	ATP synthesis by oxidative phosphorylation
chloroplasts (in plant cells)	ATP synthesis and carbon fixation by photosynthesis
peroxisomes	oxidation of toxic molecules

4.1 Overview of endomembrane system

Membrane-enclosed organelles are thought to have arisen in evolution at least in two ways. Mitochondria and chloroplasts are unique cytoplasmic organelles as they arise only from preexisting mitochondria or chloroplasts, respectively. They are thought to have originated in a way of endosymbiosis. The similarity of their genomes to those of bacteria and the close similarity of some of their proteins to bacterial proteins strongly suggest that mitochondria and chloroplasts evolved from bacteria that were engulfed by primitive eukaryotic cells with which they initially lived in symbiosis. The nuclear membranes and the membrane of the ER, Golgi apparatus, endosomes, and lysosomes are believed to have originated by invagination of the plasma membrane. These membranes, and the organelles they enclose, are all part of the **endomembrane system** which form a dynamic, integrated network in which materials are shuttled back and forth from one part of the cell to another. In the present chapter, we will introduce the structure and functions of the endoplasmic reticulum, Golgi complex, endosomes, lysosomes, and vacuoles. Nucleus is the most prominent organelle in eukaryotic cells which stores genetic information for a cell and will be introduced in detail in a separated chapter.

 A typical mammalian cell contains up to 10,000 different kinds of proteins, a yeast cell, about 5,000. For a cell to function properly, each of its numerous proteins must be localized to the correct cellular membrane or aqueous compartment. Directing newly made proteins to their correct organelle is therefore necessary for a cell to be able to grow, divide, and function properly. The process of directing each newly made polypeptide to a particular destination is referred to as protein targeting, or **protein sorting**.

 Except a few mitochondrial and chloroplast proteins that are synthesized on ribosomes inside these organelles, virtually all proteins in the cell are encoded by nuclear DNA and synthesized on ribosomes in the cytosol. After these proteins are made, they are imported into different organelles by three mechanisms (Figure 4-2).

(1) Gated transport through nuclear pores   Proteins that function inside nucleus are transported through the nuclear pores from the cytosol. The pores function as selective gates that actively transport specific macromolecules, but also allow free diffusion of smaller molecules. Such process will be discussed in the chapter of nucleus.

(2) Transport across membranes   Most proteins of mitochondria and chloroplasts, and all

proteins of the ER and peroxisomes are transported across the organelle membrane by protein translocators located in the membrane. In this case, the transported protein molecular must usually unfold in order to snake through the membrane. This process will be discussed in detail within the current chapter in ER section.

(3) **Transport by vesicles** Proteins moving from the ER onward and from one compartment of the endomembrane system to another are transported by transport vesicles, which become loaded with a cargo of proteins from the interior space, or lumen, of one compartment, as they pinch off from its membrane. The vesicles subsequently discharge their cargo into a second compartment by fusing with its membrane. This kind of vesicular transport will be introduced in the last part of the current chapter.

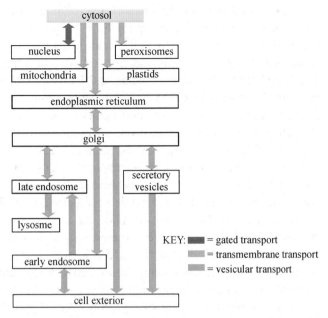

Figure 4-2　Three mechanisms that proteins are imported into organelles (From: Alberts B, et al. Molecular Biology of the Cell. 2008)

The protein **localization** is determined by its **sorting signal** that directs the protein to the organelle. The typical sorting signal is a continuous stretch of amino acid sequence of about 15 ~ 60 amino acids long. The properties of signal sequences that direct proteins from the cytosol to organelles are shown in table 4-2. The signal sequences of mitochondria, chloroplast and endoplasmic reticulum proteins are usually located in the N-terminus and are often removed from the finished protein once the sorting decision has been executed. The signal sequence that directs a protein from the cytosol into the nucleus is called **nuclear localization signal** (**NLS**). Typically it consists of one or two short sequences containing several positively charged lysines or arginines. It is located on the internal of a protein and will not be removed after transportation.

Table 4-2　Some typical signal sequences

target organelle	signal location	signal removal	nature of signal
ER	N-terminal	(+)	"Core" of 6 ~ 12 mostly hydrophobic amino acids, often preceded by one or more basic amino acids
ER retention	C-terminal	(−)	KDEL
mitochondrion	N-terminal	(+)	3 ~ 5 nonconsecutive Arg or Lys residues, often with Ser and Thr, no Glu or Asp residues
chloroplast	N-terminal	(+)	no common sequence motifs; generally rich in Ser, Thr, and small hydrophobic amino acid residues and poor in Glu and Asp residues
peroxisome	C-terminal	(−)	usually Ser-Lys-Leu at extreme C-terminus
nucleus	internal	(−)	one cluster of 5 basic amino acids, or two smaller clusters of basic residues separated by 10 amino acids

Early studies using the electron microscope provided biologists with a detailed portrait of the structure of cells, but gave little insight into the functions of the components they were observing. In order to isolate and investigate the functions of cytoplasmic organelles, techniques to break up cells and isolate particular types of organelles were pioneered in the 1950s and 1960s by Albert Claude and Christian De Duve. In these techniques, cells are ruptured by homogenization and the cytoplasmic membranes become fragmented and the fractured edges of the membrane fragments fuse to form spherical vesicles less than 100 nm in diameter. Vesicles derived from different organelles have different properties and can be separated from one another by an approach called subcellular fractionation. Membranous vesicles derived from the endomembrane system (primarily the endoplasmic reticulum and Golgi complex) form a heterogeneous collection of similar sized vesicles referred to as **microsomes**. A rapid (and crude) preparation of the microsomal fraction of a cell is depicted in figure 4-3. Once isolated, the biochemical composition of various fractions can be determined. Furthermore, researchers can probe the capabilities of these crude subcellular preparations and obtained a wealth of information about complex processes that were impossible to study using intact cells. For example, vesicles derived from different parts of the Golgi complex contained enzymes that added different sugars to the end of a growing carbohydrate chain of a glycoprotein or glycolipid. Such specific enzyme could be isolated from the microsomal fraction and then used as an antigen to prepare antibodies. The antibodies could then be linked to fluorophore or gold particles that could be visualized under the fluorescent or electron microscope, and therefore the localization of the enzyme in the membrane compartment could be determined. These studies reveal the role of the Golgi complex in the stepwise assembly of complex carbohydrates. In the past few years, sophisticated proteomic technology has been used to identify the proteins present in cell fractions. Once a particular organelle has been isolated, the proteins can be extracted, separated, and identified by mass spectrometry. Hundreds of proteins can be identified simultaneously, providing a comprehensive molecular portrait of any organelle that can be prepared in a relatively pure state.

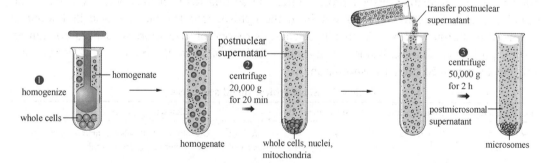

Figure 4-3 A rapid preparation of the microsomal fraction of a cell (From: Karp G, et al. Cell and Molecular Biology. 2009)

4.2 Endoplasmic reticulum

Endoplasmic reticulum (ER) comprises a system of membranes that enclose a space, or lumen, which is separated from the surrounding cytosol and forms a network of tubules, vesicles and sacs (**cisternae**) that are interconnected and often extends throughout most of the cell (Figure 4-4). The main functions of ER are the synthesis of many membrane lipids and proteins. Large areas of ER have ribosomes attached to the cytosolic surface and are designated **rough ER** (**RER**). The RER is the starting point of the biosynthetic pathway and is continuous with the outer membrane of the nuclear envelope and is typically composed of a network of flattened sacs. The attached ribosomes are actively synthesizing proteins that are delivered into the ER lumen or inserted into ER membrane. The **smooth ER** (**SER**) is scanty in most cells but is highly developed for performing particular functions in others. The membranous elements of SER are highly curved and tubular, forming an interconnecting system of pipelines curving through the cytoplasm lacking ribosome association. SER is the site of steroid hormone synthesis in cells of the adrenal gland. A variety of organic molecules, including alcohol, are detoxified by SER in liver cells. In many eukaryotic cells the SER also serves as Ca^{2+} sequesters from the cytosol which are involved in the rapid response to many extracellular signals.

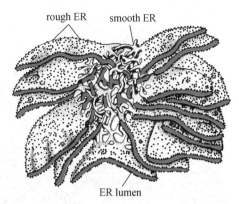

Figure 4-4   Rough ER and smooth ER
(From: Alberts B, et al. Molecular Biology of the Cell. 2008)

4.2.1 RER and protein synthesis

Certain polypeptides are synthesized on ribosomes attached to the cytosolic surface of the RER membranes. These include: (a) Soluble proteins that reside within compartments of the endomembrane system; (b) Secreted proteins; (c) Integral membrane proteins. The ER serves as the starting point of such biosynthetic pathways: ①Water-soluble proteins are completely translocated across the ER membrane and released into the ER lumen. Once inside the ER or in the ER membrane, they will be ferried by transport vesicles from organelle to organelle and, in some cases, secreted from organelle to the cell exterior. ②Prospective transmembrane proteins are only partly translocated across the ER membrane and destined to ER itself, Golgi apparatus, endosomes, lysosomes, as well as cell plasma membrane.

(1) **Synthesis of water-soluble proteins on RER**   Synthesis of all proteins begins on "free" ribosomes in the cytosol. What drives the translocation of ER-targeted proteins in the ER lumen or on ER membrane during protein synthesis? In the early 1970s, Günter Blobel, David Sabatini and Bernhard Dobberstein of Rockefeller university, first proposed, and then demonstrated, that the sequence of amino acids in the N-terminal portion of secreted proteins (**signal sequence**), which is the first part to emerge from the ribosome during protein synthesis, directs the polypeptide and ribosome to the ER membrane (known as the **signal hypothesis**) (Figure 4-5). The polypeptide moves into the cisternal space of the ER through a protein-lined, aqueous channel (so called **translocator**) in the ER membrane. It was proposed that the polypeptide moves through the membrane as it is being synthesized. A signal peptidase closely associated with the translocator clips off the signal sequence during translation, and the mature protein is released into the lumen of the ER immediately after synthesis. This proposal has been proved by a large body of experimental evidence and Blobel's original concept that proteins contain built-in "address codes" has been shown to apply in principle to virtually all types of protein trafficking pathways throughout the cell.

The ER signal sequence located at the N-terminus of the polypeptide varies greatly in

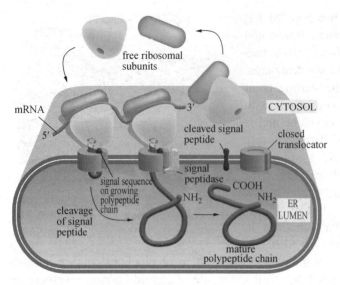

Figure 4-5 The signal hypothesis (From: Alberts B, et al. Molecular Biology of the Cell. 2008)

amino acid sequence. However, each ER signal sequence contains eight or more nonpolar hydrophobic amino acids at its center. After the ER signal sequence is translated on "free" ribosomes in the cytosol, it is recognized by a **signal-recognition particle (SRP)**, a complex consisting of six different polypeptide chains bound to a single small 7S RNA molecule. The crystal structure revealed that the signal-sequence-binding site of the SRP protein forms a large hydrophobic pocket lined by methionines. Since methionines have unbranched, flexible side chains, the pocket is sufficiently plastic to accommodate hydrophobic signal sequences of different sequences, sizes, and shapes. Binding of an SRP to the signal sequence of the polypeptide chain slows down protein synthesis until the ribosome and its bound SRP bind to an SRP receptor, an integral membrane protein complex embedded in the rough ER membrane. After binding to its receptor, the SRP is released and protein synthesis recommences, with the polypeptide now being threaded into the lumen of the ER through a translocation channel in the ER membrane (Figure 4-6).

The transient pause of protein translation caused by SRP binding presumably gives the ribosome enough time to bind to the ER membrane before completion of the polypeptide chain, thereby ensuring that proteins, especially those secreted and lysosomal hydrolases, are not released into the cytosol before reaching their destination. In contrast to the post-translational import of proteins into mitochondria and chloroplasts, SRP-mediated translation does not require chaperone proteins to keep the newly synthesized protein unfolded during translocation to the ER membrane.

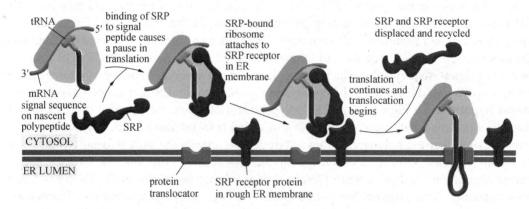

Figure 4-6 ER signal sequences and SRP direct ribosomes to the ER membrane (From: Alberts B, et al. Molecular Biology of the Cell. 2008)

Chapter 4 Endomembrane system, protein sorting and vesicle transport

In addition to directing proteins to the ER via SRP recognition, the signal sequence can be recognized by a specific site inside the translocation channel and trigger the opening of the pore in the protein translocator, where it serves as a start-transfer signal (or **start-transfer peptide**). Such dual recognition may help to ensure that only appropriate proteins enter the lumen of the ER. The signal peptide remains bound to the channel while the rest of the protein chain slides through the membrane as a large loop. At some stage during translocation, the signal sequence is cleaved off by a signal peptidase located on the luminal side of the ER membrane; the signal peptide is then released from the translocation channel and rapidly degraded into amino acids. Once the C-terminus of the protein has passed through the membrane, the protein is released into the ER lumen (Figure 4-7).

The core of the translocator is Sec61 complex containing three subunits that are highly conserved from bacteria to eukaryotic cells. The X-ray crystal structure of Sec61 complex showed the α helices contributed by the largest subunit form a central pore through which the polypeptide chain traverses the membrane. The pore is gated by a short helix to keep the translocator closed when it is idle. The short helix

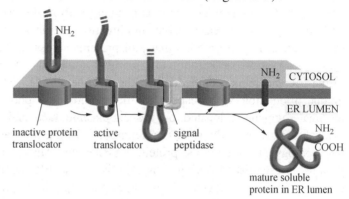

Figure 4-7 **Translocation of a water-soluble protein across the ER membrane** (From: Alberts B, et al. Molecular Biology of the Cell. 2008)

move aside to open the pore when it is engaged in passing a polypeptide chain. Thus, the pore is a dynamic gated structure that opens only transiently when a polypeptide chain traverses the membrane so that there is no unwanted passage of calcium and other ions between the cytosol and the ER lumen. In eukaryotic cells, Sec61 complexes may provide binding sites for the ribosome and other accessory proteins that help polypeptide chains fold as they enter the ER. According to one view, the bound ribosome forms a tight seal with the translocator, so that no molecules can escape from the ER lumen. Alternatively, the pore in the translocator could form a tight-fitting structure around the translocating peptide that prevents the escape of other molecules. The structure of the Sec61 complex suggests that the pore can also open along a seam on its side to allow lateral access of a translocating peptide chain into the hydrophobic core of the membrane. Such process is important both for the release of a cleaved signal peptide into the membrane and for the integration of membrane proteins into the bilayer membrane.

As we have seen, translocation of proteins into mitochondria, chloroplasts, and peroxisomes occurs post-translationally, after the protein has been made and released into the cytosol, whereas translocation across the ER membrane usually occurs during translation (**co-translation**). Such co-translational transfer process creates two spatially separate populations of ribosomes, membrane-bound and free ribosomes. They are structurally and functionally identical but differ only in the proteins they are making at any given time. Many ribosomes can bind to a single mRNA molecule and form a polyribosome directed to the ER membrane by the signal sequences on multiple growing polypeptide chains. When the translation is finished, the individual ribosomes can return to the cytosol whereas the mRNA remains attached to the ER membrane by a changing population of ribosomes that are transiently held at the membrane by the translocator.

It was found recently that some completely synthesized proteins are first released into the cytosol and then imported into the ER as well, demonstrating that translocation does not always require ongoing translation. Such post-translational translocation is especially common across the yeast ER membrane. To function in such process, completely synthesized proteins need to bind to chaperone proteins in cytosol to prevent folding, and the translocator needs accessory proteins to feed the polypeptide chain into the pore and drive translocation. These accessory proteins span the ER membrane and use a small domain on the lumenal side of the ER membrane to deposit an Hsp70-like chaperone protein (called BiP) onto the polypeptide chain as it emerges from the pore into the ER lumen. Cycles of BiP binding and release drive unidirectional translocation in a similar way as the mitochondrial Hsp70 proteins that pull proteins across mitochondrial membranes.

(2) Synthesis of transmembrane proteins on RER Transmembrane proteins translocate to the ER and remain embedded in the ER membrane. Each transmembrane protein has a unique orientation with respect to the membrane's phospholipid bilayer. Either the N-terminal segment is located on the exoplasmic face and the C-terminal segment on the cytosolic face, or in the reverse orientation. During the transport to their final destinations, the topology of a membrane protein is preserved. Thus, the orientation of these membrane proteins is established during biosynthesis on the ER membrane. Transmembrane proteins contain one to several membrane-spanning hydrophobic segments, which form **transmembrane α helices** anchoring the protein in the phospholipid bilayer.

1) **Translocation of single-pass transmembrane proteins**: There are three ways in which single-pass transmembrane proteins become inserted into the ER. In the simplest case as for the insulin receptors, the N-terminal signal sequence initiates translocation of single-pass transmembrane proteins to ER similarly as described in the water-soluble protein translocation. However, the transfer process is paused when it meets an additional hydrophobic segment in the polypeptide chain (**stop-transfer signal**). After the ER signal sequence has been released from the translocator and has been cleaved off, this stop-transfer signal is released laterally from the translocation channel and drifts into the plane of the lipid bilayer, where it forms an α-helical membrane-spanning segment that anchors the protein in the membrane. As a result, the translocated protein ends up as a transmembrane protein with the N-terminus on the luminal side of the lipid bilayer and the C-terminus on the cytosolic side (Figure 4-8).

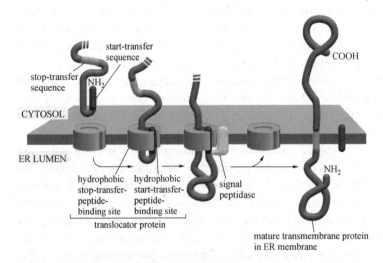

Figure 4-8 Integration of a single-pass transmembrane protein into the ER membrane (From: Alberts B, et al. Molecular Biology of the Cell. 2008)

In the other two cases, the start-transfer signal sequence is an internal hydrophobic sequence, rather than at the N-terminal end of the protein. It functions as both an ER signal sequence and membrane-anchor sequence. This **internal start-transfer sequence** directs insertion of the nascent polypeptide chain into the ER membrane. After release from the translocator, the internal start-transfer sequence remains in the lipid bilayer as a single membrane-spanning α helix (Figure 4-9). Internal start-transfer sequences can bind to the translocation apparatus in either of two orientations which in turn determines the orientation of transmembrane proteins. The orientation of the start-transfer sequence depends on

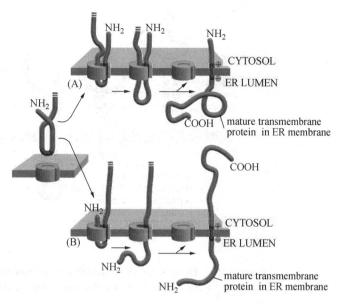

Figure 4-9 Integration of a singlepass transmembrane protein with an internal signal sequence into the ER membrane (From: Alberts B, et al. Molecular Biology of the Cell. 2008)

the distribution of nearby charged amino acids. The negatively charged end of the signal sequence inserted adjacent to the ER lumen whereas the positively charged end pointed to the cytosol.

2) **Translocation of multipass transmembrane proteins**: Many important proteins, such as ion pumps, ion channels, and transporters, span the membrane multiple times. Each membrane-spanning α helix in these multipass transmembrane proteins is used as signal sequences to start and stop the protein transfer. Combinations of start-transfer and stop-transfer signals determine the topology of multipass transmembrane proteins. In this case, hydrophobic signal sequences are thought to work in pairs: an internal start-transfer sequence serves to initiate translocation, which continues until a stop-transfer sequence is reached; the two hydrophobic sequences are then released into the bilayer, where they remain as membrane-spanning α helices (Figure 4-10).

In complex multipass proteins, in which many hydrophobic α helices span the bilayer, additional pairs of stop and start sequences come into play: one sequence reinitiates translocation further down the polypeptide chain, and the other stops translocation and causes polypeptide release, and so on for subsequent starts and stops. Thus, multipass membrane proteins are stitched into the lipid bilayer as they are being synthesized (Figure 4-11).

Whether a given hydrophobic signal sequence functions as a start-transfer or stop-transfer sequence must depend on its location in a polypeptide chain. The differences between start-transfer and stop-transfer sequences come from their relative order in the growing polypeptide chain. It seems that the SRP begins scanning an unfolded polypeptide chain for hydrophobic segments from its N-terminus toward the C-terminus. By recognizing the first appropriate hydrophobic segment to emerge from the ribosome, the SRP sets the "reading frame" for membrane integration: after the SRP initiates translocation, the translocator recognizes the next appropriate hydrophobic segment in the direction of transfer

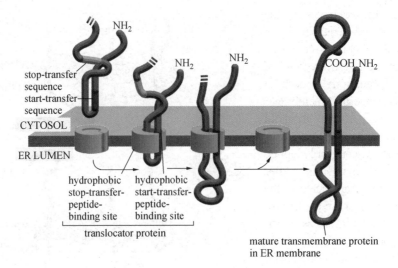

Figure 4-10　Integration of a doublepass transmembrane protein into the ER membrane (From: Alberts B, et al. Molecular Biology of the Cell. 2008)

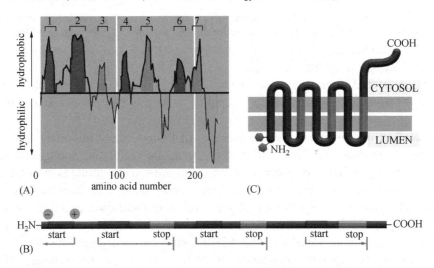

Figure 4-11　Integration of a multipass transmembrane protein rhodopsin into the ER membrane (From: Alberts B, et al. Molecular Biology of the Cell. 2008)

as a stop-transfer sequence. A similar scanning process continues until all of the hydrophobic regions in the protein have been inserted into the membrane.

(3) **Newly synthesized proteins are covalently modified in the ER**　The RER is a major protein processing place To meet its obligations, the RER lumen is packed with molecular chaperones that recognize and bind to unfolded or misfolded proteins and give them the opportunity to attain their correct three-dimensional structures.

1) **Formation of disulfide bonds**: Disulfide bonds are formed by the oxidation of pairs of cysteine side chain on the newly synthesized proteins, a reaction catalyzed by an enzyme, protein disulfide isomerase (PDI), that resides in the ER lumen. The disulfide bonds help to stabilize the tertiary and quaternary structure of those proteins. Disulfide bonds do not form in the cytosol, because of the reducing environment there. Soluble cytosolic proteins, synthesized on free ribosomes, lack disulfide bonds and utilize other interactions to stabilize

their structures.

2) **Glycosylation**: Another important modification of proteins in the ER is the protein **glycosylation**. Oligosaccharide side chains linked to amino (NH_2) group of an asparagine in a protein (*N*-linked) are by far the most common type of linkage found on glycoproteins. It is the major form of protein glycosylation in the ER. Less frequently, oligosaccharides are linked to the hydroxyl group on the side chain of a serine, threonine, or hydroxylysine amino acid, called *O*-linked glycosylation, which are carried out mainly in the Golgi apparatus. In *N*-linked glycosylation, a preformed, branched oligosaccharide containing a total of 14 sugars is transferred to the NH_2 group of an asparagine side chain on the protein immediately after the target asparagine emerges in the ER lumen during protein translocation (Figure 4-12). This ubiquitous 14-residue high-mannose precursor of *N*-linked oligosaccharides is originally formed on dolichol, a lipid that is firmly embedded in the ER membrane and acts as a carrier for the oligosacharide. Only asparagine residues in the tripeptide sequences Asn-X-Ser and Asn-X-Thr (where X is any amino acid except proline) are the glycosylation sites. The addition of the 14-sugar oligosaccharide to proteins in the ER is only the

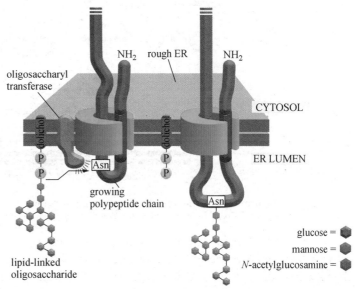

Figure 4-12 *N*-linked protein glycosylation in the ER (From: Alberts B, et al. Molecular Biology of the Cell. 2008)

first step in a series of further modifications. Despite their initial similarity, the *N*-linked oligosaccharides on mature glycoproteins are remarkably diverse. Modifications to *N*-linked oligosaccharides are completed in the Golgi complex.

The oligosaccharides on proteins serve various functions, depending on the protein. They can protect the protein from degradation, hold the protein in the ER until it is properly folded, or help guide it to the appropriate organelle by serving as a transport signal for packaging the protein into appropriate transport vesicles. When displayed on the cell surface, oligosaccharides form part of carbohydrate layer and can function in the recognition of one cell by another.

Oligosaccharides are used as tags to mark the state of protein folding. Two ER chaperone proteins, calnexin and calreticulin which require Ca^{2+} for their activities, bind to oligosaccharides on incompletely folded proteins and retain them in the ER. Like other chaperones, they prevent incompletely folded proteins from becoming irreversibly aggregated. Calnexin and calreticulin recognize *N*-linked oligosaccharides that contain a single terminal glucose, and therefore bind proteins only after two of the three glucoses on the precursor oligosaccharide have been removed by ER glucosidases. When the third glucose has been removed, the protein dissociates from its chaperone and can leave the ER. Another ER enzyme, a glucosyl transferase that keeps adding a glucose to those oligosaccharides that are attached to unfolded proteins. Thus, an unfolded protein undergoes continuous cycles of glucose removal (by glucosidase) and glucose addition (by glycosyl transferase), maintaining an affinity for calnexin and calreticulin until it has achieved its fully folded state (Figure 4-13).

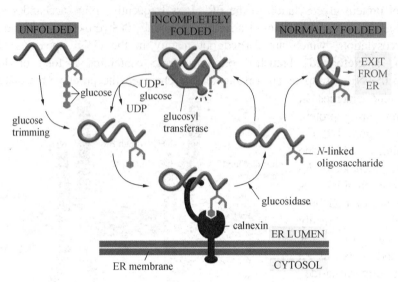

Figure 4-13 The role of *N*-linked glycosylation in ER protein folding
(From: Alberts B, et al. Molecular Biology of the Cell. 2008)

Despite all the help from chaperones, many protein molecules translocated into the ER fail to achieve their properly folded or oligomeric state. Such proteins are exported from the ER back into the cytosol and are degraded. Chaperone proteins are probably necessary to keep the polypeptide chain in an unfolded state prior to and during transport. A source of energy and a translocator might be required. The *N*-linked oligosaccharides, which serve as timers that measure how long a protein has spent in the ER. The slow trimming of a particular mannose on the core-oligosaccharide tree by an enzyme (a mannosidase) in the ER is thought to create a new oligosaccharide structure that the retrotranslocation apparatus recognizes. Proteins that fold and exit from the ER faster than the action of the mannosidase would therefore escape degradation. Once a misfolded protein has been retrotranslocated into the cytosol, an N-glycanase removes all its oligosaccharide chains. The deglycosylated polypeptide is rapidly ubiquitylated by ER-bound ubiquitin-conjugating enzymes and is then fed into proteasomes and degraded.

3) **Glycosylphosphatidyl-inositol anchor of membrane proteins**: Several cytosolic enzymes catalyze the covalent addition of a single fatty acid chain or prenyl group to selected proteins. The attached lipids help to direct these proteins to cell membranes. A related process is catalyzed by ER enzymes, which covalently attach a glycosylphosphatidyl-inositol (GPI) anchor to the C-terminus of some membrane proteins destined for the plasma membrane. This linkage forms in the lumen of the ER, where, at the same time, the transmembrane segment of the protein is cleaved off. A large number of plasma membrane proteins are modified in this way and they are attached to the exterior of the plasma membrane by their GPI anchors. In principle, they can be released from cells in soluble form in response to signals that activate a specific phospholipase in the plasma membrane.

4.2.2 Lipid bilayer assembly in the ER

The ER membrane synthesizes nearly all of the major classes of lipids, including both phospholipids and cholesterol. There are two major exceptions: sphingomyelin and glycolipid synthesis begin in the ER and are completed in Golgi; some unique lipids in mitochondrial and chloroplast membranes are synthesized by enzymes residing in those organelles. The major phospholipid made in ER is phosphatidylcholine, which can be formed from choline, two fatty acids, and glycerol phosphate. Three steps are catalyzed by enzymes in the ER membrane that have their active sites facing the cytosol, where all of the required metabolites are found. Therefore, phospholipid synthesis occurs exclusively in the cytosolic leaflet of the ER membrane. Some of these lipid molecules are later flipped into the opposite leaflet through the action of flippases.

The ER also produces cholesterol and ceramide. Ceramide is made by condensing the amino acid serine with a fatty acid to form the amino alcohol sphingosine, a second fatty acid is then added to form ceramide. The ceramide is exported to the Golgi apparatus, where it serves as a precursor for the synthesis of two types of lipids: oligosaccharide chains are added to form glycosphingolipids and phosphocholine head groups are transferred from phosphatidylcholine to other ceramide molecules to form sphingomyelin. Thus, both glycolipids and sphingomyelin are produced relatively late in the process of membrane synthesis. Because they are produced by enzymes exposed to the Golgi lumen, they are found exclusively in the noncytosolic leaflet of the lipid bilayers.

The two phospholipid leaflets in the cellular membranes are asymmetric. This asymmetry is established initially in the endoplasmic reticulum and maintained as membrane carriers bud from one compartment and fuse to the next. As a result, domains located at the cytosolic surface of the ER membrane can be identified on the cytosolic surface of transport vesicles, the cytosolic surface of Golgi cisternae, and the cytoplasmic surface of the plasma membrane. Similarly, domains situated at the luminal surface of the ER membrane are found at the exoplasmic surface of the plasma membrane. In fact, the lumen of the ER and other compartments of the secretory pathway is a lot like the extracellular space. They all have high calcium concentration and abundance of proteins with disulfide bonds and carbohydrate chains.

The membranes of different organelles have markedly different lipid composition, which indicates that changes take place as membrane flows through the cell. Several factors may contribute to such changes. First, most membranous organelles contain different enzymes that modify lipids already present within a membrane. Second, when vesicles bud from a compartment, some types of lipids may be preferentially included within the membrane of the forming vesicle, while others may be excluded. Third, cells contain phospholipid-transfer proteins that can bind and transport phospholipids through the aqueous cytosol from one membrane compartment to another. These enzymes facilitate the movement of specific phospholipids from the ER to other organelles. This is especially important in delivering lipid to mitochondria and chloroplasts, which are not part of the normal flow of membrane along the biosynthetic pathway.

4.2.3 Other functions in the ER

ER is the start site for biosynthesis in the cell. Some parts of RER are devoid of ribosomes and function as the exit sites where the first transport vesicles in the biosynthesis pathway are formed. The first step in vesicular transport begins from ER to Golgi complex as COP II coated vesicles, which will be discussed later in this chapter.

The SER is extensively developed in cells from skeletal muscle, kidney tubules, and steroid-producing endocrine glands. SER functions as the synthesis site of steroid hormones in the endocrine cells of the gonad and adrenal cortex. In liver, detoxification of a wide variety of organic compounds, including barbiturates and ethanol, is carried out by a system of oxygentransferring enzymes (oxygenases), including the cytochrome P450 family. There is no substrate specificity for these enzymes which are able to oxidize thousands of different hydrophobic compounds and convert them into more hydrophilic, more readily excreted derivatives. Cytochrome P450s metabolize many prescribed medications, and genetic variation in these enzymes among humans may explain differences from one person to the next in the effectiveness and side effects of many drugs.

Another function of ER is sequestering calcium ions from the cytoplasm of cells. SER is abundant in skeletal and cardiac muscle cells (known as the sarcoplasmic reticulum in muscle cells). A Ca^{2+}-ATPase resides on the ER membrane and pumps Ca^{2+} from cytosol into the ER lumen. A high concentration of Ca^{2+}-binding proteins in the ER facilitaes Ca^{2+} storage. The release of Ca^{2+} from ER triggers contraction of muscle cells.

4.3 Golgi apparatus

Golgi apparatus is constituted of a series of flattened membrane vesicles or sacs, called **cisternea**, surrounded by a number of more or less spherical membrane vesicles. Each Golgi stack has two distinct faces: an entry, or *cis* face and an exit, or *trans* face (Figure 4-14). The *cis* face is adjacent to the ER, while the *trans* face points toward the plasma membrane. The *cis* face is composed of an interconnected network of tubules referred to as the *cis* Golgi network (CGN). The CGN is thought to function primarily as a sorting station that distinguishes between proteins to be shipped back to the ER and those that are allowed to proceed to the next Golgi station. The Golgi cisternae can be divided into *cis*, medial, and *trans* cisternae. The *trans*-most face of Golgi contains a distinct network of tubules and vesicles called the *trans* Golgi network (TGN). The TGN is a sorting station where proteins are segregated into different types of vesicles heading either to the plasma membrane or to various intracellular destinations.

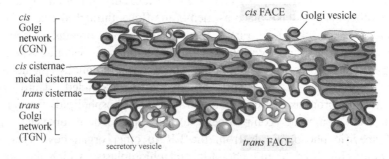

Figure 4-14 Structure of Golgi apparatus (From: Alberts B, et al. Molecular Biology of the Cell. 2008)

Golgi apparatus lies on the exit route from the ER and is a cell structure to further process the proteins synthesized in the ER. It receives proteins and lipids from the ER, modifies them, and then dispatches them to other destinations in the cell. Golgi is a major site of carbohydrate synthesis. Cells make many polysaccharides in the Golgi apparatus, including the pectin and hemicellulose of the cell wall in plants and most of the glycosaminoglycans of the extracellular matrix in animals. A large proportion of the carbohydrates that Golgi makes are attached as oligosaccharide side chains to the proteins and lipids that the ER sends to Golgi. Most proteins and lipids, once receiving their appropriate oligosaccharides in the Golgi apparatus, are recognized for targeting into the transport vesicles going to their destinations.

(1) **Glycosylation in Golgi apparatus** Many of the *N*-linked oligosaccharides that are added to proteins in the ER undergo further modifications in the Golgi apparatus. In contrast to the glycosylation events that occur in the ER, which assemble a single core oligosaccharide, the glycosylation steps in the Golgi complex can be quite varied, producing carbohydrate domains of remarkable sequence diversity. As newly synthesized glycoproteins from ER pass through the *cis* and medial cisternae of the Golgi stack, most of the mannose (Man) residues are removed from the core oligosaccharides, while a variety of sugars, such as N-acetylglucosamine (GlcNAC), galactose (Gal), fucose, and sialic acid (NANA), are added to the oligosaccharide by specific glycosyltransferases to yield a finished *N*-linked oligosaccharide. There is a clear correlation between the position of an enzyme in the chain of processing events and its localization in the Golgi stack: enzymes that act early are found in cisternae close to the *cis* face, while enzymes that act late are found in cisternae near the *trans* face (Figure 4-15).

Except that many proteins are modified by *N*-linked oligosaccharides, some proteins are glycosylated by *O*-linked oligosaccharides in Golgi apparatus. *O*-linked oligosaccharides are bound to the hydroxyl groups in certain serine, threonine, or hydroxyl lysine residues. They are generally short, often containing only one to four sugar residues. *O*-linked oligosaccharides are formed by the sequential addition of sugars in the Golgi. Addition is catalyzed by various glycosyltransferases that are specific for the donor sugar nucleotide and acceptor molecule. The Golgi apparatus confers the heaviest *O*-linked glycosylation on proteoglycan core proteins to produce proteoglycans. This process in-

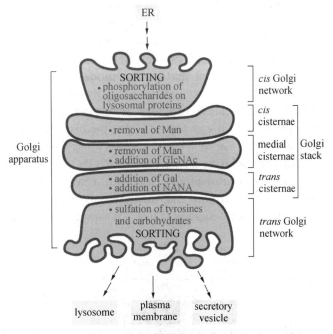

Figure 4-15 Oligosaccharide processing in Golgi apparatus
(From: Alberts B, et al. Molecular Biology of the Cell. 2008)

volves the polymerization of one or more glycosaminoglycan chains via a xylose link onto serines on a core protein. Many proteoglycans are secreted and become components of the extracellular matrix, while others remain anchored to the extracellular face of the plasma membrane. Some proteoglycans form a major component of slimy materials, such as the mucus that is secreted to form a protective coating on the surface of many epithelia. The sugars incorporated into glycosaminoglycans are heavily sulfated in the Golgi apparatus immediately after these polymers are made, thus adding a significant portion of their characteristically large negative charge. Some tyrosines in proteins also become sulfated shortly before they exit from the Golgi apparatus.

The vast abundance of glycoproteins and the complicated pathways to synthesize them suggest that the oligosaccharides on glycoproteins and glycosphingolipids have very important functions. First, it has a direct role in making folding intermediates more soluble, thereby preventing their aggregation. Second, the sequential modifications of the N-linked oligosaccharide establish a "glyco-code" that marks the progression of protein folding and mediates the binding of the protein to chaperones (discussed in the ER part) in guiding ER-to-Golgi transport. Third, the presence of oligosaccharides tends to make a glycoprotein more resistant to digestion by proteolytic enzymes. Glycosylation can also have important regulatory roles in cell signaling. The recognition of sugar chains by lectins in the extracellular space is important in many developmental processes and in cell-cell recognition.

(2) Material transport through the Golgi apparatus   How materials move through the various compartments of the Golgi complex has long been investigated. However, it is still uncertain how the Golgi apparatus achieves and maintains its polarized structure and how molecules move from one cisterna to another. Up until the mid-1980s, it was generally accepted that Golgi cisternae were transient structures and Golgi cisternae formed at the cis face of the stack by fusion of membranous carriers from the ER and ERGIC (endoplasmic reticulum Golgi intermediate compartment) and that each cisterna physically moved from the cis to the trans end of the stack, changing in composition as it progressed. This is known as the **cisternal maturation model** (Figure 4-16A). In such model, each cisterna "matures" into the next cisterna along the stack. From the mid-1980s to the mid-1990s, an alternate model called **vesicular transport model** arises (Figure 4-16B). In vesicular transport model, it was proposed that the cisternae of a Golgi stack remain in place as stable compartments. Cargo (i.e., secretory, lysosomal, and membrane proteins) is shuttled through the Golgi stack, from the CGN to the TGN, in vesicles that bud from one membrane compartment and fuse with a neighboring compartment farther along the stack. The vesicular transport and the cisternal maturation models are not mutually exclusive. Indeed, evidence suggests that transport may occur by a combination of the two

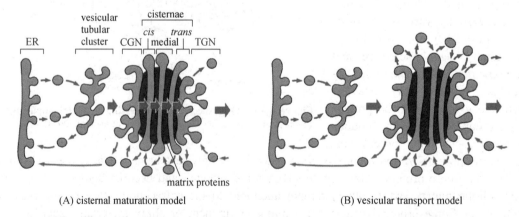

Figure 4-16   The dynamic transport through Golgi apparatus (From: Alberts B, et al. Molecular Biology of the Cell. 2008)

mechanisms, in which some cargo is moved forward (anterograde) or backward (retrograde) rapidly in transport vesicles, whereas other cargo is moved forward more slowly as the Golgi apparatus constantly renews itself through cisternal maturation.

4.4 Lysosomes and peroxisomes

4.4.1 Lysosomes

Lysosomes are found in all eukaryotic cells as tiny organelles lined by a single layer of membrane. The appearance of lysosomes in electron micrographs is quit divergent and varies in size, shape and electron density. Lysosomes are formed off the membrane of the *trans* Golgi and contain digestive enzymes that degrade worn-out organelles, as well as macromolecules and particles taken into the cell by endocytosis.

The enzymes of a lysosome share an important property: all have their optimal activity at an acid pH and thus are acid hydrolases. The pH optimum of these enzymes is suited to the low pH of the lysosomal compartment, which is approximately 4.6. The high internal proton concentration is maintained by a proton pump (an H^+-ATPase) present in the membrane. The membrane of the lysosome normally keeps these destructive enzymes out of the cytosol (whose pH is about 7.2), but the acid dependence of the enzymes protects the contents of the cytosol against damage even if some leakage should occur. Lysosomal membranes contain a variety of highly glycosylated integral proteins whose carbohydrate chains are thought to form a protective layer that shields the membrane from attack by the enclosed enzymes. The lysosomal membrane contains transport proteins that allow the final products of the digestion of macromolecules, such as amino acids, sugars, and nucleotides, to be transported to the cytosol, from where they can be either excreted or utilized by the cell.

The specialized digestive enzymes and membrane proteins of the lysosome are synthesized in the ER and transported through the Golgi apparatus to the *trans* Golgi network. *N*-linked oligosaccharides are added during transport to target lysosomal enzymes to lysosomes and to prevent their secretion. The addition and initial processing of *N*-linked oligosaccharide precursor in the rough ER is the same for lysosomal enzymes as for membrane and secretory proteins. In the *cis* Golgi, one or more mannose residues in the resulting *N*-linked oligosaccharides become phosphorylated to obtain a mannose 6-phosphate (M6P) residue (Figure 4-17). The M6P residues then bind to mannose 6-phosphate receptors, which are found primarily in the *trans* Golgi network. Vesicles containing the M6P receptors and bound lysosomal enzymes bud from the *trans* Golgi network and then fuse with a sorting vesicle, an organelle often termed the **late endosome**. The bound lysosomal enzymes are released within late endosomes. Also, a phosphatase within late endosomes generally removes the phosphate from lysosomal enzymes, preventing their rebinding to the M6P receptor. Two types of vesicles bud from late endosomes. One type containing lysosomal enzymes but not the M6P receptor fuses with lysosomes, delivering the lysosomal enzyme to their final destination. The other type of vesicle recycles the M6P receptor back to the *trans* Golgi network or, on occasion, to the cell surface.

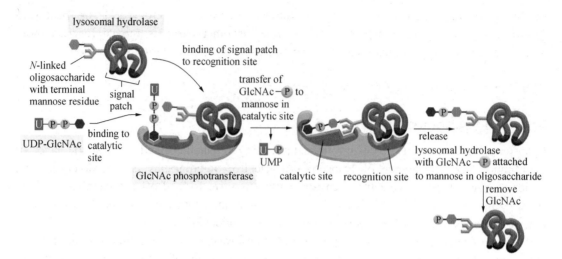

Figure 4-17 M6P pathway for targeting lysosomal enzymes to lysosomes (From: Alberts B, et al. Molecular Biology of the Cell. 2008)

Depending on their source, materials follow different paths to enter lysosomes. The extracellular particles are taken up into **phagosomes**, which then fuse with lysosomes, and the extracellular fluid and macromolecules are taken up into smaller endocytic vesicles, which deliver their contents to lysosomes via endosomes. But cells have an additional pathway for supplying materials to lysosomes, this pathway is used for degrading obsolete parts of the cell itself. In electron micrographs of liver cells, for example, one often sees lysosomes digesting mitochondria, as well as other organelles. The process seems to begin with the enclosure of the organelle by a double membrane, creating an **autophagosome**, which then fuses with lysosomes.

The dysfunction of lysosomes might lead to severe diseases. The deficiency of a single lysosomal enzyme and the corresponding accumulation of undegraded substrate in lysosomes lead to so called **lysosomal storage disorders**. Such as Pompe disease caused by the absence of a lysosomal enzyme, α-glucosidase, could lead to undigested glycogen accumulating in lysosomes, swelling of the organelles and irreversible damage to the cells and tissues. Tay-Sachs disease, which results from a deficiency of the enzyme β-N-hexosaminidase A, an enzyme that degrades the ganglioside GM2, is characterized during infancy with progressive mental and motor retardation, as well as skeletal, cardiac, and respiratory abnormalities. In I-cell disease, deficiency of an enzyme (N-acetylglucosamine phosphotransferase) required for mannose phosphorylation of lysosomal enzymes in the Golgi complex leads to the loss of an "address" for delivery of these proteins to lysosomes. Many cells in these patients contain lysosomes that are bloated with undegraded materials. Gaucher's disease, a deficiency of the lysosomal enzyme glucocerebrosidase, is characterized by accumulated large quantities of glucocerebroside lipids in the lysosomes of patient macrophages, causing spleen enlargement and anemia. Gaucher's disease can be alleviated by enzyme replacement therapy. Enzyme replacement therapy for several other lysosomal storage diseases has either been approved or is being investigated in clinical trials. However, many of these diseases affect the central nervous system, which is unable to take up circulating enzymes because of the blood-brain barrier.

In most plant cells, a single membrane-bound, fluid-filled central **vacuole**, which occupies as much as 90 percent of the cell volume, belongs to lysosomes. Plant vacuoles carry out a wide spectrum of essential functions. It functions as a temporary storage for many solutes and macromolecules, including ions, sugars, amino acids, proteins, and polysaccharides. Vacuoles may also store a host of toxic compounds, such as cyanide-containing glycosides and glucosinolates, which are part of an arsenal of chemical weapons. Other toxic compounds are simply the by-products of metabolic reactions. Plants lack the type of excretory systems found in animals, they utilize their vacuoles to isolate these by-products from the rest of the cell. The membrane that bounds the vacuole contains a number of active transport systems that pump ions into the vacuolar compartment. Such active transport leads to high ion concentration in vacuoles so that water enters the vacuole by osmosis. Hydrostatic pressure exerted by the vacuole not only provides mechanical support for the soft tissues of a plant, it also stretches the cell wall during cell growth.

4.4.2 Peroxisomes

Peroxisomes are small organelles enclosed by a single membrane. They contain enzymes used in a variety of oxidative reactions that break down lipids and destroy toxic molecules. Unlike mitochondria and chloroplasts, peroxisomes lack DNA and ribosomes and are lined by a single layer membrane. Thus, all peroxisomal proteins are encoded by nuclear genes, synthesized on ribosomes free in the cytosol, and then incorporated into pre-existing peroxisomes. Most peroxisomal membrane and matrix proteins are incorporated into the organelle post-translationally, although some of them enter the peroxisome membrane via the ER. The oxidative enzymes, such as catalase and urate oxidase, are folded in the cytosol and incorporated as a folded protein. As peroxisomes are enlarged by addition of protein (and lipid), they eventually divide, forming new ones, similar to mitochondria and chloroplasts (Figure 4-18).

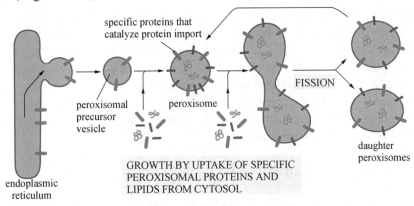

Figure 4-18 Model for peroxisome generation and proliferation (From: Alberts B, et al. Molecular Biology of the Cell. 2008)

Many peroxisomal matrix proteins, such as catalase and fatty acyl CoA oxidase, utilize a C-terminal Ser-Lys-Leu (**SKL**) **targeting sequence** that is not cleaved after import. Such proteins bind to the cytosolic receptor PTS1R, which escorts the imported protein to receptor and transporter proteins on the peroxisomal membrane. After transported across the membrane, the imported protein dissociates with PTS1R in the lumen, and PTS1R returns to cytosol to pick up another peroxisome-destined protein. Other peroxisomal matrix proteins, such as thiolase, are synthesized as precursors with an N-terminal uptake-targeting sequence. Proteins with this type of signal sequence bind to a cytosolic receptor protein named PTS2R. Like PTS1R, PTS2R escorts the precursor protein to the receptor on the membrane. Following import of such proteins, the N-terminal targeting sequence is cleaved.

Like mitochondria, peroxisomes are major sites of oxygen utilization. However, oxidative reactions performed in peroxisomes are not coupled to ATP formation. In peroxisomes, molecular oxygen is used to remove hydrogen atoms from specific organic substrates (designated as R) in an oxidation reaction that produces hydrogen peroxide (H_2O_2): $RH_2 + O_2 \rightarrow R + H_2O_2$. Catalase uses the H_2O_2 generated by other enzymes in the organelle to oxidize a variety of other substrates—including phenols, formic acid, formaldehyde, and alcohol — by the "peroxidation" reaction: $H_2O_2 + R'H_2 \rightarrow R' + 2H_2O$. This type of oxidation reaction is particularly important in liver and kidney cells, where the peroxisomes detoxify various toxic molecules that enter the bloodstream. About 25% of the ethanol we drink is oxidized to acetaldehyde in this way. In addition, when excess H_2O_2 accumulates in the cell, catalase converts it to H_2O through the reaction: $2H_2O_2 \rightarrow 2H_2O + O_2$.

Another major function of the oxidation reactions performed in peroxisomes is the breakdown of fatty acid molecules through β oxidation. Thus the fatty acids are converted to acetyl CoA in peroxisomes and then exported to the cytosol for reuse in biosynthetic reactions. In mammalian cells, β oxidation occurs in both mitochondria and peroxisomes; in yeast and plant cells, however, this essential reaction occurs exclusively in peroxisomes. An essential biosynthetic function of animal peroxisomes is to catalyze the first reactions in the formation of plasmalogens, which are the most abundant class of phospholipids in myelin. Plasmalogen deficiencies cause profound abnormalities in the myelination of nerve cell axons, which is why many peroxisomal disorders lead to neurological disease.

Peroxisomes are also important in plants. Plant peroxisomes present in leaves participate in photorespiration. The other type of peroxisome is present in germinating seeds, where it converts the fatty acids stored in seed lipids into the sugars needed for the growth of the young plant. In a process called glyoxylate cycle, two molecules of acetyl CoA produced by fatty acid breakdown in the peroxisome are used to make succinic acid, which then leaves the peroxisome and is converted into glucose in the cytosol. The glyoxylate cycle does not occur in animal cells, and animals are therefore unable to convert the fatty acids in fats into carbohydrates.

4.5 Molecular mechanisms of vesicular transport

As to eukaryotic cells, the traffic of molecules between different membrane-enclosed compartments within a cell, or between the interior of the cell and its surroundings is highly organized by **vesicular transport**. In the outward **biosynthetic-secretory pathway** (**exocytosis**), molecules are packaged into **transport vesicles**, leading from ER, through Golgi apparatus, plasma membrane and to the cell surface. At the Golgi apparatus, a side branch budding of vesicles transport molecules to endosomes and then lysosomes. The major inward **endocytic pathway** (**endocytosis**) transports materials uptake at the cell plasma membrane, through endosomes, to lysosomes. (Figure 4-19).

Chapter 4　Endomembrane system, protein sorting and vesicle transport

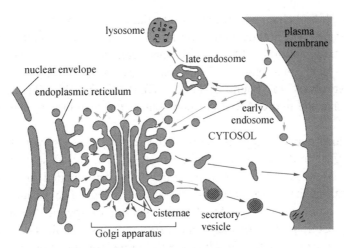

Figure 4-19　Exocytosis and endocytosis pathways (From: Alberts B, et al. Molecular Biology of the Cell. 2008)

4.5.1　Three types of coated transport vesicles

Transport vesicles have a distinctive protein coat (**coated vesicle**) on their cytosolic surface when budding from membranes. Based on the coat protein, transport vesicles are classified as three major types: **Clathrin**, **COP Ⅰ** and **COP Ⅱ** coated vesicles (Table 4-3). COP Ⅰ - coated vesicles are involved in transporting molecules from Golgi apparatus to the ER and between the Golgi cisternae. Vesicles with a COP Ⅱ coat transport proteins from the ER to the *cis* Golgi apparatus. Clathrin-coated vesicles bud from the Golgi apparatus on the outward secretory pathway and from the plasma membrane on the inward endocytic pathway.

Table 4-3　Some types of coated vesicles

type	GTPase	coat Protein	origin	destination
clathrin	Arf	clathrin, adaptin2	plasma membrane	endosomes
clathrin	Arf	clathrin, adaptin1	*trans* Golgi	endosomes
clathrin	Arf	clathrin, adaptin3	Golgi	lysosomes, melanosome, or platelet vesicles
COP Ⅰ	Arf	COP proteins	*cis* Golgi	ER
COP Ⅰ	Arf	COP proteins	later Golgi cisterna	earlier Golgi cisterna
COP Ⅱ	Sar1	Sec23/Sec24 and Sec13/Sec31 complexes, Sec16, Sec12	ER	*cis* Golgi

Different coat protein assembly requires a family of small monomeric GTP-binding proteins, so called **coat-recruitment GTPases**. They include the Arf proteins, which are responsible for both COP Ⅰ coat assembly and clathrin coat assembly at Golgi membranes, and the Sar1 protein, which is responsible for COP Ⅱ coat assembly at the ER membrane (Table 4-3). Coat-recruitment GTPases are usually found in high concentration in the cytosol in an inactive GDP-bound state. When a COP Ⅱ -coated vesicle is to bud from the ER membrane, a specific Sar1-GEF embedded in the ER membrane binds to cytosolic Sar1, causing the Sar1 to exchange its GDP with GTP. In its GTP-bound state, the Sar1 protein exposes an amphiphilic helix, which inserts into the cytoplasmic leaflet of the lipid bilayer of the ER membrane. The tightly bound Sar1 now recruits coat protein subunits to the ER membrane to initiate budding. Other GEFs and coat-recruitment GTPases operate in a similar way on other membranes. The coat-recruitment GTPases also regulate coat disassembly. Hydrolysis of bound GTP to GDP causes the GTPase to change its conformation so that the hydrophobic tail pops out of the membrane, causing vesicle's coat to disassemble.

In order to load correct cargo proteins in the correct transport vesicles, discrimination among potential membrane and soluble cargo proteins is required so that only those cargo proteins that should advance to the next compartment are loaded and those that should remain as residents in the donor compartment are excluded. The vesicle coat functions in selecting specific proteins as cargo by directly binding to specific sequences, or sorting signals, in the cytosolic portion of membrane cargo proteins (Table 4-4). The polymerized coat thus acts as an affinity matrix to cluster selected membrane cargo proteins into forming vesicle buds. Soluble proteins within the lumen of parent organelles can in turn be selected by binding to the luminal domains of certain membrane cargo proteins, which act as receptors for luminal cargo proteins.

Table 4-4 Known sorting signals that direct proteins to specific transport vesicles

signal sequence*	proteins with signal	signal receptor	vesicles
Lys-Asp-Glu-Leu (KDEL)	ER-resident luminal proteins	KDEL receptor in *cis* Golgi membrane	COP I
Lys-Lys-X-X (KKXX)	ER-resident membrane proteins (cytosolic domain)	COP I α and β subunits	COP I
Di-acidic (e.g., Asp-X-Glu)	cargo membrane proteins in ER (cytosolic domain)	COP II Sec24 subunit	COP II
mannose 6-phosphate (M6P)	soluble lysosomal enzymes after processing in *cis* Golgi	M6P receptor in *trans* Golgi membrane	clathrin/AP1
	secreted lysosomal enzymes	M6P receptor in plasma membrane	clathrin/AP2
Asn-Pro-X-Tyr (NPXY)	LDL receptor in the plasma membrane (cytosolic domain)	AP2 complex	clathrin/AP2
Tyr-X-X-Φ (YXXΦ)	membrane proteins in *trans* Golgi (cytosolic domain)	AP1 (μ_1 subunit)	clathrin/AP1
	plasma membrane proteins (cytosolic domain)	AP2 (μ_2 subunit)	clathrin/AP2
Leu-Leu(LL)	plasma membrane proteins (cytosolic domain)	AP2 complexes	clathrin/AP2

* X: any amino acid; Φ: hydrophobic amino acid. Single-letter amino acid abbreviations are in parentheses.

(1) **Formation of Clathrin-coated Vesicles** Clathrin-coated vesicles are the best-studied vesicles. The formation of this transport vesicle requires various adapter proteins (**adaptin**) as well as **clathrin**, and their final budding off requires a small GTP-binding protein, called **dynamin**.

Each clathrin consists of three large and three small polypeptide chains that together form a three-legged structure called a triskelion. During vesicle formation, the transported molecules carry specific transport signals that are recognized by cargo receptors in the compartment membrane. Adaptins bind to the cargo receptors via a four amino acid motif on the receptors, Tyr-X-X-Φ (X for any amino acid, and Φ for hydrophobic amino acid such as Phe, Leu or Met), to help selecting cargo molecules. There are several types of adaptins that are specific for a different set of cargo receptors to form distinct clathrin coated vesicles (Figure 4-20). On the other hand, adaptins bind and secure the clathrin coat to the vesicle membrane. The clathrins assemble into a basketlike network on the cytosolic surface of the membrane to shape it into a **clathrin-coated pit**. Dynamin assembles as a ring around the neck of each deeply invaginated coated pit and contracts by consuming the energy generated from GTP hydrolysis until the vesicle pinches off. After budding from its parent organelle, the vesicle sheds its coat, allowing its membrane to interact directly with the membrane to which it will fuse.

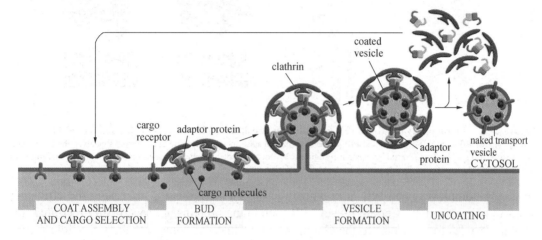

Figure 4-20 **Formation of clathrin coated vesicles** (From: Alberts B, et al. Molecular Biology of the Cell. 2008)

The inositol phospholipids of the vesicle membrane also play important role in vesicle formation. The adaptor proteins found in clathrin coats also bind to phosphoinositides (PIPs), which are used as molecular markers of compartment identity and direct when and where coats assemble in the cell. Various types of PIPs can be produced through rapid cycles of phosphorylation and dephosphorylation at the 3', 4', and 5' positions of their inositol sugar head groups by the distinct sets of PI and PIP kinases and PIP phosphatases located on different organelles in the endocytic and biosynthetic-secretory pathways. As a consequence, the distribution of PIPs varies from organelle to organelle, and even within a continuous membrane from one region to another, thereby defining specialized membrane domains. Adaptor proteins contain domains that bind with high specificity to the head groups of particular PIPs to help regulate vesicle formation and other steps in membrane transport (Figure 4-21).

(2) Formation of COP II vesicles   The first step in the secretory pathway, from the ER to Golgi, is mediated by COP II vesicles. Formation of COP II vesicles is triggered by a guanine nucleotide — exchange factor, Sec12-mediated GDP/GTP exchange on Sar1 GTPase. The GTP-bound Sar1 is recruited to the ER membrane followed by binding of a complex of Sec23 and Sec24 proteins. A second complex comprising Sec13 and Sec31 proteins then binds to the previous complexes to complete the coat structure. Sec16, a large fibrous protein, is bound to the cytosolic surface of the ER, interacts with the Sec13/31 and

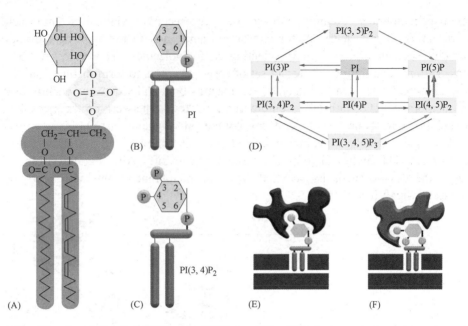

Figure 4-21 Phosphatidylinositol（PI）and phosphoinositides（PIPs）（A，B）The structure of PI；（C）A variety type of PIPs；（D）Several PI and PIP kinases and phosphatase localized to different organelles catalyze the production of particular PIPs；（E,F）PIP head groups are recognized by adaptor protein domains that discriminate between the different forms（From：Alberts B，et al. Molecular Biology of the Cell. 2008）

Sec23/24 complexes, and acts to organize the other coat proteins, increasing the efficiency of coat polymerization. Certain integral ER membrane proteins are specifically recruited into COP II vesicles for transport to the Golgi. Many of these proteins contain a di-acidic sorting signal（such as Asp-X-Glu）in the cytosolic segments. This sorting signal binds to the Sec24 subunit of the COP II coat and is essential for the selective export of certain membrane proteins from the ER. These vesicles bud from specialized regions of the ER, called ER exit sites, whose membrane lacks bound ribosomes.

(3) Formation of COP I vesicles Start of formation of COP I vesicles requires the membrane recruitment of small GTP-binding protein Arf1. COP I vesicles mediate retrograde transport within the Golgi and from the Golgi to the ER and transport escaped proteins back to the ER. Proteins that normally reside in the ER contain retrieval signals at their C-termini which can be captured by specific receptors in COP I -coated vesicles when the proteins escape from ER via COP II vesicles to the Golgi apparatus. Soluble resident proteins of the ER lumen（such as protein disulfide isomerase and the molecular chaperones that facilitate folding）typically possess the retrieval signal "lys-asp-glu-leu"（KDEL）. These proteins are recognized and returned to the ER by the KDEL receptor, an integral membrane protein that shuttles between the *cis* Golgi and the ER compartments. Membrane proteins that reside in the ER also have a retrieval signal with two closely linked basic residues, most commonly "lys-lys-X-X"（KKXX, where X is any residue）, that binds to the COP I coat, facilitating their return to the ER.

4.5.2 Targeting vesicles to a particular compartment

After a transport vesicle buds from a membrane, it is actively transported by motor proteins that move along cytoskeleton fibers to its destination. Once a transport vesicle has reached its target, it must recognize and dock with the organelle. The vesicle membrane fuses with the target membrane and unload the vesicle's cargo. Targeting vesicles to a particular compartment requires specific interactions between different membranes. Two types of proteins are crucial in these processes: Rab proteins direct the vesicle to specific spots on the correct target membrane and then SNARE proteins mediate the fusion of the lipid membrane.

(1) **Rabs** Like coat-recruitment GTPases, Rabs belong to a family of small G proteins and associate with membranes by a lipid anchor when bound GTP. Over 60 different Rab genes were identified in humans, which constitute the most diverse group of proteins involved in membrane trafficking. Different Rabs associated with different membrane compartments give each compartment a unique surface identity. The GTP-bound Rabs play a key role in vesicle targeting by recruiting specific cytosolic tethering proteins to specific membrane surfaces. Each Rab appears to bind to a specific Rab effector, a typically long coiled-coil protein, associated with the target membrane.

The Rab effectors vary greatly in structure. Some are motor proteins that propel vesicles along cell skeleton to their target membrane. Others are tethering proteins containing long threadlike domains that link two membranes. Rab effectors can also interact with SNAREs coupling membrane tethering to fusion (Figure 4-22).

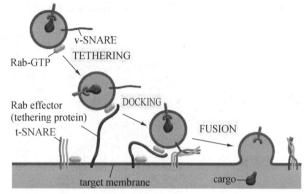

Figure 4-22 Tethering of a vesicle to a target membrane by Rabs and SNAREs (From: Alberts B, et al. Molecular Biology of the Cell. 2008)

Rab proteins assembly on a membrane with their effectors resulting in the formation of large specialized membrane patches. Rab5, for instance, is anchored to the membrane of endosomes upon GTP binding to mediate the capture of clathrin-coated vesicles budding from plasma membrane. Simultanously, Rab5 activate PI3-kinase to convert PI to PI(3)P which in turn binds some Rab effectors and help to establish functionally distinct membrane domains within a continuous membrane. Different Rab proteins help to create multiple specialized membrane domains. Rab5 receives incoming vesicles from plasma membranes, Rab11 and Rab4 domains in the same membrane are thought to organize the budding of recycling vesicles that return proteins from endosomes to the plasma membrane.

(2) **SNAREs** SNAREs are transmembrane proteins response for the fusion of transport vesicles to target membrane. Uncoating of transport vesicles exposes specific v-SNARE proteins on the surface of each type of vesicle. Each v-SNARE recognizes and binds to the complementary t-SNARE proteins on the membrane of target vesicles or organelles. Each organelle and each type of transport vesicle is believed to carry a unique SNARE, and the interactions between complementary SNAREs help ensuring that transport vesicles fuse only with the correct membrane (Figure 4-22).

Once a transport vesicle has recognized its target membrane and docked there, the vesicle has to fuse with the membrane to deliver its cargo. Fusion not only delivers the contents of the vesicle into the interior of the target organelle, it also adds the vesicle membrane to the membrane of the organelle. Membrane fusion does not always follow immediately after docking, however, it often awaits a specific molecular signal. Although the SNARE proteins themselves are thought to play a central role in the fusion process, some other proteins, probably Rabs, are also involved membrane fusion.

4.6 Secretory pathways

Newly made proteins, lipids, and carbohydrates are delivered from the ER, via the Golgi apparatus, to the cell surface by **transport vesicles** that fuse with the plasma membrane in a process called **exocytosis** (Figure 4-23). Each molecule that travels along this route passes through a fixed sequence of membrane-enclosed compartments and is often modified en route. During this process, polypeptides in the membrane and lumen of the ER undergo five principal modifications before they reach their final destinations: formation of disulfide bonds; proper folding; addition and processing of carbohydrates; specific proteolytic cleavages; and assembly into multimeric proteins.

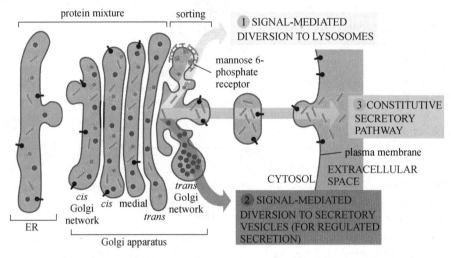

Figure 4-23 The regulated and constitutive pathways of exocytosis in secretory cells
(From: Alberts B, et al. Molecular Biology of the Cell. 2008)

4.6.1 Constitutive exocytosis

There are two exocytosis pathways that proteins and lipids from the *trans* Golgi network are transported to plasma membrane and outside cells. All eukaryotic cells have the **constitutive exocytosis** pathway that a steady stream of vesicles bud from the *trans* Golgi network and fuse with the plasma membrane. This constitutive exocytosis pathway operates continually and supplies newly made lipids and proteins to the plasma membrane, it is the pathway for plasma membrane growth when cells enlarge before dividing. The constitutive pathway also carries proteins to the cell surface to be released to the outside, a process called **secretion**. Some of the released proteins adhere to the cell surface, where they become peripheral proteins of the plasma membrane, some are incorporated into the extracellular matrix or diffuse into the extracellular fluid to nourish or to signal other cells. Because entry into this nonselective pathway does not require a particular signal sequence, it is sometimes referred to as the **default pathway**. In an unpolarized cell such as a white blood cell, it seems any protein in the lumen of the Golgi apparatus is automatically carried by the constitutive pathway to the cell surface unless it is specifically returned to the ER, retained as a resident protein in Golgi or selected for the pathways that lead to regulated secretion or to lysosomes.

4.6.2 Regulated exocytosis

In addition to the constitutive exocytosis pathway, which operates continually in all eukaryotic cells, there is a **regulated exocytosis pathway**, which operates only in cells that are specialized for secretion. Specialized secretory cells produce large quantities of particular products, such as hormones, mucus, or digestive enzymes, which are stored in **secretory vesicles** for later release. These vesicles bud off from the *trans* Golgi network and accumulate near the plasma membrane. There they wait for the extracellular signal that will stimulate them to fuse with the plasma membrane and release their contents to the cell exterior. An increase in blood glucose, for example, signals cells in the pancreas to secrete the hormone insulin. The signal is often a chemical messenger, such as a hormone, that binds to receptors on the cell surface. The resulting activation of the receptors generates intracellular signals, often including a transient increase in the concentration of free Ca^{2+} in the cytosol. In nerve terminals, the initial signal for exocytosis is usually an electrical excitation triggered by a chemical transmitter binding to receptors elsewhere on the same cell surface. When the action potential reaches the nerve terminals, it causes an influx of Ca^{2+} through voltage-gated Ca^{2+} channels. The binding of Ca^{2+} ions to specific sensors then triggers the secretory vesicles (called synaptic vesicles) to fuse with the plasma membrane and release their contents to the extracellular space.

4.6.3 Secretory vesicles

Secretory vesicles bud from the *trans* Golgi network. Proteins destined for secretory vesicles are sorted and packaged in the *trans* Golgi network. Proteins that travel by this pathway have special surface properties that cause them to aggregate with one another under the ionic conditions (acidic pH and high Ca^{2+}) that prevail in the *trans* Golgi network. The aggregated proteins are recognized by an unknown mechanism and packaged into secretory vesicles, which pinch off from the network. Proteins secreted by the constitutive pathway do not aggregate and are therefore carried automatically to the plasma membrane by the transport vesicles of the constitutive pathway. Selective aggregation has another function: it allows secretory proteins to be packaged into secretory vesicles at concentrations much higher than the concentration of the unaggregated protein in the Golgi lumen. This increase in concentration can reach up to 200 fold, enabling secretory cells to release large amounts of the protein promptly when triggered to do so.

Secretory and membrane proteins become concentrated as they move from the ER through the Golgi apparatus because of an extensive retrograde retrieval process mediated by COP I -coated transport vesicles. Membrane recycling is important for returning Golgi components to the Golgi apparatus, as well as for concentrating the contents of secretory vesicles. The vesicles that mediate this retrieval originate as clathrin-coated buds on the surface of immature secretory vesicles.

Besides the concentration of secretory proteins during secretory vesicles mature, proteins are often proteolytically processed. Many polypeptide hormones and neuropeptides, as well as many secreted hydrolytic enzymes, are synthesized as inactive protein precursors which require proteolysis to liberate the active molecules from these precursors. The cleavages begin in the *trans* Golgi network, and they continue in the secretory vesicles and sometimes in the extracellular fluid after secretion has occurred.

When a secretory vesicle fuses with the plasma membrane, its contents are discharged from the cell by exocytosis, and its membrane becomes part of the plasma membrane. Very rapidly, membrane components are removed from the surface by endocytosis almost as fast as they are added by exocytosis. After their removal from the plasma membrane, the proteins of the secretory vesicle membrane are either recycled or shuttled to lysosomes for degradation. Control of membrane traffic thus has a major role in maintaining the composition of the various membranes of the cell. To maintain each membrane enclosed compartment in the secretory and endocytotic pathways at a constant size, the balance between the outward and inward flows of membrane needs to be precisely regulated.

Most cells in tissues are polarized and have two or more distinct plasma membrane domains that are the targets of different types of vesicles. Differences between plasma membrane domains can be established in two ways: First, membrane components could be delivered to all regions of the cell surface indiscriminately but then be selectively stabilized in some locations and selectively eliminated in others. Alternatively, deliveries can be specifically directed to the appropriate membrane domain. Epithelial cells, for example, frequently secrete one set of products-such as digestive enzymes or mucus in cells lining the gut-at their apical surface and another set of products-such as components of the basal lamina-at their basolateral surface. By examining polarized epithelial cells in culture, it has been found that proteins from the ER destined for different domains travel together until they reach the TGN. Here they are separated and dispatched in secretory or transport vesicles to the appropriate plasma membrane domain. The apical plasma membrane of most epithelial cells is greatly enriched in glycosphingolipids. Similarly, plasma membrane proteins that are linked to the lipid bilayer by GPI anchor are found predominantly in the apical plasma membrane. GPI-anchored proteins are thought to be directed to the apical membrane because they associate with glycosphingolipids in lipid rafts that form in the membrane of the TGN. Having selected a unique set of cargo molecules, the rafts then bud from the *trans* Golgi network into transport vesicles destined for the apical plasma membrane. Thus, similar to the selective partitioning of some membrane proteins into the specialized lipid domains in caveolae at the plasma membrane, lipid domains may also participate in protein sorting in the TGN. Membrane proteins destined for delivery to the basolateral membrane contain sorting signals in their cytosolic tails. Such signals are recognized by coat proteins that package them into appropriate transport vesicles in the TGN. The same basolateral signals that are recognized in the TGN also function in endosomes to redirect the proteins back to the basolateral plasma membrane after they have been endocytosed.

4.6.4 Transport from ER to Golgi

The first step in the secretory pathway, from the ER to Golgi, is mediated by COP II vesicles, which is a highly selective and regulated process. Several strategies are used to ensure that only proteins targeted to other organelles are transported. One aspect of ER quality control is the retention of ER "resident" proteins that are made in the ER and destined to function there. The KDEL sequence at the C-terminus of ER soluble protein or KKXX sequence on ER membrane proteins is necessary and sufficient for their retention in the ER. If proteins with these sequences escaped to Golgi apparatus, their corresponding receptor in COP I -coated vesicles acts mainly to retrieve the escaped proteins back to the ER through vesicle transport.

Another aspect of ER quality control is the retention of unassembled or misfolded proteins. Proteins that fold up incorrectly, and dimeric or multimeric proteins that fail to assemble properly, are actively retained in the ER by binding to chaperone proteins, such as Bip and Calnexin, that reside there. Interaction with chaperones holds the proteins in the ER until proper folding occurs; otherwise, the proteins are dragged to the cytosol and ultimately degraded. Antibody molecules, for example, are composed of four polypeptide chains that assemble into the complete antibody molecule in the ER. Partially assembled antibodies are retained in the ER until all four polypeptide chains have assembled; any antibody molecule that fails to assemble properly is ultimately degraded.

4.7 Endocytic pathways

Eukaryotic cells are continually taking up fluid, as well as large and small molecules, by the process of **endocytosis**. Specialized cells are also able to internalize large particles and even other cells. The material to be ingested is progressively enclosed by a small portion of the plasma membrane, which first buds inward and then pinches off to form an intracellular endocytic vesicle, called **endosomes**. Endosomes transport their contents in a series of steps to a lysosome, which subsequently digests the materials. The mebabolites generated by digestion are transferred directly out of the lysosome into the cytosol, where they can be used by the cell. In other instances, however, endosomes are used by the cell to transport various substances between different portions of the external cell membrane.

Two main types of endocytosis are distinguished on the basis of the size of the endocytic vesicles formed. **Pinocytosis** (cellular drinking) involves the ingestion of fluid and molecules via small vesicles (<150 nm in diameter). **Phagocytosis** (cellular eating) involves the ingestion of large particles, such as microorganisms and cell debris, via large vesicles called phagosomes (generally >250 nm in diameter). Whereas all eukaryotic cells are continually ingesting fluid and molecules by pinocytosis, large particles are ingested mainly by specialized phagocytic cells. In this final section we trace the endocytic pathway from the plasma membrane to lysosomes.

4.7.1 Phagocytosis and pinocytosis

The most dramatic form of endocytosis, phagocytosis, was first observed more than a hundred years ago. In protozoa, phagocytosis is a form of feeding: microorganisms ingest large particles, such as bacteria, by taking them up into phagosomes; these phagosomes then fuse with lysosomes, where the food particles are digested. Few cells in multicellular organisms are able to ingest large particles efficiently. In the animal gut, for example, large particles of food have to be broken down to individual molecules by extracellular enzymes before they can be taken up by the absorptive cells lining the gut. Nevertheless, phagocytosis is important in most animals for purposes other than nutrition. Phagocytic cells — including macrophages, which are widely distributed in tissues, and some other white blood cells — defend us against infection by ingesting invading microorganisms. To be taken up by a macrophage or other white blood cell, particles must first bind to the phagocytic cell surface and activate one of a variety of surface receptors. Some of these receptors recognize antibodies, the proteins that protect us against infection by binding to the surface of microorganisms. Binding of antibody-coated bacteria to these receptors induces the phagocytic cell to extend sheetlike projections of the plasma membrane, called pseudopods, that engulf the bacterium and fuse at their tips to form a phagosome. This compartment then fuses with a lysosome and the microbe is digested. Phagocytic cells also play an important part in scavenging dead and damaged cells and cellular debris.

Eukaryotic cells continually ingest bits of their plasma membrane in the form of small pinocytic vesicles that are later returned to the cell surface. The rate at which plasma membrane is internalized by pinocytosis varies from cell type to cell type, but it is usually surprisingly large. Pinocytosis is mainly carried out by the clathrin-coated pits and vesicles that we discussed earlier. After they pinch off from the plasma membrane, clathrin-coated vesicles rapidly shed their coat and fuse with an endosome. Extracellular fluid is trapped in the coated pits as it invaginates to form a coated vesicle, and so substances dissolved in the extracellular fluid are internalized and delivered to endosomes. This fluid intake is generally balanced by fluid loss during exocytosis.

4.7.2 Receptor-mediated endocytosis

Pinocytosis, as just described, is indiscriminate. The endocytic vesicles simply trap any molecules that happen to be present in the extracellular fluid and carry them into the cell. In most animal cells, however, pinocytosis via clathrin-coated vesicles also provides an efficient pathway for taking up specific macromolecules from the extracellular fluid. The macromolecules bind to complementary receptors on the cell surface and enter the cell as receptor-macromolecule complexes in clathrin-coated vesicles. This process, called **receptor-mediated endocytosis**, provides a selective concentrating mechanism that increases the efficiency of internalization of particular macromolecules more than 1,000 fold compared with ordinary pinocytosis, so that even minor components of the extracellular fluid can be taken up in large amounts without taking in a correspondingly large volume of extracellular fluid. An important example of receptor-mediated endocytosis is the ability of animal cells to take up the cholesterol they need to make new membrane.

Cholesterol is extremely insoluble and is transported in the blood-stream bound to protein in the form of particles called low-density lipoproteins, or LDL. The LDL binds to receptors located on cell surfaces, and the receptor-LDL complexes are ingested by receptor-mediated endocytosis and delivered to endosomes. The interior of endosomes is more acid than the surrounding cytosol or the extracellular fluid, and in this acidic environment the LDL dissociates from its receptor: the receptors are returned in transport vesicles to the plasma membrane for reuse, while the LDL is delivered to lysosomes. In the lysosomes, the LDL is broken down by hydrolytic enzymes; the cholesterol is released and escapes into the cytosol, where it is available for new membrane synthesis. The LDL receptors on the cell surface are continually internalized and recycled, whether they are occupied by LDL or not (Figure 4-24).

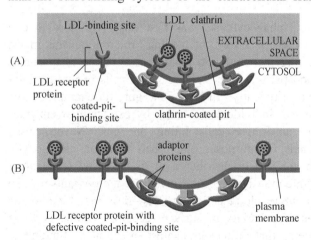

Figure 4-24 The LDL receptor-mediated endocytosis
(From: Alberts B, et al. Molecular Biology of the Cell. 2008)

Receptor-mediated endocytosis is also used to take up many other essential metabolites, such as vitamin B_{12} and iron, that cells cannot take up by the processes of membrane transport. Vitamin B_{12} and iron are both required, for example, for the synthesis of hemoglobin, which is the major protein in red blood cells; these metabolites enter immature red blood cells as a complex with protein. Many cell-surface receptors that bind extracellular signaling molecules are also ingested by this pathway: some are recycled to the plasma membrane for reuse, whereas others are degraded in lysosomes. Unfortunately, receptor-mediated endocytosis can also be exploited by viruses: the influenza virus and HIV, which causes AIDS, gain entry into cells in this way.

4.7.3 Endocytosed macromolecules are sorted in endosomes

Extracellular material taken up by pinocytosis is rapidly transferred to endosomes. The endosomal compartment can be visualized by an electron microscope when living cells are incubated in fluid containing an electron-dense marker. It reveals itself to be a complex set of connected membrane tubes and larger vesicles. The marker molecules appear first in **early endosomes**, just beneath the plasma membrane; 5~15 minutes later they show up in **late endosomes**, near the nucleus. Early endosomes mature gradually into late endosomes as the vesicles within them fuse, either with one another or with a pre-existing late endosome. The interior of the endosome compartment is kept acidic by an ATP-driven H^+ (proton) pump in the endosomal membrane that pumps H^+ into the endosome lumen from the cytosol.

The endosomal compartment acts as the main sorting station in the inward endocytic pathway, just as the *trans* Golgi network who serves this function in the outward secretory pathway. The acidic environment of the endosome plays a crucial part in the sorting process by causing many receptors to release their bound cargo. The routes taken by receptors once they have entered an endosome differ according to the type of receptor: ①most are returned to the same plasma membrane domain from which they came, as is the case for the LDL receptor discussed earlier; ②some travel to lysosomes, where they are degraded; ③some proceed to a different domain of the plasma membrane, thereby transferring their bound cargo molecules from one extracellulare space to another, a process called **transcytosis**.

Cargo molecules that remain bound to their receptors share the fate of their receptors. Those that dissociate from their receptors in the endosome are doomed to destruction in lysosomes along with most of the contents of the endosome lumen. It remains uncertain how molecules move from endosomes to lysosomes. One possibility is that they are carried in transport vesicles; another is that endosomes gradually convert into lysosomes.

Summary

Eukaryotic cells are highly compartmented. These membrane-enclosed compartments or organelles perform different functions. Proteins synthesized at the ribosomes are selectively delivered to various compartments. The directing of newly made proteins to their correct organelle is determined by sorting signal sequence(s), a continuous stretch of amino acid sequence on proteins. Proteins destined to mitochondrion, chloroplast, peroxisome, and the interior of the nucleus are delivered directly from the cytosol. For others, including the Golgi apparatus, lysosomes, endosomes, and the nuclear membranes, proteins and lipids are delivered through transport vesicles via the ER, which is itself a major site of lipid and protein synthesis. The transport pathways mediated by transport vesicles extend outward from the ER to the plasma membrane, and inward from the plasma membrane to lysosomes, and thus provide routes of communication between the interior of the cell and its surroundings. As proteins and lipids are transported outward along these pathways, many of them undergo various types of chemical modification, such as the formation of disulfide bonds, and N-linked or O-linked glycosylation. Cells ingest fluid, molecules, and sometimes even particles, by endocytosis. Receptor-mediated endocytosis also provides a selective concentrating mechanism to take up macromolecules. Much of the endocytotic material is delivered to endosomes and then to lysosomes. Most of the components of the endocytic vesicle membrane, however, are recycled in transport vesicles back to the plasma membrane for reuse.

Questions
1. What are the main functions of membrane-enclosed organelles in a typical eukaryotic cell?
2. How proteins synthesized at the ribosomes are transported to various organelles?
3. How do the transport vesicles specifically select cargo molecules?
4. What is the difference between the N-linked and O-linked glycosylation?
5. How is the cholesterol taken by cells?

（张　儒　杨仲南　张　森）

Chapter 5

Mitochondria and chloroplasts

The basic energy source of living things comes from the energy of sunlight, but organisms cannot use sunlight energy directly unless it was transformed into chemical energy. This type of energy transformation takes place in chloroplasts of plant cells and on the photosynthetic lamellae of blue-green algae. **Chloroplasts** via photosynthesis change sunlight energy into chemical energy that is stored in organic substances. Animals and other organisms without chloroplasts acquire energy from breaking down organic nutrition from plants. **Mitochondrion** is this type of organelle that can decompose organic substances and turn chemical energy into ATP that cells can use directly. Evidently, mitochondria and chloroplasts are energy-producing organelles in cells.

5.1 Mitochondria and oxidative phosphorylation

Mitochondria are present in nearly all eucaryotic cells — in plants, animals, and most eucaryotic microorganisms — and it is in these organelles that most of a cell's ATP is produced. Without them, present-day eucaryotes would be dependent on the relatively inefficient process of glycolysis for all of their ATP production. So, mitochondria are generally called "powerhouse" in a cell.

5.1.1 Structure of mitochondria

Mitochondria occupy a substantial portion of the cytoplasmic volume of the eucaryotic cells, and they have been essential for the evolution of complex animals. Mitochondria are usually depicted as stiff, elongated cylinders with a diameter of $0.5 \sim 1$ μm, similar in size and shape to bacteria, although these attributes can vary depending on the cell type. In a living cell, mitochondria have the characteristics of polymorphism, variability, mobility and adaptivity. Their number varies dramatically in different cell types, and can change with the energy needs of the cell. In skeletal muscle cell, for example, the number of mitochondria may increase five-to ten-fold due to mitochondrial growth and division that occurs if the muscle has been repeatedly stimulated to contract. Their locations inside the cell are not fixed. As they move about in the cytoplasm, they often seem to be associated with microtubules, which can determine the unique orientation and distribution of mitochondria in different types of cells.

Each mitochondrion is bounded by two highly specialized membranes — one wrapped around the other — that play a vital part in its activities. The outer and the inner mitochondrial membranes create two mitochondrial compartments: a large internal space called the matrix (or inner chamber) and the much narrower intermembrane space (Figure 5-1).

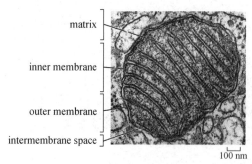

Figure 5-1 The structure of a mitochondrion

The outer membrane contains many molecules of transport protein called porin molecules, which form wide aqueous channels through the lipid bilayer. As a result, the outer membrane is like a sieve that is permeable to all molecules of 5,000 Daltons or less, including small proteins. This makes the intermembrane space chemically equivalent to the cytosol with respect to the small molecules it contains.

The inner membrane is highly specialized, its lipid bilayer contains a high proportion of the "double" phospholipid cardioolipin, which has four fatty acids rather than two and may help to make it especially impermeable to the passage of ions and most small molecules, except where a path is provided by membrane transport proteins. The mitochondrial matrix therefore contains only molecules that can be selectively transported into the matrix across the inner membrane, and its contents are highly specialized.

The inner mitochondrial membrane is the site of electron transport and proton pumping, and it contains the ATP synthase. Most of the proteins embedded in the inner mitochondrial membrane are components of the electron-transport chains required for oxidative phosphorylation. This membrane has a distinctive lipid composition and also contains a variety of transport proteins that allow the entry of selected small molecules, such as pyruvate and fatty acids, into the **matrix**.

The inner membrane is usually highly convoluted, forming a series of infoldings, known as cristae, that project into the matrix space to increase the surface area of the inner membrane(Figure 5-1). These folds provide a large surface on which ATP synthesis can take place. For example, the number of cristae is three times greater in a mitochondrion of a cardiac muscle cell than in a mitochondrion of a liver cell, presumably because of the greater demand for ATP in heart cells.

5.1.2 Biochemical compositions of mitochondria

Main chemical components of mitochondria are proteins and lipids. Proteins can be classified into two types: soluble proteins and insoluble proteins. Soluble proteins are mainly enzymes and proteins located on the outside of the membrane — peripheral proteins; while insoluble proteins are proteins embedded inside the membrane — integral proteins and structural proteins. The lipids are largely phospholipids. The chemical components of the outer and inner membranes are different in their compositions of protein and lipid. The outer membrane is composed of about half protein and half lipid; the inner membrane is about 80 percent protein and 20 percent lipid — a higher proportion of protein than occurs in other cellular membranes. In the matrix there are enzymes of the citric acid cycle and other metabolic pathways.

5.1.3 Electron transport and oxidative phosphorylation

The metabolism of food molecules is completed in the mitochondria. Mitochondria can use both pyruvate and fatty acids as fuel. Pyruvate comes mainly from glucose and other sugars, and fatty acids come from fats. Both of these fuel molecules are transported across the inner mitochondrial membrane and then converted to the vital metabolic intermediate acetyl CoA by enzymes located in the mitochondrial matrix. The acetyl groups in acetyl CoA are then oxidized in the matrix via the citric acid cycle. The cycle converts the carbon atoms in acetyl CoA to CO_2, which is released from the cell as a waste product. Most importantly, the cycle generates high-energy electrons, carried by the activated carrier molecules NADH and $FADH_2$. These high-energy electrons are then transferred to the inner mitochondrial membrane, where they enter the **electron-transport chain**; the loss of electrons regenerates the NAD^+ and FAD that are needed for continued oxidative metabolism. Electron transport along the chain now begins. The entire sequence of reactions is outlined in figure 5-2.

(1) **Electron-transport chain** (**respiratory chain**) The electron-transport chain that carries out **oxidative phosphorylation** is present in many copies in the inner mitochondrial membrane. Most of the proteins involved in the mitochondrial electron-transport chain are grouped into three large respiratory enzyme complexes, each containing multiple individual proteins. Each complex includes transmembrane proteins that hold the entire protein complex firmly in the inner mitochondrial membrane.

The three respiratory enzyme complexes are: ① The NADH dehydrogenase complex (generally known as complex Ⅰ) is the largest of the respiratory enzyme complexes, con-

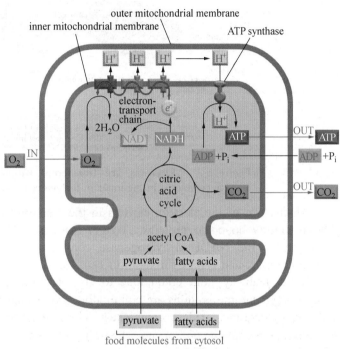

Figure 5-2 A summary of energy-generating metabolism in mitochondria

taining more than 40 polypeptide chains. ②The cytochrome b-c_1 complex contains at least 11 different polypeptide chains and functions as a dimmer. ③The cytochrome oxidase complex also functions as a dimmer, each monomer contains 13 different polypeptide chains, including two cytochromes and two copper atoms. Each contains metal ions and other chemical groups that form a pathway for the passage of electrons through the complex. The respiratory complexes are the sites of proton pumping, and each can be thought of as a protein machine that pumps protons across the membrane as electrons are transferred through it.

The process of electron transport begins when the hydride ion (H^-) is removed from NADH (to generate NAD^+) and is converted into a proton and two high-energy electrons: $H^- \rightarrow H^+ + 2e^-$ (Figure 5-3). This reaction is catalyzed by the first of the respiratory enzyme complexes, the NADH dehydrogenase, which accepts the electrons. The electrons are then passed along the chain to each of the other enzyme complexes in turn. Within each of the three respiratory enzyme complexes, the electrons move mainly between metal atoms that are tightly bound to the proteins, travelling by skipping from one metal ion to the next.

In contrast, electrons are carried between the different respiratory complexes by molecules that diffuse along the lipid bilayer, picking up electrons from one complex and delivering them to another in an orderly sequence. Ubiquinone, a small hydrophobic molecule that dissolves in the lipid bilayer, is the only carrier that is not part of a protein. Ubiquinone picks up electrons from the NADH dehydrogenase complex and delivers them to the cytochrome b-c_1 complex (Figure 5-3).

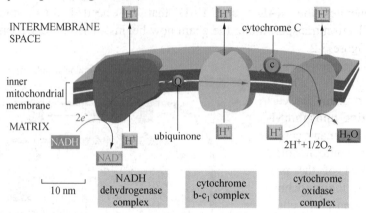

Figure 5-3　Electrons are transferred through three respiratory enzyme complexes in the inner mitochondrial membrane

All the rest of the electron carriers in the electron-transport chain are small molecules that are tightly bound to proteins. To get from NADH to ubiquinone, the electrons are passed inside the NADH dehydrogenase complex between a flavin group bound to the protein complex and a set of iron-sulfur centers.

In the pathway from ubiquinone to O_2, iron atoms in heme groups that are tightly bound to cytochrome proteins are commonly used as electron carriers, as in the cytochrome b-c_1 and cytochrome oxidase complexes. The cytochromes constitute a family of colored proteins; each contains one or more heme groups whose iron atom changes from the ferric (Fe^{3+}) to the ferrous (Fe^{2+}) state whenever it accepts an electron. Cytochrome C shuttles electrons between the cytochrome b-c_1 complex and the cytochrome oxidase complex.

At the very end of the respiratory chain, just before oxygen, the electron carriers are either iron atoms in heme groups or copper atoms that are tightly bound to the complex in the cytochrome oxidase complex. For cytochrome oxidase, the energetically favorable reaction is the addition of electrons to O_2 at the enzyme's active site. The structure of this large enzyme complex has recently been determined by X-ray crystallography, as illustrated in Figure 5-4. Cytochrome oxidase receives electrons from cytochrome c, thus oxidizing it, and donates these electrons to oxygen. It is here that nearly all of the oxygen we breathe is used, serving as the final repository for the electrons that NADH donated at the start of the electron-transport chain.

(2) The redox potential determines the electron transfer along the chain　In biochemical reactions, any electrons removed from one molecule are always passed to another, so that whenever one molecule is oxidized, another is reduced. Like any other chemical reaction, the tendency of such oxidation-reduction reactions, or redox reactions, to proceed spontaneously depends on the relative affinities of the two molecules for electrons. The tendency to transfer electrons from any redox pair depends on the redox potential — the voltage difference between two redox pairs. Electrons will move

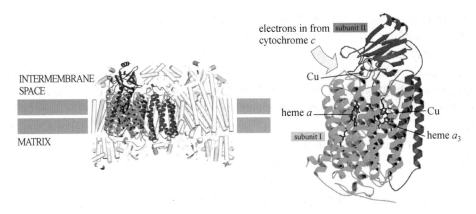

Figure 5-4 The molecular structure of cytochrome oxidase

spontaneously from a redox pair like NADH/NAD$^+$ with a low redox potential (a low affinity for electrons) to a redox pair like O_2/H_2O with a high redox potential (a high affinity for electrons). Thus NADH is a good molecule to donate electrons to the respiratory chain, while O_2 is well suited to act as the "sink" for electrons at the end of the pathway. The difference in redox potential is a direct measure of the standard free-energy change for the transfer of an electron from one molecule to another (Figure 5-5).

(3) **Protons are moved with the transfer of electrons**   Whenever a molecule is reduced by acquiring an electron, the electron (e^-) brings with it a negative charge. In many cases, this charge is rapidly neutralized by the addition of a proton (H^+) from water, so that the net effect of the reduction is to transfer an entire hydrogen atom, $H^+ + e^-$. Similarly, when a molecule is oxidized, the hydrogen atom can be readily dissociated into its constituent electron and proton, allowing the electron to be transferred separately to a molecule that accepts electrons, while the proton is passed to the water. Therefore, in a membrane in which electrons are being passed along an electron-transport chain, it is a relatively simple matter,

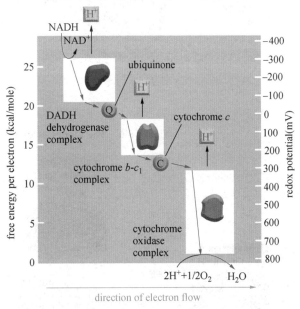

Figure 5-5 Redox potential increases along the mitochondrial electron-transport chain

in principle, to pump protons from one side of the membrane to another. All that is required is that the electron carrier would be arranged in the membrane in a way that causes it to pick up a proton from one side of the membrane when it accepts an electron, while releasing the proton on the other side of the membrane as the electron is passed on to the next carrier molecule in the chain.

(4) **The mechanism of H^+ pumping**   Some respiratory enzyme complexes pump one H^+ per electron across the inner mitochondrial membrane, whereas others pump two. The

detailed mechanism by which electron transport is coupled to H^+ pumping is different for the three different respiratory enzyme complexes. In the cytochrome b-c_1 complex, the quinones clearly have a role. As mentioned previously, a quinone picks up a H^+ from the aqueous medium along with each electron it carries and liberates it when it releases the electron. Since ubiquinone is freely mobile in the lipid bilayer, it can accept electrons near the inside surface of the membrane and donate them to the cytochrome b-c_1 complex near the outside surface, thereby transfering one H^+ across the bilayer for every electron transported. However, two protons are pumped per electron in the cytochrome b-c_1 complex. The complicated series of electron transfers that make this possible are still being worked out at the atomic level, aided by the complete structure of the cytochrome b-c_1 complex determined by X-ray crystallography (Figure 5-6).

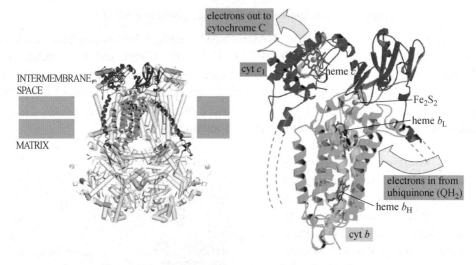

Figure 5-6 The atomic structure of cytochrome b-c_1

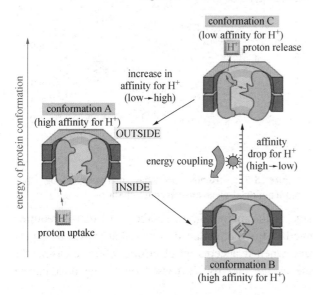

Figure 5-7 A general model for H^+ pumping

For both the NADH dehydrogenase complex and the cytochrome oxidase complex, it seems likely that electron transport drives sequential allosteric changes in protein conformation by altering the redox state of the components, which in turn cause the protein to pump H^+ across the mitochondrial inner membrane. This type of H^+ pumping requires at least three distinct conformations for the pump protein, a general mechanism is presented in Figure 5-7.

A schematic view of the reaction catalyzed by cytochrome oxidase is presented in Figure 5-8. In brief, four electrons from cytochrome c and four protons from the aqueous environment are added to each O_2 molecule in the reaction $4e^- + 4H^+ + O_2 \rightarrow 2H_2O$. In addition, four more protons are pumped across the membrane during electron transfer, building up the electrochemical proton gradient. Proton pumping is caused by allosteric changes in the conformation of the protein, which are driven by energy derived from electron transport.

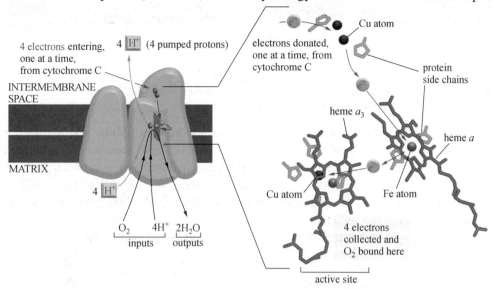

Figure 5-8 Cytochrome oxidase consumes nearly all the oxygen we breathe

(5) **Electron transport generates a proton gradient across the inner membrane** Each of the respiratory enzyme complexes couples the energy to an uptake of protons from water ($H_2O \rightarrow H^+ + OH^-$) in the mitochondrial matrix, accompanied by the release of protons on the other side of the inner membrane into the intermembrane space. As a result, the energetically favorable flow of electrons along the electron-transport chain pumps protons across the inner membrane out of the matrix, creating an electrochemical proton gradient across the inner mitochondrial membrane (Figure 5-9).

The active pumping of protons thus has two major consequences: ①It generates a pH gradient across the inner mitochondrial membrane, with the pH higher in the matrix than in the intermembrane space, where the pH is generally close to 7. ②It generates a voltage gradient (membrane potential) across the inner mitochondrial membrane, with the inside negative and the outside positive as a result of a net outflow of positive ions-protons.

Because protons are positively charged, they will move more readily across a membrane if the membrane has an excess of negative electrical charges on the other side. In the case of the inner mitochondrial membrane, the pH gradient and membrane potential work together to create a steep electrochemical proton gradient that makes it energetically very favorable for H^+ to flow back into the mitochondrial matrix. The membrane potential adds to the driving force pulling H^+ back across the membrane; hence this potential increases the amount of energy stored in the proton gradient (Figure 5-9).

(6) **The proton gradient drives ATP synthesis** The electrochemical proton gradient across the inner mitochondrial membrane can be used to drive ATP synthesis in the process of **oxidative phosphorylation** — a process involves both the consumption of O_2 and the synthesis of ATP through the addition of a phosphate group to ADP (Figure 5-10). This is made possible by a large membrane-bound enzyme called **ATP synthase**. This enzyme

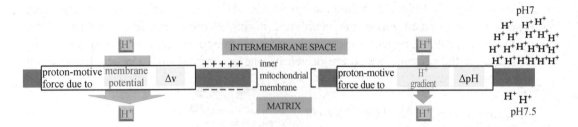

Figure 5-9 The two components of the electrochemical proton gradient. The total proton-motive force across the inner mitochondrial membrane consists of large force due to the membrane potential (ΔV) and a smaller force due to the H$^+$ gradient (ΔpH). Both forces combine to drive H$^+$ into the matrix.

creates a hydrophilic pathway across the inner mitochondrial membrane that allows protons to flow down their electrochemical gradient. As these ions thread their way through the ATP synthase, they are used to drive the energetically unfavorable reaction between ADP and Pi that makes ATP. The ATP synthase is of ancient origin; the same enzyme occurs in the mitochondria of animal cells, the chloroplasts of plants and algae, and in the plasma membrane of bacteria and archaea.

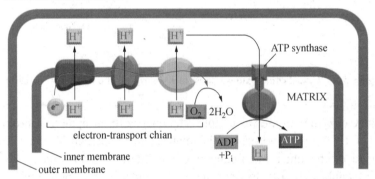

Figure 5-10 The general mechanism of oxidative phosphorylation

The structure of ATP synthase is shown in Figure 5-11. It is a large, multisubunit protein. A large enzymatic portion, shaped like a lollipop head and composed of 6 subunits, projects on the matrix side of the inner mitochondrial membrane and is attached through a thinner multisubunit "stalk" to a transmembrane proteins that form a "stator". The stator is in contact with a "rotor" composed of a ring of 10 to 14 identical transmembrane protein subunits. As protons pass through a narrow channel formed at the stator-rotor contact, their movement causes the rotor ring to spin rapidly within the head, inducing the head to make ATP. The synthase essentially acts as an

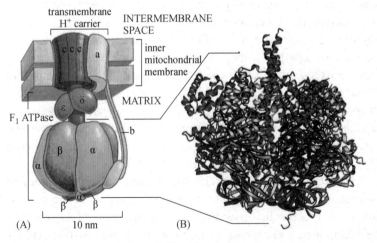

Figure 5-11 ATP synthase

energy-generating molecular motor, converting the energy of proton flow down a gradient into the mechanical energy of two sets of proteins rubbing against one another — rotating stalk proteins pushing against stationary head proteins. The changes in protein conformation driven by the rotating stalk subsequently convert this mechanical energy into the chemical bond energy needed to generate ATP. This marvelous device is capable of producing more than 100 molecules of ATP per second, calling for about three protons pass through the synthase to make each molecule of ATP. Through the above processes proton gradients across the inner mitochondrial membrane produce most of the ATP in cells.

(7) **The proton gradient drives coupled transport across the inner membrane** The synthesis of ATP is not the only process driven by the electrochemical proton gradient. In mitochondria, many charged small molecules, such as pyruvate, ADP, and Pi, are pumped into the matrix from the cytosol, while others, such as ATP, must be moved in the opposite direction. Transporters that bind these molecules can couple their transport to the energetically favorable flow of H^+ into the mitochondrial matrix. Thus, for example, pyruvate and inorganic phosphate (Pi) are co-transported inward with H^+ as the H^+ moves into the matrix.

In contrast, ADP is co-transported with ATP in opposite directions by a single carrier protein. As an ATP molecule has one more negative charge than ADP, each nucleotide exchange results in the moving of a total of one negative charge out of the mitochondrion. The ADP-ATP co-transport is therefore driven by the voltage difference across the membrane (Figure 5-12). All in all, a typical ATP molecule in the human body shuttles out of a mitochondrion and back into it (as ADP) for recharging more than once per minute, keeping the concentration of ATP in the cell about 10 times higher than that of ADP.

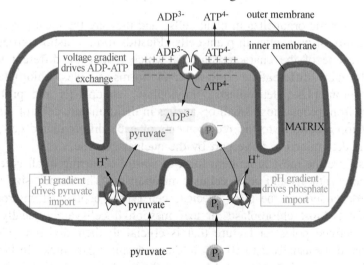

Figure 5-12 Some of the active transport processes driven by the electrochemical proton gradient across the inner mitochondrial membrane

(8) **The efficiency of respiration** During the biological oxidations — the total amount of energy generated and stored in the phosphate bonds of ATP, the efficiency with which oxidation energy is converted into ATP bond energy is often greater than 40%. This is considerably better than the efficiency of most nonbiological energy-conversion devices. If cells worked only with the efficiency of an electric motor or a gasoline engine

(10% ~20%), an organism would have to eat voraciously in order to maintain itself.

Oxidation produces huge amounts of free energy, which can be utilized efficiently only in small bits. Biological oxidative pathways involve many intermediates, each differing only slightly from its predecessor. The energy released is thereby parceled out into small packets that can be efficiently converted to high-energy bonds in useful molecules, such as ATP and NADH, by means of coupled reactions.

5.2 Chloroplasts and photosynthesis

The living organisms on Earth rely on solar energy to survival, and **photosynthesis** is the only biological way to harvest this energy. Photosynthesis is a process used by plants and other organisms to capture the solar energy to the photolysis of water. Nearly all of the organic material and energy required for surviving is from photosynthesis. Therefore, photosynthesis has a very important significance for the human beings, and is regarded as the most important chemical reaction on Earth.

In plants, photosynthesis is carried out in the **chloroplast**. Chloroplasts perform photosynthesis during the daylight hours and thereby produce ATP and NADPH, which are used to convert CO_2 into sugars inside the chloroplast. The sugars produced are exported to the surrounding cytosol, where they are used as fuel to make ATP and as starting materials for many of the other organic molecules that the plant cell needs. Sugar is also exported to all those cells in the plant that lack chloroplasts. As is the case for animal cells, most of the ATP present in the cytosol of plant cells is made by the oxidation of sugars and fats in mitochondria.

5.2.1 Morphology and structure of chloroplasts

Chloroplasts are specific organelles in plant cells, and they are the most prominent members of the plastid family of organelles in plant cells. Plastids form a distinct group of organelles in plants and are one of the characteristics by which plants are different to animals. In plants, plastids may differentiate into several forms besides chloroplast, including chromoplasts, gerontoplast, leucoplasts, amyloplast, elaioplastis and proteinoplast. All plastids develop from proplastids, small organelles in the immature cells of plant meristems. Proplastids develop according to the requirements of each differentiated cell, and the type that is present is determined in large part by the nuclear genome.

Chloroplasts are the most prominent form of plastid occurring in all green plant tissues and perform photosynthesis and other cellular metabolism. The shape, size and number of chloroplast are dependent on the plant species, cell types, ecological environment and the physiological state. Most chloroplasts in leaf mesophyll cells are typically ellipsoidal in shape but with defined poles, a feature that is crucial to their division. Chloroplasts are observable as flat discs usually 2 to 10 μm in diameter and 1 μm thick. In land plants, they are, in general, 5 μm in diameter and 2 to 3 μm thick. A typical parenchyma cell contains about 50 to 200 chloroplasts in higher plants, which accounts for 40% of the cytoplasmic volume; while algae plants usually contain a huge chloroplasts and its shape displays a ribbon, spiral and star which are dependent on the shape of the cells.

Like all plastids, they are bounded by a double envelope membrane, that is, a highly permeable outer membrane and a much less permeable inter membrane — in which membrane transport proteins are embedded, and a narrow intermembrane space in between (Figure 5-13). The inner membrane surrounds a large space called the **stroma**, which, analogous to the mitochondrial matrix, contains various metabolic enzymes. Besides the inner and outer membranes of the envelope, chloroplasts have a third internal membrane system, called the **thylakoid** membrane. The thylakoid membrane forms a network of flattened discs called thylakoids, which are frequently arranged in stacks called **granum** (Figure 5-13). The space inside each thylakoid is thought to be connected with that of other thylakoids, thereby defining a continuous third internal compartment that is separated from the stroma by the thylakoid membrane (Figure 5-14). Need to point out that the light-capturing systems, the electron-transport chains, and ATP synthases are all incorporated in the thylakoid membrane, a third membrane that forms a set of flattened disclike sacs, the thylakoids (Figure 5-14).

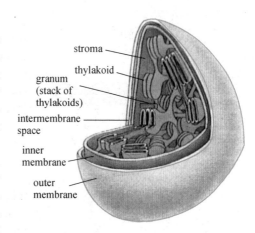

Figure 5-13 Structure of a chloroplast

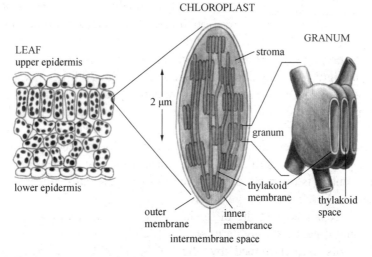

Figure 5-14 A chloroplast contains a third internal compartment

5.2.2 Photosynthesis of chloroplast

Photosynthesis is a process that plants, alage, and prokaryotes directly use light energy to synthesize organic compounds. The process of photosynthesis encompasses a complex series of chemical reactions that involve light absorption, energy conversion, electron transfer, carbon dioxide fixation and water photolysis. Most photosynthetic organisms convert light energy into stable chemical products, according to the following reaction equation:

$$2H_2A + CO_2 \rightarrow (CH_2O) + 2A + H_2O$$

As shown in the equation, photosynthesis is a biological oxidation-reduction process. CO_2 is the electron acceptor, and H_2A (for example, H_2O or H_2S) is any reduced compound that can serve as the electron donor. CH_2O represents the carbohydrate generated by the reduction, and A represents the product formed by oxidation of H2A (for example, O_2 or S). During the 50s and 60s of 20th century, Danial I. Arnon and his colleagues showed that isolated chloroplast can convert CO_2 to carbohydrate in the light. These experiments also documented that the photosynthetic process involved two stages: **light**

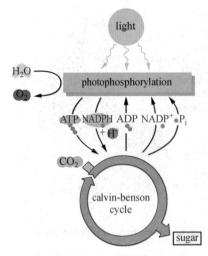

Figure 5-15 Overview of the two stages in the photosynthesis process

reactions and dark reaction or carbon dioxide fixation (Figure 5-15). Photosynthesis is a very complex reaction process, and based on current information photosynthesis can be divided into two stages: the light-dependent process and the light-independent process. The first process, the light dependent process (also called the "light reactions"), requires the direct energy of light to make energy carrier molecules — ATP and NADPH that are used in the second process. Water is split in this process, releasing oxygen gas. The second process, the light independent process (also called the "dark reactions"), occurs when the products of the light reaction — the ATP and the NADPH — serve as the source of energy and reducing power respectively, to drive the conversion of CO_2 to carbohydrate. The details will be discussed in the following text.

(1) **Light reactions (photosynthetic electron-transfer reactions)**

1) **The nature of light**: One of the major advances of the 20th century physics was that finding that, Light, a form of electromagnetic radiation, has properties of both waves and particles. According to the description of the quantum mechanics, radiant energy is visualized as being a stream of energy-carrying particles called quanta. The energy of a photon of a particular wavelength can be described by the following equation: $E = h v / \lambda$ (λ means that wavelength that is defined as the distance from peak to peak).

The energy content is inversely proportional to its wavelength: longer wavelengths have less energy than do shorter ones (Figure 5-16).

White light is separated into the different colors of light when it passes through a prism. The order of colors is determined by the wavelength of light. Thus, not all colors of light have equal energy. Visible light is one small part of the electromagnetic spectrum. The longer the wavelength of visible light, the more red parts of the color. Likewise the shorter wavelengths are towards the violet side of the spectrum. Wavelengths longer than red are referred to as infrared, while those shorter than violet are ultraviolet.

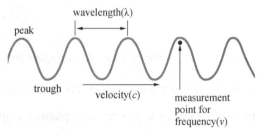

Figure 5-16 Wavelength and other aspects of the wave nature of light

Wave properties of light include the bending of the wave path when passing from one material (medium) into another (i.e., the prism, rainbows, pencil in a glass-of-water, etc.). The particle properties are demonstrated by the photoelectric effect. When we consider events at the level of a single molecule — such as the absorption of light by a molecule of chlorophyll, we have to picture light as being composed of discrete packets of energy called photons. Light of different colors is distinguished by photons of different energy, with longer wavelengths corresponding to lower energies. Thus photons of red light have a lower energy than photons of green light.

2) **Chlorophyll and accessory pigments**: For light energy to be used by any system, the light must first be absorbed. Almost all photosynthetic organisms contain chlorophyll or a related pigment. Most photosynthetic organisms contain chlorophyll, one form of the light-absorbing pigment. **Chlorophyll** is a complex molecule and it contains tetrapyrrole ring-like structure, which is structurally similar to and produced through the same metabolic pathway as other porphyrin pigments such as heme. At the center of the chlorin ring is a magnesium ion which coordinated to the four modified pyrrole rings. Additionally, chlorophyll also contains a long hyfrocarbon tail that makes the molecules hydrophobic (Figure 5-17).

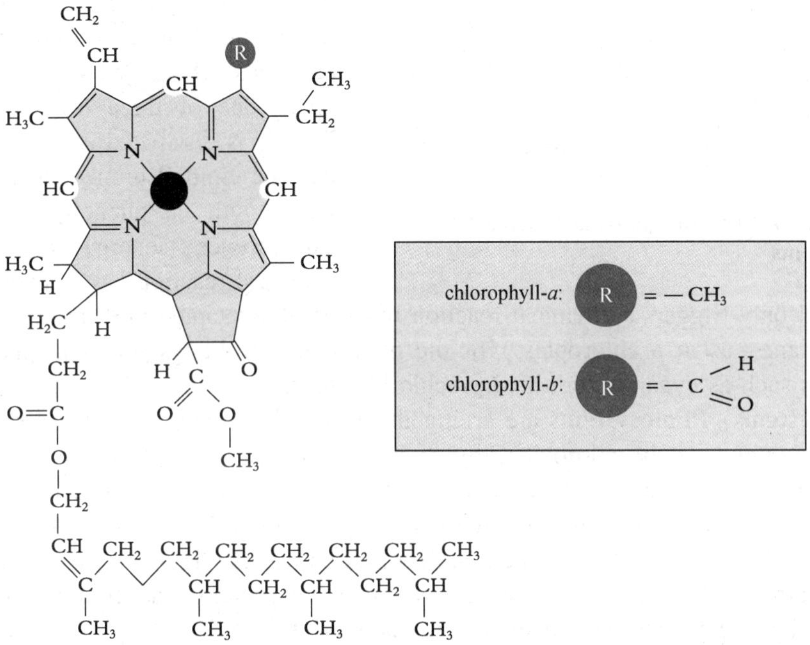

Figure 5-17 The molecular structure of chlorophylls

All photosynthetic organisms (plants, certain protistans, prochlorobacteria, and cyanobacteria) have chlorophylla. Among organisms that perform oxygenic photosynthesis, only land plants, green algae, and a few groups of cyanobacteria (prochlorophytes) possess chlorophyllb. Chlorophylla and chlorophyllb have different spectrophotometry characteristic, and in diethyl ether, chlorophylla has approximate absorbance maxima of 430 nm and 662 nm, while chlorophyllb has approximate maxima of 453 nm and 642 nm (Figure 5-18). **Carotenoids** are tetraterpenoid organic pigments derived from eight isoprene units and they include the carotenes and xanthophylls. Carotenoids, which are responsible for the orange-yellow colors observed in the leaves of plants, absorb light which wavelength is between 400 and 500 nm (Figure 5-18), a range in which absorption by chlorophylla is relatively weak. Thus, carotenoids play a minor role as accessory light-harvesting pigments absorbing and transferring light energy to chlorophyll molecules, but they play important structural role in the assembly of light-harvesting complexes and have an essential function in protecting the photosynthetic apparatus from photooxidative damage. Carotenoids also help prevent singlet oxygen formation in chloroplast, since they can accept excitation energy from triplet chlorophyll. Moreover, recent studies also have implicated that zeaxanthin in a similar process with singlet chlorophyll, the excited form of chlorophyll involved in energy transfer and photochemistry. There is another group of photosynthetic pigments the

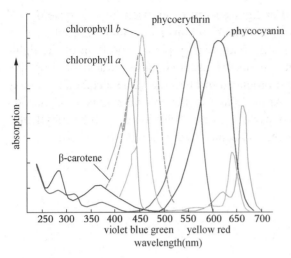

Figure 5-18 Absorption spectrum of several plant pigments

phycobilins. Phycobilins are water-soluble linear tetrapyrroles which also derives from chlorophyll biosynthesis pathway, but lacks a phytol tail. **Phycobline** is organized into phycobilisomes with proteins and exists in red algae and cyanobacteria.

The action spectrum of photosynthesis is the relative effectiveness of different wavelengths of light at generating electrons. If a pigment absorbs light energy, one of three things will occur. Energy is dissipated as heat; the energy may be emitted immediately as a longer wavelength, a phenomenon known as fluorescence; energy may trigger a chemical reaction, as in photosynthesis. Chlorophyll only triggers a chemical reaction when it is associated with proteins embedded in a membrane (as in a chloroplast) or the membrane infoldings found in photosynthetic prokaryotes such as cyanobacteria and prochlorobacteria.

3) **Photosystems**: **Photosystems** are arrangements of light-absorbing chlorophyll and large multiprotein complexes in chloroplast thylakoid membranes of plant and in the membranes of photosynthetic bacteria. A photosystem consists of two closely linked components: an antenna complex and a photochemical reaction center (Figure 5-19). The former consists of proteins bound to a large set of pigment molecules that capture light energy and feed it to the reaction centre, and the latter consists of a complex of proteins and chlorophyll molecules that enable light energy to be converted into chemical energy.

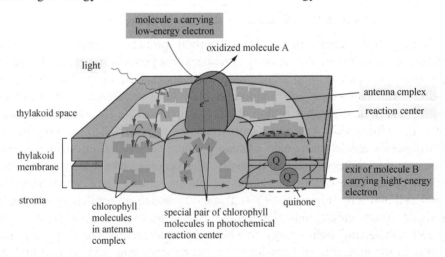

Figure 5-19 A photosystem contains a reaction center and an antenna

The antenna complex is important for capturing light energy, and it consists of light-harvesting complexes. A light-harvesting complex is a complex of proteins and photosynthetic pigments and surrounds a photosynthetic reaction center to focus energy, attained from photons absorbed by the pigment, toward the reaction center using Förster resonance energy transfer. The antenna complex in plants also contains accessory another pigments called carotenoids, which protect the chlorophylls from oxidation and can help the antenna complex. When a chlorophyll molecule in the antenna is excited by light, the excitation energy may also be transferred from one molecular to another molecule nearby by resonance energy transfer until it reaches a special pair of chlorophyll molecules in the reaction centre. Despite that the antenna complex exists in plant and algae, there are some different in components. A phycobilisome is a light-harvesting protein complex only presenting in sea water, such as glaucocystophyta and red algae, and is structured like a real antenna. Phycobilisomes are protein complexes associated with thylakoid membranes, and they include stacks of chromophorylated proteins, the phycobiliproteins, and their associated linker polypeptides. Each phycobilisome consists of a core made of allophycocyanin, from which several outwardly oriented rods made of stacked disks of phycocyanin and (if present) phycoerythrin(s) or phycoerythrocyanin. The spectral property of phycobiliproteins is mainly determined by their prosthetic groups, which are linear tetrapyrroles known as phycobilins including phycocyanobilin, phycoerythrobilin, phycourobilin and phycobiliviolin.

The photochemical reaction center is a transmembrane complex of several proteins, pigments and other co-factors assembled together to perform the primary energy conversion reactions of photosynthesis. The special pair of chlorophyll molecules in the reaction center acts as an irreversible trap for an excited electron, as these chlorophylls are positioned to pass a high-energy electron to a precisely positioned neighboring molecule in the same protein complex, creating a charge separation by moving the energized electron rapidly away from the chlorophylls, which are transferred by reaction center is much more stable.

The chlorophyll molecule in the reaction center loses an electron, it becomes positively charged. As illustrated in figure 5-20A, this chlorophyll rapidly regains an electron from an adjacent electron donor (orange) to return to its unexcited, uncharged state. Then, in slower reactions, the electron donor has its missing electron replaced with an electron removed from water, and the high-energy electron that was generated by the excited chlorophyll is transferred to the electron transport chain (Figure 5-20B).

Although these species are separated by billions of years of evolution, the reaction centers are homologous for all photosynthetic species. Green plants and algae have two different types of reaction centres that are part of larger supercomplexes known as photosystem Ⅰ (PS Ⅰ) and photosystem Ⅱ (PS Ⅱ). PSII, called water-plastoquinone oxidoreductase, located in the thylakoid membrane of plants, algae, and cyanobacteria, is the first protein complex in the light-dependent reactions. In cyanobacteria and green plants, PS Ⅱ is composed of about 20 subunits as well as other accessory, light-harvesting proteins. Each PS Ⅱ contains at least 99 cofactors which include 35 chlorophyll a, 12 beta-carotene, two pheophytin, three plastoquinone, two heme, bicarbonate, 25 lipid, the Mn_4CaO_5 cluster (including chloride ion), and one non heme Fe^{2+} and two putative Ca^{2+} ion per monomer. Currently, several PS Ⅱ crystal structure have been defined. PS Ⅱ captures light photons to energize electrons that are then transferred through a variety of coenzymes and cofactors to reduce plastoquinone to plastoquinol. The energized electrons are replaced by oxidizing water to form hydrogen ions and molecular oxygen. By obtaining these electrons from water, PS Ⅱ provides the electrons for all of photosynthesis to occur. PS Ⅰ is an integral membrane protein complex that was discovered in the 1950s and it is the

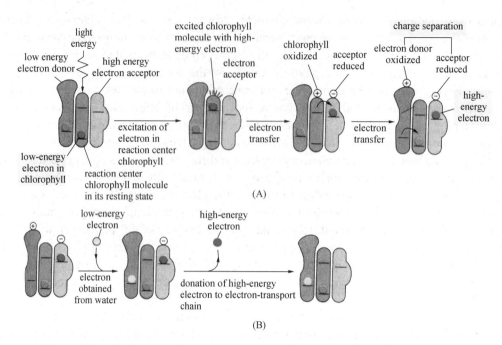

Figure 5-20 Light energy is harvested by a reaction center chlorophyll molecule

second photosystem in the photosynthetic light reactions of algae, plants, and some bacteria. The PS I system comprises more than 110 co-factors, significantly more than PS II. These various components have a wide range of functions. PS I uses light energy to mediate electron transfer from plastocyanin to ferredoxin.

4) Electron-transport chain and photophosphorylation: Photosynthetic **phosphorylation** is the process of phosphate group transfer into ADP to synthesize energy rich ATP molecule making use of light as external energy source. Photosynthesis in plants and cyanobacteria has at least two photophosphorylation ways that is noncyclic photophosphorylation and cyclic photophosphorylation. Noncyclic photophosphorylation produces both ATP and NADPH directly, while cyclic photophosphorylation only produces ATP but no reduced NADP(NADPH). In the non-cyclic reaction, light energy is captured in the light-harvesting antenna complexes of photosystem II by chlorophyll and other accessory pigments. When a chlorophyll molecule at the core of the photosystemII reaction center obtains sufficient excitation energy from the adjacent antenna pigments, an electron is transferred to the primary electron-acceptor molecule, pheophytin, through a process called photo-induced charge separation. These electrons are shuttled through a Z-scheme electron transport chain, that initially functions to generate a chemiosmotic potential across the membrane. An ATP synthase enzyme uses the chemiosmotic potential to make ATP during photophosphorylation, whereas NADPH is a product of the terminal redox reaction in the Z-scheme (Figure 5-21). The electron enters a chlorophyll molecule in photosystem I. The electron is excited due to the light absorbed by the photosystem. A second electron carrier accepts the electron, which again is passed down lowering energies of electron acceptors. The energy created by the electron acceptors is used to move hydrogen ions across the thylakoid membrane into the lumen. The electron is used to reduce the co-enzyme NADP, which has functions in the light-independent reaction.

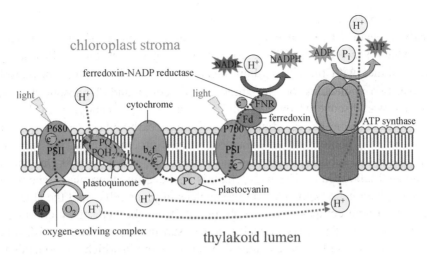

Figure 5-21 During photosynthesis electrons travel down an electron transport chain in the thylakoid membrane

In cyclic electron flow, the electron begins in a pigment complex called photosystem I, passes from the primary acceptor to ferredoxin, then to cytochrome b_6-f (a similar complex to that found in mitochondria), and then to plastocyanin before returning to chlorophyll. This transport chain produces a proton-motive force, pumping H^+ ions across the thylakoid membrane, which produces a concentration gradient that can be used to power ATP synthase during chemiosmosis. This cyclic photophosphorylation produces neither O_2 nor NADPH. Unlike non-cyclic photophosphorylation, $NADP^+$ does not accept the electrons; they are instead sent back to photosystem I (Figure 5-22).

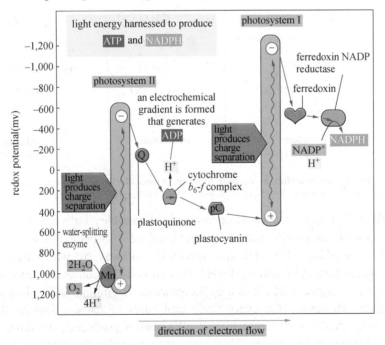

Figure 5-22 The components in the electron-transport chain have different redox potentials

(2) Dark reactions (carbon-fixation reactions)

1) The calvin cycle (C-3 pathway): The central reaction of photosynthetic carbon fixation, in which a molecule of CO_2 is converted to organic carbon, is illustrated in figure 5-23. CO_2 combines with the five-carbon sugar derivative ribulose 1,5-bisphosphate plus water to give two molecules of the three-carbon compound 3-phosphoglycerate. This reaction, which was discovered by Melvin Calvin in 1948, is catalyzed in the chloroplast stroma by a large enzyme called ribulose bisphosphate carboxylase (rubisco). In plants, algae, cyanobacteria, phototropic and chemoautotropic proteobacteria, the rubisco enzyme usually consists of two subunit, called the large chain (L, about 55,000 Da) and the small chain (S, about 13,000 Da). The large-chain is encoded by chloroplast genome in plants, while the small-chain gene is encoded by the nucleus of plant cells, and the small chains are imported to the stromal compartment of chloroplasts from the cytosol by crossing the outer chloroplast membrane. The enzymatically active substrate (ribulose 1,5-bisphosphate) binding sites are located in the large chains that form dimers in which amino acids from each large chain contribute to the binding sites. A total of eight large-chains (= 4 dimers) and eight small chains assemble into a larger complex of about 540,000 Da. In some proteobacteria and dinoflagellates, enzymes consisting of only large subunits have been found. Since this enzyme works extremely sluggishly compared with most other enzymes (processing about three molecules of substrate per second compared with 1,000 molecules per second for a typical enzyme), many enzyme molecules are needed. RuBisCO often represents more than 50% of the total chloroplast protein, and it is widely claimed to be the most abundant protein on earth.

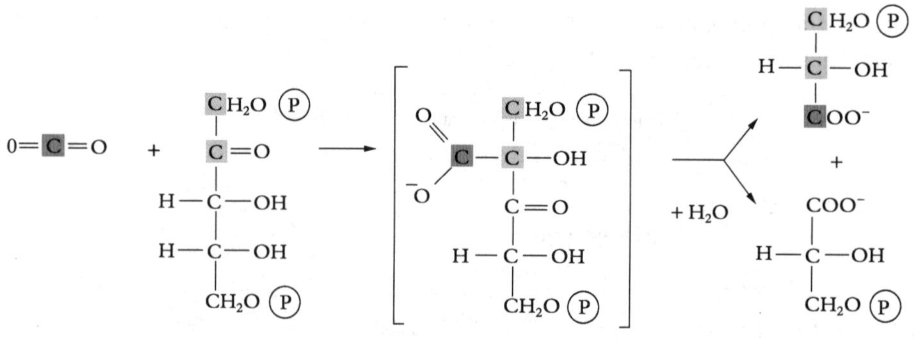

Figure 5-23 The initial reaction in carbon fixation. Carton dioxide is converted into organic carbon, is catalyzed in the chloroplast stroma by the abundant enzyme ribulose bisphosphate carboxylase.

This initial CO_2 fixing reaction is energetically favorable, but only when it receives a continuous supply of the energy-rich compound ribulose 1,5-bisphosphate, to which each molecule of CO_2 is added. The elaborate metabolic pathway by which this compound is regenerated requires both ATP and NADPH. This carbon-fixation cycle (also called Calvin cycle) is outlined in figure 5-24; it is a cyclic process, beginning and ending with ribulose 1,5-bisphosphate. However, for every three molecules of carbon dioxide that enter the cycle, one new molecule of glyceraldehyde 3-phosphate is produced, the three-carbon sugar that is the net product of the cycle. This sugar then provides the starting material for the synthesis of many other sugars and organic molecules.

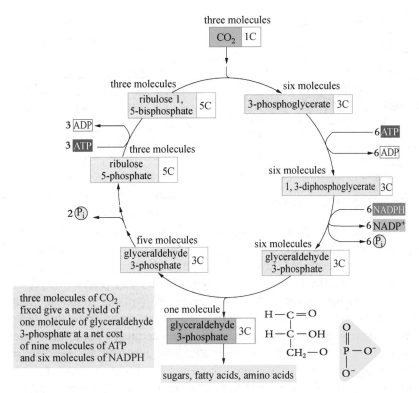

Figure 5-24　The carbon-fixation cycle forms organic molecules from CO_2 and H_2O

In the carbon-fixation cycle, three molecules of ATP and two molecules of NADPH are consumed for each CO_2 molecule converted into carbohydrate. Thus both phosphate bond energy (as ATP) and reducing power (as NADPH) are required for the formation of sugar molecules from CO_2 and H_2O.

2) **C-4 pathway and CAM pathway**: In most plants, carbon fixation occurs when CO_2 reacts with a five-carbon compound called RuBisCO (ribulose 1, 5-bisphosphate) as described above. The product splits immediately to form a pair of three-carbon compounds, and therefore this pathway is called the C-3 pathway. Further reaction leads to the production of a sugar (glyceraldehyde-3-phosphate) and the regeneration of RuBisCO. However, RuBisCO is not very efficient at grabbing CO_2, and it has an even worse problem.

When the concentration of CO_2 in the air inside the leaf falls too low, RuBisCO starts combining oxygen instead. The ultimate consequence of this process, called photorespiration (Figure 5-25), is that sugar is burned up rather than being created. Photorespiration becomes a significant problem for plants during hot, dry days, when they must keep their stomates closed to prevent water loss.

Some plants have developed a preliminary step to the Calvin cycle (C-3 pathway), which is known as C-4. While most C-fixation begins with RuBP, C-4 begins with a new molecule, phosphoenolpyruvate (PEP), a 3-C chemical that is converted into oxaloacetic acid (OAA, a 4-C chemical) when carbon dioxide is combined with PEP. The OAA is converted to malic acid and then transported from the mesophyll cell into the bundle-sheath cell, where OAA is broken down into PEP plus carbon dioxide. The carbon dioxide then enters the Calvin cycle, with PEP returning to the mesophyll cell. The resulting sugars are

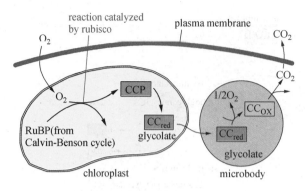

Figure 5-25 Photorespiration

now adjacent to the leaf veins and can readily be transported throughout the plant.

The capture of carbon dioxide by PEP is mediated by the enzyme PEP carboxylase, which has a stronger affinity for carbon dioxide than does RuBP carboxylase. When carbon dioxide levels decline below the threshold for RuBP carboxylase, RuBP is catalyzed with oxygen instead of carbon dioxide, which forms glycolic acid, a chemical that can be broken down by photorespiration, producing neither NADH nor ATP, in effect dismantling the Calvin cycle (Figure 5-26). C-4 plants, which often grow close together, have had to adjust to decreased levels of carbon dioxide by raising the carbon dioxide concentration in certain cells to prevent photorespiration. C-4 plants evolved in the tropics and are adapted to higher temperatures than are the C-3 plants found at higher latitudes. Common C-4 plants include crabgrass, corn, and sugar cane. Note that OAA and malic acid also have functions in other processes, thus the chemicals would have been present in all plants, leading scientists to hypothesize that C-4 mechanisms evolved several times independently in response to a similar environmental condition, a type of evolution known as convergent evolution. By concentrating CO_2 in the bundle sheath cells, C-4 plants promote the efficient operation of the Calvin-Benson cycle and minimize photorespiration. C-4 plants include corn, sugar cane, and many other tropical grasses. Crassulacean acid metabolism, also known as CAM photosynthesis, is another carbon fixation pathway that evolved in some plants as an adaptation to acid conditions. CAM plants initially grab CO_2 to PEP and form OAA. CAM plants fix carbon at night rather than during the day, and store the OAA in large vacuoles within the cell (Figure 5-27). This allows them to have their stomates open in the cool of the evening, avoiding water loss, and to use the CO_2 for the Calvin-Benson cycle during the day, when it can be driven by the sun's energy. CAM plants are more common than C-4 plants and include cacti and a wide variety of other succulent plants.

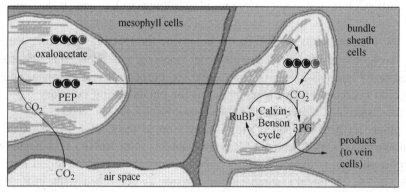

Figure 5-26 C-4 photosynthsis involves the separation of carbon fixation and carbohydrate systhesis in space and time

3) **Carbon fixation in chloroplasts generates sucrose and starch**: Much of the glyceraldehyde 3-phosphate produced in chloroplasts is moved out of the chloroplast into the cytosol. Some of it enters the glycolytic pathway, where it is converted to pyruvate that is then used to produce ATP by oxidative phosphorylation in plant cell mitochondria. The glyceraldehyde 3-phosphate is also converted into many other metabolites, including the disaccharide sucrose. Sucrose is the major form in which sugar is transported between plant cells; just as glucose is transported in the blood of animals, sucrose is exported from the leaves via the vascular bundle to provide carbohydrate to the rest of the plant.

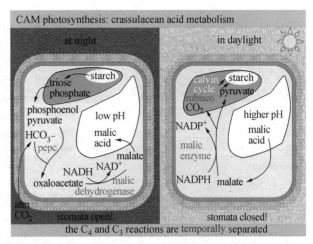

Figure 5-27 CAM photosynthesis (From: http://plantphys.info/plant_physiology/c4cam.shtml)

The glyceraldehyde 3-phosphate that remains in the chloroplast is mainly converted to starch in the stroma. Like glycogen in animal cells, starch is a large polymer of glucose that serves as a carbohydrate reservoir. The production of starch is regulated so that it is synthesized and stored as large grains in the chloroplast stroma during periods of excess photosynthetic capacity. At night, starch is broken down to sugars to help support the metabolic needs of the plant. Starch forms an important part of the diet of all animals that eat plants.

(3) Chloroplasts also perform other crucial biosyntheses It is well known that the chloroplast performs photosynthesis. Moreover, chloroplast also synthesizes amino acids, fatty acids, and the lipid components of their own membranes. The reduction of nitrite (NO_2^-) to ammonia (NH_3), an essential step in the incorporation of nitrogen into organic compounds, also occurs in chloroplasts. Moreover, chloroplasts are only one of several types of related organelles (plastids) that play a variety of roles in plant cells.

5.3 The origins of chloroplasts and mitochondria

Most people believed that chloroplasts and mitochondria evolved from bacteria that were engulfed by ancestral eucaryotic cells more than a billion years ago. As a relic of this evolutionary past, both types of organelles contain their own genomes, as well as their own biosynthetic machinery for making RNA and organelle proteins. The way that mitochondria and chloroplasts reproduce — through the growth and division of pre-existing organelles — provides additional evidence of their bacterial ancestry (Figure 5-28).

The growth and proliferation of mitochondria and chloroplasts are complicated, as their component proteins are encoded by two separate genetic systems, one in the organelle and one in the cell nucleus. In the case of the mitochondrion, most of the original bacterial genes have become transposed to the cell nucleus, leaving only relatively few genes inside the organelle itself. Animal mitochondria in fact contain a uniquely simple genetic system: the human mitochondrial genome, for example, contains only 16,569 nucleotide pairs of DNA encoding 37 genes. The vast majority of mitochondrial proteins-including those

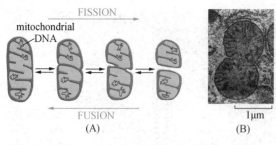

Figure 5-28　A mitochondrion divides like a bacterium

needed to make the mitochondrion's RNA polymerase and ribosomal proteins, and all of the enzymes of its citric acid cycle — are produced from nuclear genes, and these proteins must therefore be imported into the mitochondria by specialized protein translocases of the outer (TOM, translocale of the outer membrane) and inner (TIM) membrane from the cytosol, where they are made.

Like mitochondria, chloroplasts also contain their own genome. More than 100 chloroplast genomes have now been sequenced. The genomes of even distantly related plants (such as tobacco and liverwort) are nearly identical, and even those of green algae are closely related. Chloroplast genome is circle and most organism chloroplast genomes contain a pair inverted repeat (IRA and IRB) which are separated by long and short single copy regions (LSC and SSC). A typical chloroplast genome structure is shown in figure 5-29.

In a general, plastids possess a genome (plastome) of between 120 and 160 kb that encodes between 120 and 135 genes [see the Organelle Genome Megasequencing Program (http://megasun.bch.umontreal.ca/ogmp), for a complete set of available genomes]. These chloroplast genes are categorized into three groups according to their functions of these gene products. This first group is the photosynthesis-related genes which include photosynthetic polypeptides, for example several for the four main thylakoid complexes, NADH dehydrogenase genes; the second group is these genes which are responsible for genetic machinery such as eubacterial-type ribosomal and transfer RNAs; and the third group is the genes which products is biogenesis, such as, accd, ycf 3. Many plastome genes are organized in operons which is similar with the E. coli. Plastids are highly polyploid, with plastomes probably arranged as concatenated, long, linear molecules rather than small circles.

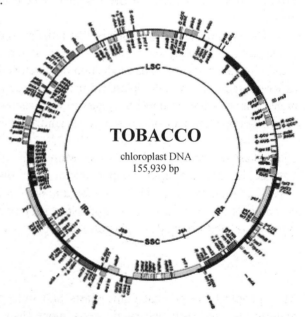

Figure 5-29　The tabacco chloroplast genome structure
(From: Wakasugi T, et al. 2001)

Summary

Mitochondria are enclosed by two concentric membranes, the innermost of which encloses the mitochondrial matrix. The matrix space contains many enzymes, including those of the citric acid cycle. These enzymes produce large amounts of NADH and $FADH_2$ from the oxidation of acetyl CoA. In the inner mitochondrial membrane, high-energy electrons donated by NADH and FADH$_2$ pass along an electron-transport chain — the respiratory chain — eventually combining with molecular oxygen in an energetically favorable reaction.

Much of the energy released by electron transfers along the respiratory chain is harnessed to pump H$^+$ out of the matrix into the intermembrane space, thereby creating a transmembrane electrochemical proton (H$^+$) gradient. The proton pumping is carried out by three large respiratory enzyme complexes embedded in the inner membrane. The resulting electrochemical proton gradient across the inner mitochondrial membrane is harnessed to make ATP when H$^+$ ions flow back into the matrix through ATP synthase, an enzyme located in the inner mitochondrial membrane.

In photosynthesis in chloroplasts and photosynthetic bacteria, high energy electrons are generated when sunlight is absorbed by chlorophyll; this energy is captured by protein complexes known as photosystems, which are located in the thylakoid membranes of chloroplasts.

Electron-transport chains associated with photosystems transfer electrons from water to NADP$^+$ to form NADPH, with the concomitant production of an electrochemical proton gradient across the thylakoid membrane. Molecular oxygen is generated as a by-product. The proton gradient is used by an ATP synthase embedded in the membrane to generate ATP.

The ATP and the NADPH made by photosynthesis are used within the chloroplast to drive the carbon-fixation cycle in the chloroplast stroma, thereby producing carbohydrate from CO_2.

Both mitochondria and chloroplasts are thought to have evolved from bacteria that were endocytosed by primitive eucaryotic cells.

Questions

1. Please explain the following terms: ATP synthase, mitochondrion, matrix, electron-transport chain, cytochrome, quinone, oxidative phosphorylation, chloroplast, stroma, photosynthesis, photosystem, reaction center, light reaction, carbon fixation.
2. Which of the following statements are correct? Explain your answers.
   A. Many, but not all, electron-transfer reactions involve metal ions.
   B. The electron-transport chain generates an electrical potential across the membrane because it moves electrons from the intermembrane space into the matrix.
   C. The electrochemical proton gradient consists of two components: a pH difference and an electrical potential.
   D. Ubiquinone and cytochrome c are both diffusible electron carriers.
   E. Plants have chloroplasts and therefore can live without mitochondria.
3. Please describe the structure of mitochondria.
4. Please describe the structure of chloroplasts.
5. Please describe the structural differences between leaves of C-3 plant and C-4 plant.
6. Explain the working mechanism of ATP synthase.

(张 森 余庆波)

Chapter 6

Cytoskeleton

The cytoskeleton is an elaborate network of protein filaments that support and spatially organize the cytoplasm of a eukaryotic cell, allowing the cell to mechanically interact with its environment and make coordinated movements. The primary building blocks of the cytoskeleton are three types of protein filaments: **microtubules**, **actin filaments** and **intermediate filaments**(Figure 6-1).

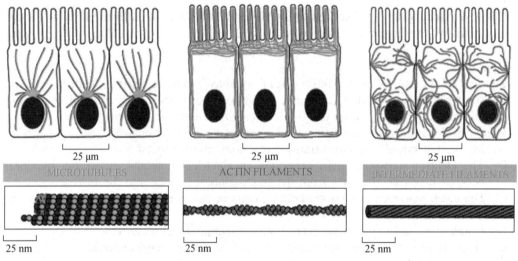

Figure 6-1　Three types of protein filaments form the cytoskeleton. The cells illustrated are epithelial cells lining the gut(From: Alberts B, et al. Essential Cell Biology. 2004)

6.1　Microtubules

6.1.1　Structure and functions of microtubules

Microtubules are long, straight and relatively stiff hollow cylindrical tubes that are ubiquitous in all eukaryotic cells. Microtubules have many functions in a cell. They are stiff enough to provide mechanical support to cells. Their cytoplasmic distribution helps to determine and maintain the cell shape. They are thought to help organizing internal cell contents. They provide a set of "tracks" for cell organelles and vesicles to move on. During mitosis, microtubules form the spindle fibers for separating chromosomes. When arranged in geometric patterns inside flagella and cilia, microtubules are used for locomotion.

A microtubule has an outer diameter of approximately 24 ~ 25 nm, a wall thickness of about 4 nm, and may extend across the length and breadth of a cell. In Figure 6-2A, the bumps at the surface of the microtubules are **microtubule-associated proteins** (MAPs). Each MAPs has its one domain attached to the side of a microtubule and another domain projects outward as a filament. The general functions of MAPs are to increase the stability of microtubules and to promote microtubule assembly. Some MAPs form cross-bridges connecting microtubules to each other to maintain their parallel alignment.

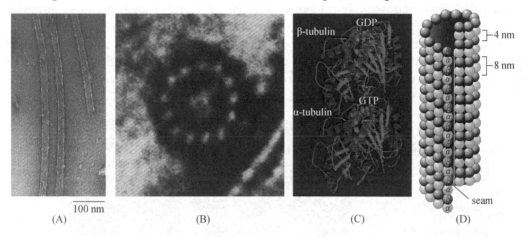

Figure 6-2 **The structure of microtubules.** (A) Electron micrograph of negative stained microtubules from brain showing the globular subunits that make up the protofilaments. The bumps at the surface of the microtubules are microtubule-associated proteins (MAPs). (B) Electron micrograph of a cross section through a microtubules of these plants cells are most abundant in a cortical zone about 100 nm thick just beneath the plasma membrane (seen in the lower right of the micrograph). (C) A ribbon model showing the three-dimensional structure of the α/β-tubulin heterodimer. Note the complementary shapes of the subunits at their interacting surfaces. The α-tubulin subunit has a bound GTP, which is not hydrolyzed and is nonexchangeable. The β-tubulin subunit has a bound GDP, which is exchanged for a GTP prior to assembly into a polymer. The plus end of the dimer is at the top. (D) Diagram of a longitudinal section of a microtubule shown in the B-lattice, which is the structure thought to occur in the cell. The wall consists of 13 protofilaments composed of α/β-tubulin heterodimers stacked in a head-to-tail arrangement. Adjacent protofilaments are not aligned in register, but are staggered about 1 nm so that the tubulin molecules form a helical array around the circumference of the microtubule. The helix is interrupted at one site where α and β subunits make lateral contacts. The produces a "seam" that runs the length of the microtubule. (From: Karp G. Cell and Molecular Biology — concepts and experiments. 2005)

The wall of a microtubule consists of globular proteins termed protofilaments which are arranged parallel to the long axis of the tubule. The cross section of a microtubule consists of 13 protofilaments aligned side by side in a circular pattern within the wall (Figure 6-2B). Each protofilament is built from dimeric protein subunits. Each **tubulin dimer** is made of one **α-tubulin** and one **β-tubulin**. These two tubulins are very similar in 3-dimensional structure and they bind tightly together by noncovalent bonding. Tubulin dimers stack tightly together again by noncovalent bonding to form the wall of microtubules (Figure 6-2D). Each tubulin dimer binds two GTP molecules, with the α-tubulin binding to one GTP irreversibly and the β-tubulin binding to another GTP reversibly, and β-tubulin hydrolyzes GTP to GDP (Figure. 6-2C). As is discussed later in the chapter, the guanine bound to β-tubulin regulates the addition of tubulin subunits at the ends of a microtubule. Because each protofilament is made of a linear chain of heterodimers with alternating α and β tubulins, it has a structural polarity with an α-tubulin at one end and a β-tubulin at the other end. All protofilaments of one microtubule have the same polarity, therefore, the entire microtubule has polarity with one end made of all β-tubulin subunits called the plus end, and the other end made of all α-tubulin subunits called the minus end. This structural polarity is crucial in microtubule formation and function, as the plus end and the minus end differ in their rate of microtubule assembly.

6.1.2 Microtubule organizing center (MTOC)

Microtubules in cells grow from a variety of specialized organizing centers called **microtubule-organizing centers (MTOCs)**. The best-studied MTOC is the **centrosome** in animal cells. The centrosome is composed of a pair of barrel-shaped centrioles surrounded by amorphous, electron-dense **pericentriolar material (PCM)**. **Centrioles** are cylindrical structures about 0.2 μm in diameter and 0.4 μm in length. Centrioles contain nine evenly spaced fibrils, each of which is made of three microtubules called A, B, C tubules. Only the A tubule is a complete microtubule connecting to the center of the centriole by a radial spoke, thus giving the cross section of the centriole a pinwheel look (Figure 6-3).

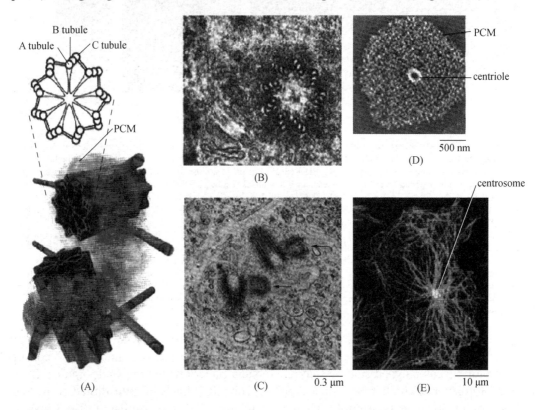

Figure 6-3 The centrosome. (A) Schematic diagram of a centrosome showing the paired centrioles, the surrounding PCM, and micrograph of a cross section of a centriole showing the pinwheel arrangement of the nine peripheral fibrils, each of which consists of one complete microtubule and two incomplete microtubules. (B) Electron micrograph of a cross section of a centriol. (C) Electron micrograph showing two pairs of centrioles. Each pair consists of a longer parental centriole and a small daughter centriole (arrow), which is undergoing elongation in this phase of the cell cycle. (D) Election micrographic reconstruction of a 1.0 mmol/L potassium iodide extracted centrosome, showing the PCM to consist of a loosely organized fibrous lattice. (E) Fluorescence micrograph of a cultured mammalian cell showing the centrosome at the center of an extensive microtubular network (From: Karp G. Cell and Molecular Biology — concepts and experiments. 2005)

All MTOCs share similar functions in all cells—they control the number of microtubules, microtubule polarity, the number of protofilaments that make up microtubule walls, as well as the time and location of microtubule assembly. All MTOCs have a common protein component called **γ-tubulin**. Approximately 80% of γ-tubulin in cells is part of a 25S complex named the γ-tubulin ring, a ring-shaped structure that serves as the starting point, or nucleation site, for the growth of one microtubule (Figure 6-4 A). The α/β-tubulin dimers add to the γ-tubulin ring in a specific direction such that the end result is that the minus end of each microtubule is embedded in the centrosomes while the plus end of each microtubule faces outward where further growth (polymerization) can occur (Figure 6-4 B).

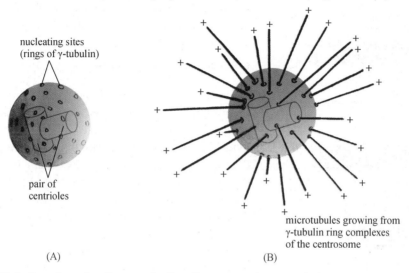

Figure 6-4 Tubulin polymerizes from nucleation sites on a centrosome. (A) Schematic drawing showing that a matrix of protein containing the γ-tubulin rings the nucleate microtubule growth. In animal cells, the centrosome contains a pair of centrioles, each made up of a cylindrical array of short microtubules. (B) A centrosome with attached microtubule is embedded in the centrosome, having grown from a nucleating ring, whereas the plus end of each microtubule is free in the cytoplasm (From: Alberts, B, et al. Essential Cell Biology. 2004)

Growing microtubules show dynamic instability: once a microtubule has been nucleated, its plus end grows outward by the addition of subunits for minutes then suddenly shrinks by losing subunits from its free end and subsequently start growing again. Thus a given microtubule oscillates unpredictably between assembly (growing/polymerization) and disassembly (shrinking/depolymerization) phases; growing and shrinking microtubules can coexist in the same region of a cell. This dynamic instability results mainly from GTP hydrolysis. As mentioned before, each tubulin dimer has one tightly bound GTP molecule that is hydrolyzed to GDP (still tightly bound) shortly after adding to a growing microtubule. After disassembly, the GDP bound to the released dimer is replaced by a new GTP. This nucleotide exchange allows the dimer to be once again ready to become a building block for polymerization.

The GTP-bound tubulin molecules pack more strongly to each other compared with GDP-bound tubulin molecules. When polymerization is progressing rapidly, tubulin molecules add to the end of microtubule faster than GTP hydrolysis rate, giving rise to a microtubule end full of GTP-tubulin subunits called GTP cap. However, occasionally, when the tubulin at the free end of the microtubule hydrolyzes its GTP to GDP before the next tubulin has been added, the end of microtubule consists of GDP bound subunits. Because GDP-tubulins bind less effectively to each other, this favors disassembly of the microtubule. Once disassembly starts, it tends to continue at a catastrophic rate. Therefore, microtubules shrink rapidly (Figure 6-5).

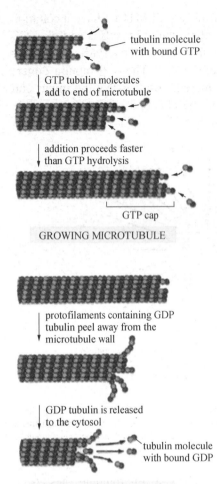

Figure 6-5 GTP hydrolysis controls the growth of microtubules. Tubulin dimers carrying GTP bind more tightly to one another than do tubulin dimers carrying GDP. Therefore, microtubules that have freshly added tubulin dimers at their end with GTP bound tend to keep growing. From time to time, however, especially when microtubule growth is slow, the subunit in this "GTP cap" will hydrolyze their GTP to GDP before fresh subunits loaded with GTP have time to bind. The GTP cap is thereby lost; the GDP-carrying subunits are less tightly bound in the polymer and are readily released from the free end, so that the microtubule begins to shrink continuously (From: Alberts B, et al. Essential Cell Biology. 2004)

In a normal cell, the MTOCs continuously explore randomly by shooting out new microtubules in all directions and retracting them. This process ceases when the plus end is somehow stabilized by attachment to another molecule or cell structure and a relatively stable link is established between that structure and the centrosome. The random exploration and selective stabilization enables MTOCs to set up a highly organized system of microtubules linking specific parts of the cell.

6.1.3 Motor proteins

Cells can modify the dynamic instability of their microtubules for specific reasons. For example, when cells enter mitosis, microtubules become more dynamic, growing and shrinking more frequently than they normally do, which allows them to disassemble rapidly and reassemble into a mitotic spindle. However, differentiated cells often suppress the dynamic instability of their microtubules by having proteins bind to the ends or along the length of microtubules to stabilize them. Stabilized microtubules can maintain the organization of the cell. Most differentiated cells are polarized; meaning that one end of the cell is structurally and functionally different from the other end of the cell. Polarized systems of microtubules determine a cell's polarity, help to position intracellular organelles in the cell and guide the intracellular movement. For example, in a nerve cell, all the plus ends of microtubules in the axon point toward the axon terminal (Figure 6-6). Nerve cells are able to transport materials such as membrane vesicles and proteins for secretion along these oriented microtubule tracks in axons from the producing site (cell body) to the target (end of the axon). Movement of particles along microtubules is a lot faster and more efficient than free diffusion.

Directed intracellular movement of cell materials along microtubules is accomplished by motor proteins. This is also the case with the **actin filaments**, which will be discussed later in the chapter. Motor proteins derive their energy from repeated cycles of ATP hydrolysis. There are two major families of motor proteins that move along cytoplasmic microtubules: kinesins and dyneins. **Kinesins** usually move toward the plus end of a microtubule (outward, away from the centrosome) while **dyneins** move toward the minus end (inward, toward the centrosome). They both have two globular

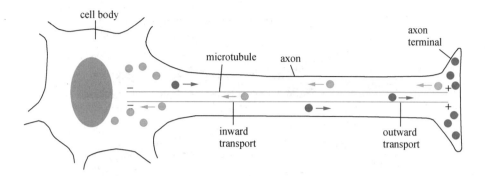

Figure 6-6 microtubules transport cargo along a nerve cell axon. In nerve cells all the microtubules in the axon point in the same direction. With their plus ends toward the axon terminal. The oriented microtubules serves as tracks for the directional transport of materials synthesized in the cell body but required at the axon terminal (such as membrane proteins required for growth.) For an axon passing from your spinal cord to a muscle in your shoulder, say, the journey takes about two days. In addition to this outward traffic of material driven by one set of motor proteins, there is inward traffic in the reverse direction driven by another set of proteins. The inward traffic carries materials ingested by the breakdown of proteins and other molecules back toward the cell body(From: Alberts B, et al. Essential Cell Biology. 2004)

ATP-binding heads and a tail (Figure 6-7).

The heads determine which microtubule to attach by interacting with microtubules of specific orientation while the tail binds stably to some specific cell materials for transport.

The heads of both kinesins and dyneins are enzymes with ATP-hydrolyzing (ATPase) activity. ATP hydrolysis provides energy for a cycle of conformation changes in the head that enables it to move along the microtubule by a cycle of binding, release, rebinding to the microtubule.

Microtubules and their associated motor proteins play an important role in positioning intracellular organelles (such as

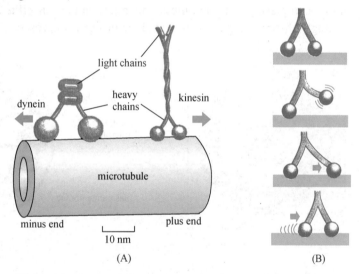

Figure 6-7 Motor proteins move along microtubules using their globular heads. (A) Kinesins and cytoplasmic dyneins are microtubule motor proteins that generally move in opposite directions along a microtubule. Each of these proteins (drawn here to scale) has two heavy chains and several smaller light chains. Each heavy chain forms a globular head that interacts with microtubules. (B) Diagram of a motor protein showing ATP-dependent "walking" along a filament(From: Alberts B, et al. Essential Cell Biology. 2004)

membranes) within a cell. For example, the normal positioning of endoplasmic reticulum relies on its membrane receptors that bind to kinesin and subsequently are pulled outward along microtubules, stretched like a net; in the case of Golgi apparatus, it is dragged by dyneins along microtubules to reach its position at the center of a cell (Figure 6-8). In this way the cell internal membrane arrangement, on which the successful function of the cell relies, are established and maintained.

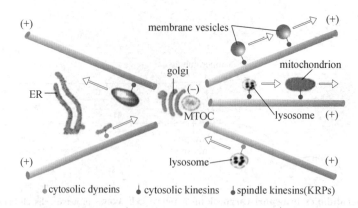

Figure 6-8 A general model for kinesin-and dynein-mediated transport in a typical cell. The array of microtubules, with their (+) ends pointing toward the cell periphery, radiates from an MTOC in the Golgi region. Kinesin-dependent anterograde transport conveys mitochondria, lysosomes, and an assortment of membrane vesicles to the endoplasmic reticulum (ER) or cell periphery. Cytosolic dynein-dependent retrograde transport conveys elements of the ER, late endosomes, and lysosomes to the cell center (From: Lodish H, et al. Molecular Cell Biology. 2000)

6.1.4 The microtubules in cilia and flagella

Unlike cytoplasmic microtubules, the microtubules in **cilia** and **flagella** are arranged in a strikingly distinctive pattern revealed by electron microscopy (Figure 6-9).

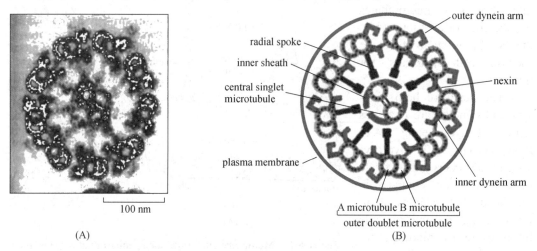

Figure 6-9 Microtubules in a cilium or in a flagellum are arranged in a "9 + 2" array. (A) Electron micrograph of a flagellum of *chlamydomonas* shown in cross section. Illustrating the distinctive "9 + 2" arrangement of microtubules. (B) Diagram of the flagellum across section. The nine outer microtubules (each a special paired structure) carry two rows of dynein molecules. The heads of these dyneins appear in this view like pairs of arms reaching toward the adjacent microtubule. In a living cilium, these dynein heads periodically make contact with the adjacent microtubule and move along it, thereby producing the force for ciliary beating. Various other links and projections shown are proteins that serve to hold the bundle of microtubules together and to convert the sliding motion produced by dyneins into bending (From: Alberts B, et al. Essential Cell Biology. 2004)

The cross section of a cilia or flagella shows a ring of nine doublet microtubules with a pair of single microtubule at the center. This "9 + 2 array" is seen in all forms of eukaryotic cilia and flagella, ranging from protozoa to human. This is a useful reminder that all living eukaryotes have evolved from a common ancestor. The movement of a cilium or a flagellum is initiated by the bending of its core as the microtubules slide against each other. The bending motion of the core is generated from ciliary dynein, a motor protein that resembles cytoplasmic dynein in both its structure and function. Ciliary dynein has its tail attached to one microtubule and its head to another microtubule to generate a sliding force between two filaments. Because the adjacent doublets are linked together, the parallel sliding movement between free microtubules is converted to a bending motion in the cilium (Figure 6-10).

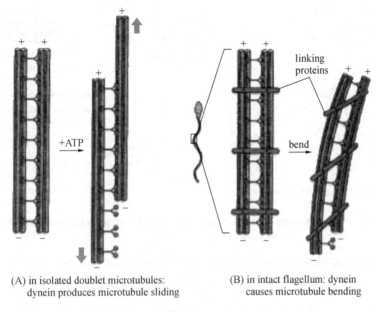

(A) in isolated doublet microtubules: dynein produces microtubule sliding

(B) in intact flagellum: dynein causes microtubule bending

Figure 6-10 The movement of dynein causes the flagellum to bend
(From: Alberts B, et al. Essential Cell Biology. 2004)

6.2 Actin filaments

Actin filaments (microfilaments) are thin and flexible filaments that are the most abundant intracellular proteins in a eukaryotic cell. They are essential for cell locomotion and intracellular motile processes. They also help to shape the cell.

6.2.1 Globular monomers and filamentous polymers

Actin exists either as G-actin, a globular monomer, or as F-actin, a filamentous polymer that is a linear chain of G-actin subunits. Small actin binding proteins such as thymosin and profilin binds to G-actin in the cytosol, preventing G-actin subunits from polymerizing into filaments. Many other actin-binding proteins, however, bind to F-actin and control the behavior of the intact filaments to form different types of stable actin filament structures (Figure 6-11). Each actin molecule contains an Mg^{2+} ion complexed with either ATP or ADP. Thus, there are four states of actin: ATP-G-actin, ADP-G-actin, ATP-F-Actin and ADP-F-actin. The more prominent forms of actin in a cell are ATP-G-actin and ADP-F-actin. The interaction between the ATP and ADP forms of actin is important in the cytoskeleton assembly.

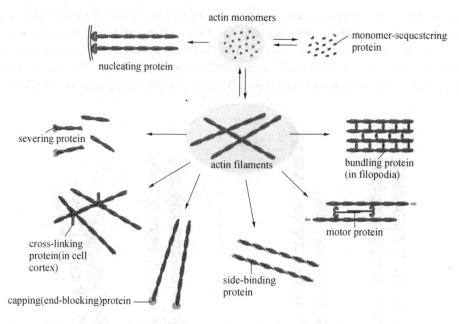

Figure 6-11 Actin-binding proteins control the behavior of actin filaments in vertebrate cells (From: Alberts B, et al. Essential Cell Biology. 2004)

X-ray crystallographic analysis shows G-actin is separated into two lobes by a deep cleft where ATP and Mg^{2+} are bound. Two lobes and a cleft make up an ATPase fold (Figure 6-12A) where the floor of the cleft acts as a hinge that allows the lobes to flex relative to each other and ATP holds together the two lobes of the actin monomer. The

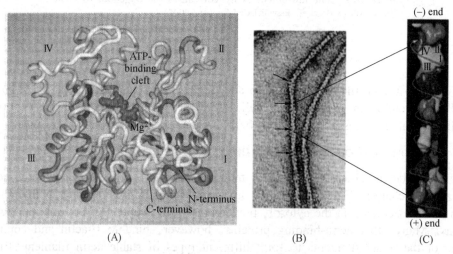

Figure 6-12 Structures of monomeric G-actin and F-actin filament. (A) Model of a β-actin monomer from a nonmuscle cell shows it to be a platelike molecule divided by acentral cleft into two approximately equal-sized lobes and four subdomains, numbered I ~ IV. ATP (red) binds at the bottom of the cleft and contacts both lobes (the center ball represents Mg^{2+}). The N- and C-termini lie in subdomain I. (B) In the electron microscope, negatively stained of beads subunits. Because of the twist, the filament appears alternately thinner (7nm diameter) and thicker (9nm diameter). (C) In one model of the arrangement of subunits in an actin filament, the subunits lie in a tight helix along the filament, as indicated by the arrow. One repeating unit consists of 28 subunits (13 turns of the helix), covering a distance of 72 nm. Only 14 subunits are shown in the figure. The ATP-binding cleft is oriented in the same direction (top) in all actin subunits in the filament. As discussed later, this end of a filament is designated the (-) end; the opposite end is the (+) end (From: Lodish H, et al. Molecular Cell Biology. 2000)

nucleotide affects the conformation of the actin molecule and stabilizes it. Without a bound nucleotide, G-actin denatures very quickly. When negatively stained by uranyl acetate for electron microscopy, F-actin appears as long twisted strings of beads whose diameter varies between 7 and 9 nm. (Figure 6-12B) From x-ray diffraction studies (Figure 6-12A), scientists have produced a model of an actin filament shown in figure 6-12C in which the subunits are organized into tightly wound helix. Each subunit is surrounded by four other subunits: one above, one below and two to one side. Each subunit corresponds to a bead seen in electron micrographs of F-actin filaments in Figure 6-12B.

6.2.2 The actin cytoskeleton is organized into bundles and networks of filaments

There are two most common organization styles of actin filaments in a cell: bundles and networks of filaments (Figure 6-13). Electron micrograph or immunofluorescence micrograph shows that bundles start from the protrusion of the cell surface membrane and continue to fan out to become a network of filaments. In bundles, the actin filaments are tightly packed in parallel arrays whereas in a network the actin filaments are loosely packed in a criss-cross pattern and often at right angles. There are two types of actin networks. One is planar or two-dimensional, like a net or a web, and is associated with the plasma membrane; the other is three-dimensional, present in the cell and provides the cytosol gel-like properties. Actin cross-linking proteins hold the filaments together in all bundles and networks. Each cross-linking protein has two actin-binding sites, with one site for each filament. Parallel actin filaments in bundles are held close together by short cross-linking protein while orthogonally oriented actin filaments in networks are tethered by long, flexible cross-linking proteins. Functionally, both bundles and networks support the plasma membrane and help to determine a cell's shape.

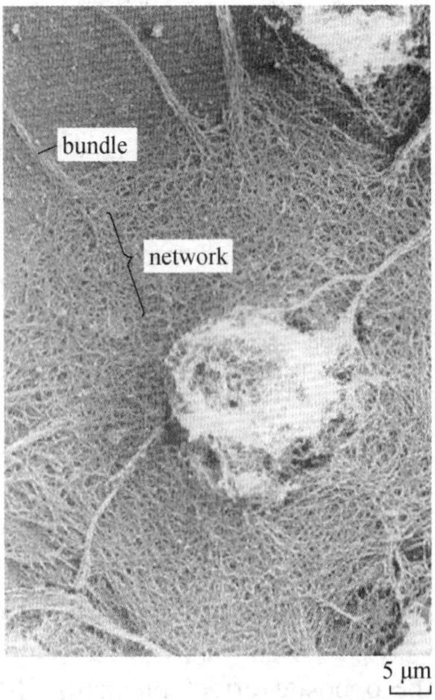

Figure 6-13 Micrograph revealing bundles and networks of actin filaments in the cytosol of a spreading platelet treated with detergent to remove the plasma membrane. Actin bundles project from the cell to form the spikelike filopodia. In the lamellar region of the cell, the actin filaments form a network that fills the cytosol. In contrast to the roughly parallel alignment of bundled filaments. The filaments in networks lie at various angles approaching 90° (From: Lodish H, et al. Molecular Cell Biology. 2000)

6.2.3 Mechanism of actin ploymerization

Actin and tubulin polymerize by similar mechanisms. *In vitro*, the addition of ions Mg^{2+}, K^+, or Na^+ to a solution of G-actin induces the polymerization of G-actin into F-actin filaments, which are indistinguishable from microfilaments (actin) isolated from cells. This process is reversible in that once the ion concentration is reduced in the solution, the F-actin filaments depolymerize into G-actin. This reversible assembly of actin lies at the core of many cell movements. The assembly of G-actin into F-actin is accompanied by the hydrolysis of ATP to ADP and Pi, although ATP hydrolysis is not necessary for polymerization to occur.

There are three sequential phases in the polymerization of actin filaments (Figure 6-14). In phase one (lag state), G-actin keeps on aggregating into short, unstable oligomers until the oligomers reach a certain length (three or four subunits), then oligomers can act as a stable seed, or nucleus. In phase two (elongation state), the nucleus rapidly elongates into a filament due to the addition of actin monomers to both of its ends until the concentration of G-actin monomers decrease enough to reach its equilibrium state with the filament. In phase three (steady state), G-actin monomers only exchange with subunits at the filament ends but there is no net change in the total mass of the filaments. The equilibrium concentration of the pool of unassembled free subunits is called the **critical concentration** (**Cc**).

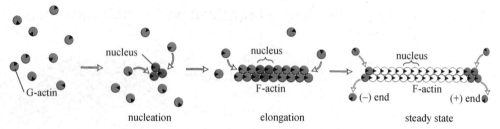

Figure 6-14 **The three phases of G-actin polymerization in *vitro*.** During the initial nucleation phase, ATP-G-actin monomers slowly form stable complexes of actin. These nuclei are more rapidly elongated in the second phase by addition of subunits to both ends of the filament. In the third phase, the ends of actin filaments are in a steady state with monomeric ATP-G-actin. After their incorporation into a filament, subunits slowly hydrolyze ATP and become stable ADP-F-actin. Note that the ATP-binding clefts (black triangles) of all the subunits are oriented in the same direction in F-actin(From: Lodish H, et al. Molecular Cell Biology. 2000)

Toxins easily perturb the equilibrium between F- and G-actins. **Cytochalasin**, a fungal alkaloid, depolymerizes actin filaments by binding to the plus end of F-actin and blocks further addition of subunits. In contrast, **latrunculin**, a toxin secreted by sponges, binds to G-actin and inhibits it from adding to a filament end. Both toxins inhibit the polymerization; therefore, with the presence of these two toxins in the bath, the actin cytoskeleton disappears and cell movements are inhibited. A third toxin, **phalloidin**, has the opposite effect on actin. It binds at the interface between F-actins, locking adjacent subunits together, and thus preventing actin filaments from depolymerization, even when G-actin is diluted below its Cc.

A cell regulates actin polymerization through several actin-binding proteins that either promote or inhibit polymerization. For example, thymosin β4 (Tβ4, an abundant cytosolic protein, binds ATP-G-actin (but not F-actin) in a 1∶1 complex. In the complex, G-actin cannot polymerize. Thus, thymosin β4 serves as a buffer for monomeric actin and inhibits actin assembly. Another cytosolic protein profilin, on the other hand, promotes actin assembly, although it also binds ATP-actin monomers in a stable 1∶1 complex. Some actin binding proteins control the lengths of actin filaments by binding to actin filaments and breaking them into shorter fragments. Another group of proteins can cap the ends of actin filaments and stabilize actin filaments.

Table 6-1 shows a list of cytosolic proteins that regulates actin polymerization. The activity of actin-binding proteins that regulates the length, location, organization, and dynamic behavior of actin filaments are in turn, regulated by extracellular signals, allowing cells to rearrange its cytoskeleton to respond to its environment. In the case of actin filaments, a group of closely related monomeric GTP-binding proteins called Rho protein family act as signaling switch molecules. Activation of different Rho proteins causes different dramatic and complex structural changes in the cell.

Table 6-1 Some cytosolic proteins that control actin polymerization

Protein	MW	Activity
cofilin	15,000	severing
gCAP39	40,000	capping [(+) end]
severin	40,000	severing, capping
gelsolin	87,000	severing, capping [(+) end]
villin	92,000	cross-linking, severing, capping
capZ	36,000(α)32,000(β)	capping [(+) end]
tropomodulin	40,000	capping [(-) end]

(From: Lodish H, et al. Molecular Cell Biology. 2000)

6.2.4 Cell movement by manipulating actin polymerization

Functionally, the cell derives its driving force for many types of movement by manipulating actin polymerization and depolymerization. For example, cell crawling, a major type of cell movement, depends on actin filaments.

Cell crawling has three steps: ① the cell pushes out at its protrusion leading edge; ② new anchorage points created at the leading edge cause the adherence between protrusions and the cell crawling surface; ③ the rest of the cell drags itself forward by traction on anchorage points (Figure 6-15).

All three steps involve actin in different ways. The first step is driven by rapid polymerization of actin filaments, which are nucleated at the plasma membrane and push out the membrane without tearing it to form exploratory, motile structures such as lamellipodia or filopodia. **Actin-related proteins (ARPs)** form a complex that promotes the formation of a two-dimensional, tree like web of actin by binding to an existing actin filament and nucleating the creation of new side growths of filaments. With the help of other actin binding proteins, this web assembles at the front end and dissembles at the rear end, pushing the lamellipodium or filopodium forward (Figure 6-16).

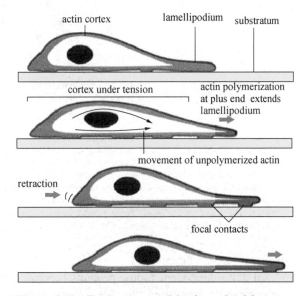

Figure 6-15 Forces generated in the actin-rich cortex move a cell forward (From: Alberts B, et al. Essential Cell Biology. 2004)

In step two, the lamellipodia and filopodia stick to a favorable patch surface through integrins, a type of transmembrane proteins embedded in their plasma membrane. Integrins adhere to the crawling surface on the extracellular side and capture actin filaments intracellularly, creating anchorage sites. In step three, cells use anchorages and internal contractions to exert a pulling force. These are based on the interaction of actin filaments with motor proteins called myosins. Although how this pulling force is generated is not clear, the mechanism of how myosin motor proteins and actin filament interaction produces

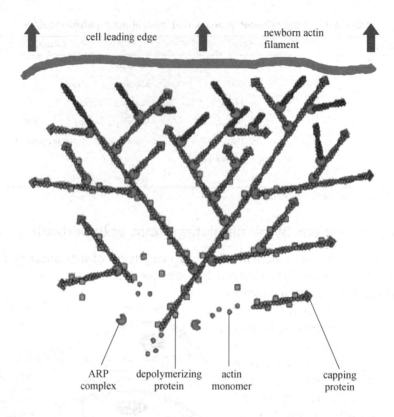

Figure 6-16 Assembly of an actin meshwork pushes forward the leading edge of a lamellipodium. Nucleation of new actin filaments is mediated by ARP(actin-related proteins) complexes at the front of the web. Newly formed filaments are thereby attached to the sides of preexisting filaments. As these filaments elongate, they push the plasma membrane forward. The actin filament plus ends will become protected by capping proteins, preventing further assembly or disassembly from the old plus ends at the front of the array. Hydrolysis of ATP bound to the polymerized actin subunits promotes depolymerization at the rear end of the actin complex by a depolymerizing protein. The spatial separation of assembly and disassmbly allows the network as a whole to move forward at a steady rate(From: Alberts B, et al. Essential Cell Biology. 2004)

movement is well understood.

All actin-dependent motor proteins belong to the myosin family. They bind to actin to form contractile structures. In fact, many cell movements, ranging from the intracellular motion of cellular components to whole cell locomotion, depend on the interaction of actin and myosin. Among these, the best-studied contractile structures are muscles.

6.2.5 Muscle contraction

(1) **Structure of muscle cell** Vertebrates and many invertebrates have two major classes of muscle—skeletal muscle and smooth muscle—each serves different functions. A typical muscle cell, muscle fiber or a myofiber, is a cylindrical structure that is large (1 ~ 40 mm in length and 10 ~ 50 μm in width) and multinucleated (containing as many as 100 nuclei). A myofiber is filled with bundles of thinner, cylindrical filaments called myofibrils that extend the length of the cell. Each myofibril consists of a repeating linear array of sarcomeres, each about 2 μm long in resting muscle. Each sarcomere displays a characteristic dark and light banding pattern, which in turn gives the muscle cell a striated appearance. The dark band (A band) is bisected by a dark region (H zone) while the light

band (I band) is bisected by a dark line (the Z disk or Z line). Each sarcomere is a segment from one Z disk to the next, consisting of two halves of an I band and an A band (Figure 6-17). Electron microscopy of stained muscle fibers reveals the banding pattern to be the result of the partial overlap of two distinct types of filaments: thin filaments composed of actin and thick filaments containing myosin II.

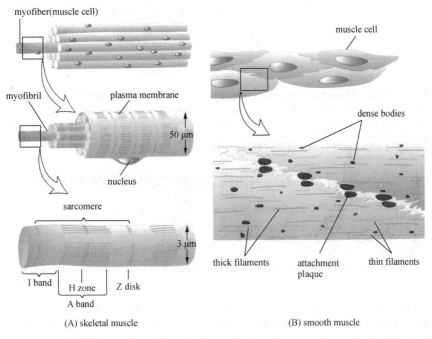

(A) skeletal muscle (B) smooth muscle

Figure 6-17 General structure of skeletal and smooth muscle. (A) Skeletal muscle tissue is composed of bundles of multinucleated muscle cell, or myofibers. Each muscle cell is packed with bundles of actin and myosin filaments, organized into myofibrils that extend the length of the cell. Packed end to end in a myofibril is a chain of sarcomeres, the functional units of contraction. The internal organization of the filaments gives skeletal muscle cells a striated appearance. (B) Smooth muscle is composed of loosely organized spindle-shaped cells that contain a single nucleus. Loose bundles of actin and myosin filaments pack the cytoplasm of smooth muscle cells. These bundles are connected to dense bodies in the cytosol and to the membrane at attachment plaques (From: Lodish H, et al. Molecular Cell Biology. 2000)

Myosin II is a dimer made of two identical myosin molecules held together by their tails. Each myosin II has two globular ATPase heads at one end and a single coiled-coil tail at the other. Myosin II molecules bind to one another through their tails and form a bipolar myosin filament in which the heads project from the sides (Figure 6-18).

(2) Actin filaments slide against myosin filament during muscle contraction A myosin filament has two sets of heads pointing in opposite directions from the middle. One set of heads binds to actin filaments in one direction and moves them one way and the other set of heads binds to actin filaments in the opposite direction and moves them accordingly (Figure 6-19). When actin and myosin are organized together in a bundle, the bundle can generate a contractile force, thus making sarcomere a contractile unit.

A contraction of a muscle cell is caused by a simultaneous shortening of all sarcomeres, which is in turn caused by the actin and myosin filaments sliding past each other without shortening (Figure 6-20). illustrates how myosin and actin interact during a cycle of muscle contraction and relaxation.

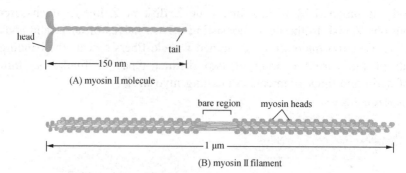

Figure 6-18　Myosin-II molecules can associate with one another to form myosin filaments. (A) A molecule of myosin-II has two globular heads and a coiled-coil tail. (B) The tails of myosin-II associate with one another to form a bipolar myosin filament in which the heads project outward from the middle in opposite direction. The bare region in the middle of the filament consists of tails only (From: Albert B. Essential Cell Biology. 2004)

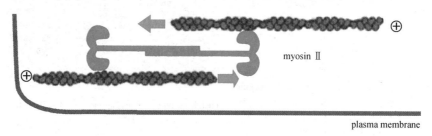

Figure 6-19　Even small bipolar filaments composed of myosin II molecules can slide actin filaments over each other, thus mediating local *shortening* of an actin filament bundle. As with myosin I, the head group of myosin II walks toward the plus end of the actin filament it contacts (From: Alberts B, et al. Essential Cell Biology. 2004)

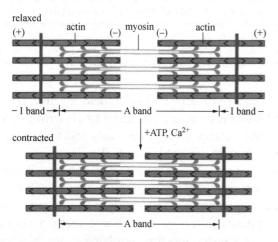

Figure 6-20　The sliding-filament model of contraction in striated muscle. The arrangement of thick myosin and thin actin filaments in the relaxed state is shown in the upper diagram. In the presence of ATP and Ca^{2+}, the myosin heads extending from the thick filaments walk toward the (+) ends of the thin filaments. Because the thin filament are anchored at the Z disks (purple), movement of myosin pulls the actin filaments toward the center of the sarcomere, shortening its length in the contracted state as shown in the lower diagram (From: Lodish H, et al. Molecular Cell Biology. 2000)

Each myosin filament has about 300 myosin heads, each of which can attach and detach from actin about five times per second. This allows the myosin and actin to slide past one another at speeds of up to 15 μm per second, which permits a sarcomere to change from fully extended state (3 μm) to contracted state (2 μm) within a tenth of a second. Therefore, when a muscle is stimulated, all of the sarcomeres of the muscle are also triggered almost instantaneously, and the entire muscle can contract extremely fast, usually within a tenth of a second.

When a muscle receives signals from the nervous system, there is a sudden rise in cytosolic Ca^{2+} in the muscle cell. The Ca^{2+} interacts with a molecular switch made of proteins such as tropomyosin and troponin that are specifically associated with actin filaments. Tropomyosin is a rigid, rod-shaped molecule that prevents the myosin heads from associating with the actin filament because it binds in the groove of the actin helix overlapping seven actin monomers. Troponin is a protein complex that includes a Ca^{2+}-sensitive protein troponin-C, which is associated with the end of a tropomyosin molecule. (Figure 6-21A) When the cytosolic Ca^{2+} rises, Ca^{2+} binds to and change the shape of troponin, causing tropomyosin molecules to shift their position slightly, allowing myosin heads to bind to actin filament and initiate contraction (Figure 6-21B). As soon as the Ca^{2+} level returns to its resting level, troponin and tropomyosin molecules recover to their original positions where they block myosin binding and end contraction.

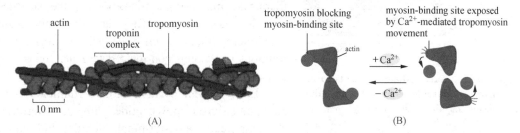

Figure 6-21 Skeletal muscle contraction is controlled by troponin. (A) A muscle thin filament showing the positions of tropomyosin and troponin along the actin filament. Every tropomyosin molecule has seven evenly spaced regions of homologous sequence, each of which is thought to bind to an actin subunit in the filament. (B) When Ca^{2+} binds to troponin, the troponin moves the tropomyosin that otherwise blocks the interaction of actin with the myosin heads (From: Alberts B, et al. Essential Cell Biology. 2004)

6.3 Intermediate filaments

The third sets of cytoskeletal fibers in eukaryotic cells are **intermediate filaments (IFs)**. IFs typically form a network throughout the cytoplasm of most animal cells. They surround the nucleus and extend out to the cell periphery. They are often interconnected with other types of filaments by thin, wispy cross-bridges that are made of a huge, elongated accessory protein called plectin that has a binding site for an intermediate filament at one end and another binding site at the other end for another type of filament depending on the plectin isoform (Figure 6-22).

6.3.1 Cytosolic intermediate filaments

IFs are often anchored to the plasma membrane at cell-cell junctions such as desmosomes where the external face of the membrane connects with that of another cell (Figure 6-23).

IFs are strong, solid, unbranched, ropelike fibers that provide strength and mechanically support the cells that are subject to physical stress such as

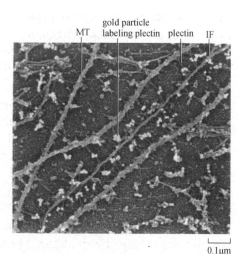

Figure 6-22 Connections between intermediate filaments and other components of the cytoskeleton. Intermediate filaments are linked to both microtubules and actin filaments by a protein called plectin. Plectin (red) links IFs (green) to MTs (orange). Gold particles (yellow) label plectin (deep etch TEM). Plectins can also bind, actin MFs (not shown). Here, IFs server as strong but elastic connectors between the different cytoskeletal filaments (From: Becker WM. The World of the Cell. 2006)

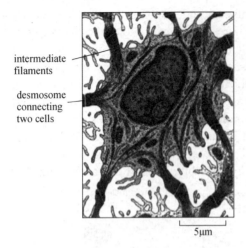

Figure 6-23 **Intermediate filaments form a strong, durable network in the cytoplasm of the cell.** Drawing from an electron micrograph of a section of epidermis showing the bundles of intermediate filaments that traverse the cytoplasm and that are inserted at desmosomes (From: Alberts B, et al. Essential Cell Biology. 2004)

neurons, muscle cells and the epithelial cells that line the body's cavities. IFs within the nucleus are called the nuclear lamina. They are a mesh of filaments that support the nuclear envelope in all eukaryotic cells. IFs were so named because in the smooth muscle cells where they were first discovered, they were found to have a diameter of about 8~12 nm, which is between that of the thin actin-containing microfilaments (about 7 nm) and thicker microtubule filaments (24 nm). Of the three types of cytoskeletal filaments, intermediate filaments are the strongest and the most stable. They are the only intact cytoskeleton filaments left when cells are treated with solutions containing non-ionic detergents and a high concentration of salts.

Each subunit of intermediate filaments is an elongated fibrous polypeptide with an N-terminal globular head, a C-terminal globular tail and a center elongated rod domain. The rod domain has an extended α-helical region that enables pairs of intermediate filament proteins to form stable dimers by wrapping around each other in a coiled-coil configuration. Because the two polypeptides are aligned parallel to each other in the same direction, the dimer has polarity with one end defined by N-termini of the polypeptides and the other by C-termini. Two such dimmers bind together side-by-side in a staggered fashion with their N-and C-termini pointing in opposite (antiparallel) direction to form a tetramer via non-covalent bonding. Then tetramers bind to each other end-to-end and side-by-side also via non-covalent bonding to form the final ropelike intermediate filament. Because the tetrameric building blocks lack polarity, so do the intermediate filaments, which is another feature that distinguishes IFs from other cytoskeletal filaments (Figure 6-24).

The central rod domains of different IFs are all similar in size and amino acid sequence, therefore, different IFs are similar in diameter and internal structure. In contrast, the globular head and tail

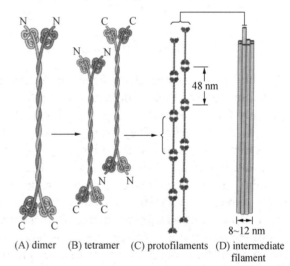

(A) dimer (B) tetramer (C) protofilaments (D) intermediate filament

Figure 6-24 **A model for intermediate filaments assembly in vitro.** (A) The starting point for assembly is a pair of IFs polypeptides. The two polypeptides are identical for all IFs except kertatin filaments, which are obligate heterodimers with one each of the type I and type II polypeptides. The two polypeptides twist around each other to form a two-chain coiled coil, with their conserved center domain aligned in parallel. (B) Two dimmers align laterally to form a tetrameric protofilament. (C) Protofilaments assemble into larger filaments by end-to-end and side-to-side alignment. (D) The fully assembled intermediate filament is thought to be eight protofilaments thick at any point (From: Becker WM. The World of the Cell. 2006)

regions of different IFs vary greatly both in size and in amino acid sequence. The head and tail regions are exposed on the surface of the filament, allowing the filament to interact with other components of the cytoplasm. Unlike microtubules and microfilaments, IFs are a heterogeneous group of structures. The polypeptides subunits of IFs can be classified into six categories based on their tissue distribution, biochemical, genetic, and immunologic criteria(Table 6-2).

Table 6-2　Properties and distribution of the major mammalian intermediate filament proteins

IF protein	sequence type	average molecular mass ($\times 10^{-3}$)	estimated number of polypeptides	primary tissue distribution
keratin (acidic)	I	40~64	>25	epithelia
keratin (basic)	II	53~67	>25	epithelia
vimentin	III	54	1	mesenchymal cells
desmin	III	53	1	muscle
glial fibrillary acidic protein (GFAP)	III	50	1	glial cells, astrocytes
peripherin	III	57	1	peripheral neurons
neurofilament proteins				neurons of central and peripheral nerves
NF-L	IV	62	1	
NF-M	IV	102	1	
NF-H	IV	110	1	
Lamin proteins				
Lamin A	V	70	1	all cell types (nuclear envelopes)
Lamin B	V	67	1	
Lamin C	V	60	1	
nestin	VI	240	1	heterogeneous

(From: Karp G, et al. Cell and Molecular Biology — concepts and experiments. 2005)

Because of the tissue specificity of intermediate filaments, animal cells from different tissues can be distinguished on the basis of the IF protein present, as determined by immunofluorescence microscopy. This intermediate filament typing serves as a diagnostic tool in medicine. IF typing is especially useful in the diagnosis of cancer, because tumor cells are known to retain the IF proteins characteristic of the tissue of origin, regardless of where the tumor occurs in the body.

6.3.2　Nuclear intermediate filaments

The nuclear envelope is supported by a meshwork of intermediate filaments. The inner nuclear membrane are organized as a two dimensional mesh called nuclear lamina which are composed of a class of intermediate filaments proteins called lamins(Figure 6-25).

Lamins are not as stable as cytoplasmic IFs. They disassemble and reassemble at each cell division when the nuclear envelope breaks down during mitosis and then re-forms in each daughter cell. Disassembly and reassembly of the nuclear lamins are controlled by phosphorylation and dephosphorylation of the lamins by protein kinases. Phosphorylation of lamins causes conformational change in lamins that weakens the binding between IF tetramers and subsequently the filaments fall apart. On the other hand, dephosphorylation of

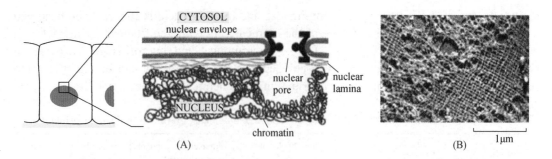

Figure 6-25 Intermediate filaments support and strengthen the nuclear envelope. (A) Schematic cross section through the nuclear envelope. The intermediate filaments of the nuclear lamina line the inner face of the nuclear envelope and are thought to provide attachment sites for the DNA-containing chromatin. (B) Electron micrograph of a portion of the nuclear lamina from a frog egg, or oocyte. The lamina is formed from a square lattice of intermediate filaments composed of lamins (Nuclear laminae from other cell types are not always as regularly organized as the one shown here) (From: Alberts B. et al. Essential Cell Biology. 2004)

lamins at the end of mitosis causes the lamins to reassemble (Figure 6-26).

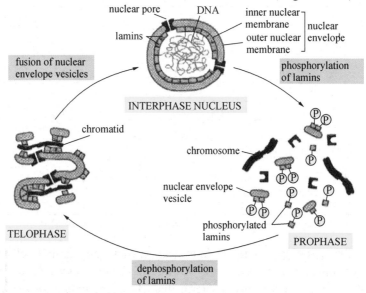

Figure 6-26 The nuclear envelope breaks down and reforms during mitosis. The phosphorylation of the lamins at prophase helps trigger the disassembly of the nuclear lamina, which in turn causes the nuclear envelope to break up into vesicles. Dephosphorylation of the lamins at telophase helps reverse the process (From: Alberts B, et al. Essential Cell Biology. 2004)

Summary

The cytoskeleton is an extensive system of fibers that are responsible for cell shape and cell motility. The cytoskeleton contains three principal types of filaments: microtubules, actin filaments and intermediate filaments. All three types of filaments undergo constant remodeling through the assembly and disassembly of their subunits. While tubulin and actin have been strongly conserved in eukaryotic evolution, the family of intermediate filaments is very diverse. There are a variety of tissue specific forms of intermediate filaments, including keratin, desmin and neurofilaments. In most animal cells, microtubules are nucleated at MTOC, whereas most actin filaments are nucleated near the plasma membrane. Through ATP hydrolysis, motor proteins convert chemical energy stored in ATP into movement along microtubules or actin filaments. They mediate the sliding of filaments relative to one another and thus help transporting membrane-enclosed organelles along filament tracks. The motor proteins that move on actin filaments are members of the myosin superfamily. During muscle contraction, actin filaments slide against myosin filaments. The motor proteins that move on microtubules are members of either the kinesin family or the dynein family.

Questions
1. Define these key terms: actin filament, myosin, cell cortex, centriole, dynein, kinesin, intermediate filament, nuclear lamina, sarcomere, tubulin.
2. Why do eukaryotic cells, especially animal cells have such large and complex cytoskeletons?
3. There are no known motor proteins that move on intermediate filaments, why do you think that is the case?
4. Which of the following changes takes place when a skeletal muscle contracts:
   A. Z discs move farther apart
   B. Actin filaments contract
   C. Myosin filament contract
   D. Sarcomeres become shorter
5. Describe the functions of microtubule.
6. Describe the structure of the sarcomere of a skeletal muscle myofibril and the changes that occur during muscle contraction.
7. Why we say intermediate filament typing is a useful diagnostic tool?

（沈大棱）

Chapter 7

Cell communication and signaling

To carry out one or more specific biological functions, cells need to communicate and work with each other to coordinate their activities. **Cell communication** is also called **cell signaling**, which is one of the most basic characteristics of the cell. Cell communication (signaling) affects every aspect of cell structure and function. An understanding of how cells communicate with each other can tie together a variety of seemingly independent cellular processes.

7.1 Signaling components

Communication between cells is mediated mainly by extracellular signal molecules. The **signaling molecules** from physical and chemical changes of the environment, or produced by a **signaling cell** can be detected by a **target cell**, which in turn responds specifically to the signaling molecule.

Single and cells in multi-cellular organisms use hundreds of signaling molecules such as cytokines, growth factors, hormones, fatty acid derivatives, neurotransmitter, and even dissolved gases. The target cells responds by means of a **receptor**, which usually embedded in the plasma membrane of the target cell and specifically binds the signal molecule and then initiates a response through one or more intracellular signaling pathways mediated by a series of signaling proteins.

7.1.1 Cell communication styles

There are huge varieties of signaling molecules and receptors used in cell communication, however, relatively few basic styles of cell communication are existed. **Contact-dependent signaling** and non-contact dependent signaling are two main styles.

Contact-dependent signaling is the most intimate and short-range cell-cell communication. This happens between the cells that make direct contact through signaling molecules anchored in the plasma membrane of the signaling cell and bind to a receptor molecule embedded in the plasma membrane of the target cell. Lateral inhibition mediated by Notch and Delta during nerve cell production in Drosophila is a good example of contact-dependent communication (Figure 7-1A). The nervous system originates in the embryo from a sheet of epithelial cells. Isolated cells in this sheet begin to specialize as neurons, while their neighbors remain non-neuronal and maintain the epithelial structure of the sheet. The signals that control this process are transmitted via direct cell-cell contacts: each future neuron delivers an inhibitory signal molecule Delta and binds to North receptor proteins on the neighboring cells. The same mechanism, mediated by essentially the same molecules, controls the detailed pattern of differentiated cell types in various other tissues, in both vertebrates and invertebrates.

Non-contact mediated signaling is also referred as **chemical signaling**. Signaling cells secrete signal molecules into the extracellular fluid, and it is the most common style of communication. Depending on the distance over which the signal acts, chemical signaling can be classified into three sub-types.

(1) Paracrine signaling (Figure 7-1B) The signaling molecules stay in the neighborhood of the signaling cells and only affect nearby target cells. But cells may also produce signals that they themselves respond to, which is **autocrine signaling**.

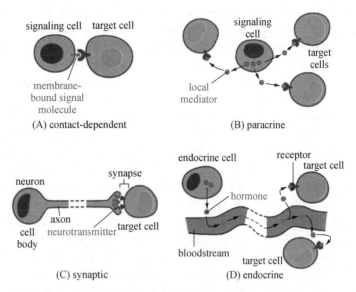

Figure 7-1 **Four forms of intercellular signaling** (From: Alberts B, et al. Molecular Biology of The Cell. 2008)

This is especially common in tumor cells in which over-produced growth factors stimulate over-growth of tumor cells themselves and adjacent non-tumor cells.

(2) Synaptic signaling It requires a special structure called the synapse between the cell originating and the cell receiving the signal to achieve rapid communication. Synaptic signaling only occurs between cells with the synapse; for example between a neuron and the muscle that is controlled by neural activity. For instance, neuronal signaling involves **neurotransmitter** release from nerve cells, then diffuse across the narrow (< 100 nm) gap between the axon-terminal membrane and membrane of the target cell in less than 1 millisecond (Figure 7-1C).

(3) Endocrine signaling (Figure 7-1D) **Hormones** are used as signaling molecules and are secreted into the bloodstream (in an animal) or the sap (in a plant) from endocrine cells. This is the most "public" style of communication in which signals are broadcasted throughout the body, reaching distant target cells. For example, the endocrine gland of the pancreas secrets the hormone insulin into the bloodstream and regulates glucose uptake in cells all over the body.

In contrast with other modes of cell signaling, **gap junctions** generally allow communication to pass in both directions symmetrically. Gap junctions are channels that allow the diffusion of ions and small molecules (up to 1,000 Da) between closely apposed cells, and their typical effect is to homogenize conditions in the communicating cells, including the propagation of electrical signals, metabolic cooperation etc. Thus, gap junctions allow neighboring cells to share signaling information.

7.1.2 Types of receptors

Detection of signals arriving from outside of the cell is usually fulfilled by the presence of specific receptors. In most cases, the receptors are transmembrane proteins on the target cell surface. When these proteins bind an extracellular signal molecule (a ligand), they become activated and generate various intracellular signals that alter the behavior of the cell.

The number of different types of receptors is even greater than the number of signaling molecules because in many cases, there are multiple receptors for a single signaling molecule. Most cell-surface receptors belong to three large families: **ion-channel-coupled receptors**, **G-protein-coupled receptors**, or **enzyme-coupled receptors** (Figure 7-2).

(1) Ion-channel-coupled receptors   Ion-channel-coupled receptors (also known as transmitter-gated ion channels) (Figure 7-2A) are linked to ion channels, and the conductance of the channels is modulated by the binding of signal molecules (agonists or antagonists). Ion-channel-coupled receptors can be found in the nervous system and other electrically excitable cells such as muscle, neuroendocrine cells etc., and often involved in the detection of neurotransmitter molecules. They function in the simplest and most direct way compared with how other cell-surface receptor types act. In most cases, **neurons** pass signals to other cells at **synapses** where a presynaptic neuron releases a chemical signal in the form of neurotransmitters. Neurotransmitters bind to ion-channel-coupled receptors on the postsynaptic target cells. These receptors alter their conformation to transiently open or close the efflux and influx of specific ions such as Na^{2+}, K^+, Ca^{2+}, or Cl^- and to allow them going across the cell membrane. The ion-flows are driven and directed by their **electrochemical gradient** (i.e. much higher concentration of ions on one side of the membrane than on the other side). Ions rushing into or out of the cell create a temporary change in the membrane electric potential due to the positive or negative nature of the ions. This conversion of chemical signals to electric signals happens within a milli-second or so. The sudden change in membrane potential can trigger or suppress a nerve impulse depending on what ion channels were involved. It may also affect the activity of other membrane protein such as voltage-gated ion channels, which in turn can trigger a nerve impulse, or stop an impulse from happening. Furthermore, this process can also change the activity of cytoplasmic proteins. For example, the opening of Ca^{2+} channels increases the intracellular concentration of Ca^{2+}, which in turn profoundly alter the activities of many proteins because Ca^{2+} is an intracellular messenger and has many substrates.

(2) G-protein-coupled receptors   The majority of cell-surface receptors are G-protein-coupled receptors (GPCRs) (Figure 7-2B), with hundreds of members identified in different organisms ranging from yeast to flowering plants and mammals. GPCRs respond to a huge variety of extracellular signaling molecules including hormones, neurotransmitters, opium derivatives, chemoattractants, odorants and tastants, as well as photons. These signaling molecules (**ligands**) can be proteins, small peptides, or derivatives of amino acids or fatty acids; and for each of them there is a different receptor or set of receptors. As the name indicated, all GPCRs interact with G proteins. G proteins are a group of proteins that were discovered and characterized in the early 1970s and so named because they all bind guanine nucleotides, either GDP or GTP. All GPCRs have a similar structure: each is made of single polypeptide chain that passes through plasma membrane seven times. That is why GPCRs are also called seven-transmembrane (7TM) receptors. A **trimeric GTP-binding protein** (**G protein**) mediates the interaction between the activated receptor and the target protein.

(3) **enzyme-coupled receptors** Like G-protein-coupled receptors, enzyme-coupled receptors (Figure 7-2C) are also transmembrane proteins that have a ligand-binding domain on the outer surface of the plasma membrane and a cytosolic domain. Unlike G-protein-coupled receptors, enzyme-coupled receptors usually have only one transmembrane segment, trespassing plasma membrane by a single α helix. Unlike G-protein-coupled receptors, the cytoplasmic domain of enzyme-coupled receptors acts as an enzyme or complex with another protein that acts as an enzyme.

However, not all extracellular signaling molecules are detected on the surface of the cell by plasma membrane receptors. Many very important signals are released by cells but the receptors for their perception are inside the target cells, either in the cytoplasm or in the nucleus. The signaling molecule has to enter the cell to bind to them. Intracellular ligand binding for these extracellular signaling molecules means that they have to move through the plasma membrane, and commonly these signals are small and hydrophobic. Various small hydrophobic signal molecules diffuse directly across the plasma membrane of the target cells and bind to intracellular receptors that are gene regulatory proteins.

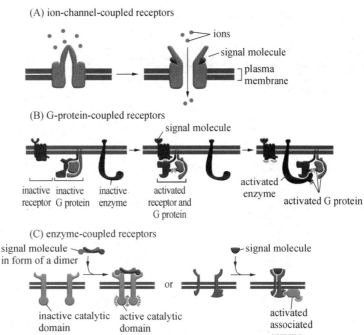

Figure 7-2 **Cell-surface receptors fall into three basic classes.** (A) An ion-channel-coupled receptor opens (or close, not shown) in response to binding of its signal molecule. (B) When a G-protein-coupled receptor binds its extracellular signal molecule, the signal is passed first to a GTP-binding protein (a G protein) that is associated with the receptor. The activated G protein then leaves the receptor and turns on a target enzyme (or ion channel, not shown) in the plasma membrane. (C) An enzyme-coupled receptor binds its extracellular signal molecule, switching on an enzyme activity at the other end of the receptor. Although many enzyme-coupled receptors have their own enzyme activity (*left*), others rely on associated enzymes (*right*) (From: Alberts B, et al., Molecular Biology of The Cell. 2008)

The signal molecules such as steroid hormones, thyroid hormones, retinoids, and vitamin D, they all act by a similar mechanism. When these signal molecules bind to their receptor proteins, they activate the receptors, which bind to DNA to regulate the transcription of specific genes. The receptors are **nuclear receptors**, that is, the ligand-activated gene regulatory proteins. Some nuclear receptor proteins may be activated by intracellular metabolites rather than by secreted signal molecules. Some receptor proteins are located in the cytosol and enter the nucleus only after ligand binding; others are bound to DNA in the nucleus even in the absence of the ligand.

7.1.3 The basic cell signaling process

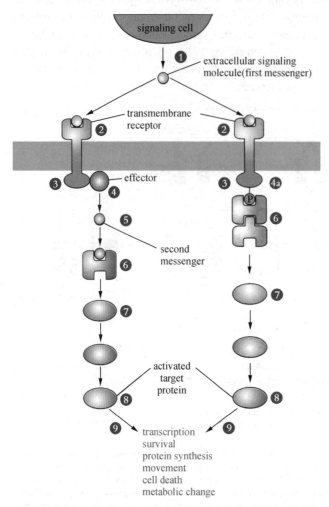

Figure 7-3 An overview of the signaling pathways by which extracellular messenger molecules can elicit intracellular responses. One signaling pathway is activated by a diffusible second messenger, and another pathway is activated by recruitment of proteins into the plasma membrane. Most signal transduction pathways refer to a combination of these two mechanisms. It should also be noted that signaling pathways are not typically linear tracks as depicted here, but are branched and interconnected to form a complex web (From: Karp G, et al. Cell and Molecular Biology — concepts and experiments. 2010)

An overview of the basic cell signaling process is described in figure 7-3. Cell signaling usually starts with extracellular messenger molecules (**first messenger**) produced by signaling cells. The first messenger reaches the target cells and binds to its receptor of the target cells. This binding causes a signal to be relayed across the membrane to the cytoplasmic domain of the receptor. Subsequently, most signals are transmitted to the cell interior using one of the two pathways. One way is through a **second messenger** (a.k.a. effector), which is a small substance that typically activates or inactivates specific proteins. The other way is through making the cytoplasmic domain a recruiting site for docking the intracellular signaling proteins. Either way, a protein positioned at the upstream of the pathway is activated, which in turn activates a series of distinct downstream proteins one by one, relaying the signal downstream.

The activation of proteins in the pathways usually involves alterations in signaling protein conformation (Figure 7-4). This is usually done by protein kinases and protein phosphatases that add (**phosphorylate**) or remove (**dephosphorylate**) phosphate groups from target proteins respectively. Approximately one third of all proteins of a cell are thought to be subjected to phosphorylation. Phosphorylation can change protein activities in many ways. It can activate or inactivate an enzyme, increase or decrease protein-protein interaction, induce a protein to move from one subcellular compartment to another, or initiate protein degradation. Many signaling proteins controlled by phosphorylation are often organized into phosphorylation cascade, which is a sequence of events where one enzyme phosphorylates another, causing a chain reaction leading to the phosphorylation of different proteins.

The other class of proteins that function by gaining and losing phosphate groups consists of GTP-binding proteins. The protein can be further classified to G proteins and **monomeric GTPases** (also called **monomeric G proteins**). Both proteins are activated by GTP-binding and inactivated by GDP-binding. The figure 7-5 illustrates the regulation of monomeric GTPases.

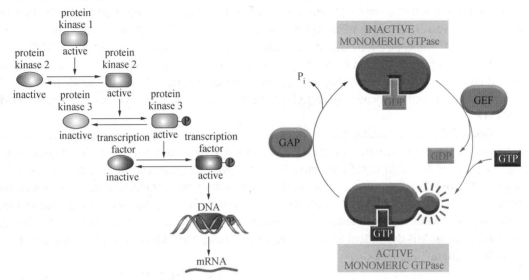

Figure 7-4　Signal transduction pathway consisting of protein kinases and protein phosphatases whose catalytic actions change the conformations, and thus change the activities, of the proteins they modify. Note that each of these activation steps in the pathway is reversed by a phosphatase (From: Karp G, et al. Cell and Molecular Biology — concepts and experiments. 2010)

Figure 7-5　The regulation of a monomeric GTPases. GTPase-activating proteins (GAPs) inactivate the protein by stimulating it to hydrolyze its bound GTP to GDP. Guanine nucleotide exchange factors (GEFs) activate the inactive protein by stimulating it to release its GDP (From: Alberts B, et al. Molecular Biology of The Cell. 2008)

Ultimately, signals reach target proteins. Depending on the type of cells and messages relayed, these target proteins may regulate gene expression, alter the activity of metabolic enzymes, reconfigure the cytoskeleton, change ion permeability, activate DNA synthesis, or even cause cell death. Virtually all cell activity is regulated by signals originated from the cell surface. The whole process, where a signal carried by the extracellular messenger molecules is relayed into changes that occur inside a cell, is called **signal transduction.**

Finally, all signaling process must be turned off in order for cells to be responsive to other messages or next wave of the same message that they may receive. This can be accomplished in many ways. Some cells produce extracellular enzymes that destroy specific extracellular messenger/signaling molecules. Some cells internalize activated receptors along with ligands (signaling molecules). Once inside the cell, the receptors and the ligands are either both degraded, or the ligands are degraded and the receptors are returned to the cell surface. Both processes decrease cells' ability to respond to subsequent stimuli. In addition, cells have enzymes in the cytosol that de-activate signaling proteins in the pathway. Some cytosolic signaling proteins are degraded and new signaling proteins are synthesized to maintain signaling capabilities.

7.2 The role of intracellular receptor: signaling of nitric oxide

Nitric Oxide (NO), an inorganic gas, was found to be an intracellular messenger in 1986. How it works is best illustrated by dissecting the mechanism of acetylcholine-induced smooth muscle cell relaxation (Figure 7-6). Endothelial cells are the flattened cells that line every blood vessel. Binding of acetylcholine to the endothelial cell membrane gives a rise in cytosolic Ca^{2+} concentration which in turn activates nitric oxide synthase (NOS), an enzyme that catalyzes the amino acid L-arginine into nitric oxide (NO). NO formed in the endothelial cell diffuses across the plasma membrane into the adjacent smooth muscle cells where it binds and activates **guanylyl cyclase**, which in turn synthesizes **cyclic GMP (cGMP)**.

One of the main cellular receptors of NO is the enzyme guanylyl cyclase, the enzyme responsible for the production of cGMP. It is usually the soluble form of the cyclase that is the receptor of NO, but some data have been reported that suggest the membrane form may also be NO regulated to some extent. cGMP is an important intracellular second messenger, which resembles **cyclic AMP (cAMP)**, a much more commonly used intracellular signaling molecule. cGMP leads to a decrease in cytosolic Ca^{2+} concentration through activation of cGMP-dependent kinase (PKG), causing the muscle cells to relax and blood vessels to dilate (Figure 7-6).

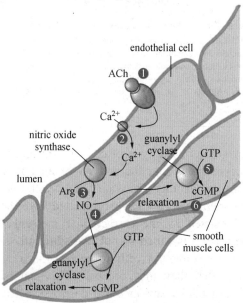

Figure 7-6 A signal transduction pathway that operates by means of NO and cyclic GMP that leads to the dilation of blood vessels (From: Karp G, et al. Cell and Molecular Biology — concepts and experiments. 2010)

The discovery of NO acting as an activator of guanylyl cyclase sheds light on the therapeutic effect of nitroglycerine, a commonly used drug in treatment of the pain of angina caused by an inadequate flow of blood to the heart muscle. In the body, nitroglycerine is metabolized into NO, which rapidly relaxes the smooth muscles lining the blood vessels of the heart, increasing blood flow into the heart. Many nerve cells also use NO to signal neighboring cells. For instance, during sexual arousal, nerve terminals in the penis release NO, which triggers the local blood-vessel dilation and penis erection.

NO acts only locally because NO is converted to nitrates and nitrites within seconds by reaction with oxygen and water outside of the cells, thus terminating the NO signaling. NO signaling is a great example of how an extracellular signal can directly activate an enzyme and alter a cell within a few seconds or minutes.

Recent research shows that NO can also function without involving production of cGMP. For example, in a post-translational modification process called S-nitrosulation, NO is added to the SH group of certain cysteine residues in a number of proteins causing the alteration of protein properties and activity.

Hemoglobin, ras, ryanodine channels and caspases are some good examples of these proteins.

7.3 Signaling through G-protein-coupled cell-surface receptors

The typical topology of GPCRs includes three loops of amino-terminus presented outside of the cell and forming the ligand-binding site; the seven α-helices that traverse the membrane are connected by loops of varying length; another three loops of carboxyl-terminus on the cytoplasmic side of the cell provide the binding sites for intracellular signaling molecules. G proteins bind to the third intracellular loop (Figure 7-7). Arrestins, an inhibitor of G protein actions, compete the same binding site of GPCR with G proteins. More and more proteins acting as molecular scaffolds that link receptors to various intracellular signaling proteins and effectors are found to bind to the carboxyl-termini of GPCRs.

Binding of a ligand to the extracellular ligand-binding site of a GPCR causes a conformational change in the extracellular ligand-binding site. This conformational change is transferred across the plasma membrane and changes the conformation of cytoplasmic loops of GPCR, which in turn increase the affinity of GPCR to a cytoplasmic G protein. Consequently, the G protein binds to the activated receptor forming a receptor-G protein complex. The G protein thus becomes active.

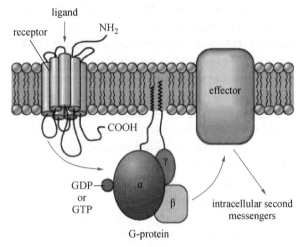

Figure 7-7 **The membrane-bound machinery for transducing signals by means of a seven transmembrane receptor and a heterotrimeric G protein.** When bound to their ligand, GPCRs interact with a trimeric G protein, which activates an effector, such as adenylyl cyclase. Note that the α and γ subunits of the G protein are linked to the membrane by lipid groups that are embedded in the lipid bilayer. (From: Karp G, et al. Cell and Molecular Biology — concepts and experiments. 2010)

There are different kinds of G proteins; each is specific for a particular set of receptors and a particular set of downstream target proteins. Most of G proteins are called **heterotrimeric G proteins** to distinguish them from small, **monomeric G proteins**, such as Ras, which are discussed later.

Heterotrimeric G proteins share similar structure. The physiologic processes mediated by heterotrimeric G proteins include glycogen breakdown, behavioral sensitization and learning, visual excitation, chemotaxis, etc. They all have three protein subunits — α, β, γ, two of which are covalently tethered to the plasma membrane short lipid tails. In idle state, the α subunit of G protein has GDP bound to it. The interaction between a G protein with its receptor activates G protein by inducing a conformation change in the α subunit of the G protein, causing it to lose some of its affinity for GDP, which releases GDP in exchange for the binding of a molecule of GTP. The activated G protein breaks into two free activated subunits: the GTP-bound α subunit and the activated βγ complex unit. Both of these two subunits can bind and interact directly with effector proteins located in the plasma membrane, which in turn may relay signals to other destinations. As long as these effector proteins have a α or a βγ subunit bound to them, they can amplify the relayed signals.

To regain sensitivity to future stimuli, the receptor, the G protein, and the effector protein must all return to their inactive state. This is a two-step process called desensitization. The first step is the phosphorylation of the cytoplasmic domain of the activated GPCR by a **G protein-coupled receptor kinase (GRK)**. GRKs are a small family of serine-threonine protein kinases. Most of them are located next to the cytoplasmic surface of the plasma membrane. Some of them that located in the cytoplasm are recruited to the membrane upon activation of certain G proteins. The conformation changes that occur for GPCRs to activate G proteins also make G proteins good substrates for GRKs. Therefore, following ligand binding, GPCRs become sensitive to phosphorylation by GRKs. The second step involves protein arrestins that have the same binding sites on GPCRs with G proteins. As a result, **arrestins** binding to the phosphorylated GPCRs prevent the further activation of G proteins. When desensitized, the cell stops responding to the stimulus even though the stimulus is still acting on the outer surface of the cell. Desensitization is an important mechanism that allows a cell to respond to changes in its environment rather than continue to be in a state of activation endlessly in the presence of an unchanging environment. The importance of desensitization can be illustrated in a disease called retinitis pigmentosa. People who suffer from this disease have mutations that interfere with phosphorylation of rhodopsin by a GRK, leading to the death of the **photoreceptor** cells in the retina causing retinal death and subsequent blindness.

Arrestins can also bind to **clathrin** molecules that are located in **clathrin-coated pits**. The bound arrestins and clathrin promotes a process called receptor internalization where phosphorylated GPCRs are taken into the cell by endocytosis. Under certain circumstances, endocytosed receptors are dephosphorylated and returned to the plasma membrane, while other times receptors are degraded in the lysosomes. When the receptors are returned, the cells remain sensitive to ligands; when receptors are degraded, the cells lose, at least temporarily, sensitivity to the ligands. Interestingly, both desensitization and receptor internalization appears to be initiated by a large variety of extracellular signals activating GRKs.

Termination of signaling via activated α subunit of the G protein goes through another type of process, which depends on the behavior of the α subunit. Following its interaction with an effector, the α subunit goes through an intrinsic GTP-hydrolyzing (GTPase) process, which hydrolyzes its bound GTP back to GDP and Pi. GTP hydrolysis is accelerated by regulators of G protein signaling (RGSs). GTP hydrolysis induces a conformational change that causes α subunit to dissociate from the effector protein and reunite with the βγ complex unit, which thus inactivates G protein. The reconstituted G protein is now ready to be reactivated by another activated receptor, thus the system is returned to a resting state.

The target proteins for activated G protein subunits are either ion channels or membrane-bound enzymes. The most popular G-protein-coupled receptor-initiated enzyme-mediated signaling pathways are **the cyclic AMP pathway** and **the inositol phospholipid pathway.**

7.3.1 G-protein regulation of ion channels

G protein-gated ion channels are associated with a specific type of G protein-coupled receptor. These ion channels are transmembrane ion channels with selectivity filters and a G protein binding site. G-proteins directly activate these ion channels using effector proteins or the G protein subunits themselves. Unlike most effectors, not all G protein-gated ion channels have their activity mediated by G_α of their corresponding G proteins. For instance, the opening of inwardly rectifying K^+ channels is mediated by the binding of $G_{\beta\gamma}$. Two sets of nerve fibers control the heartbeat in animals: one speeds up the heart rate, the other slows it down. The nerves that slow the heartbeat do so by releasing neurotransmitter acetylcholine (Ach), which binds to a G-protein-coupled receptor (muscarinic receptor) on the surface of heart muscle cells. When Ach binds to this receptor, a G protein (Gi, "i" stands for inhibitory) is activated by dissociating into a α subunit and a βγ complex. βγ complex in turn binds to the cytoplasmic side of a type of K^+ channels in the heart

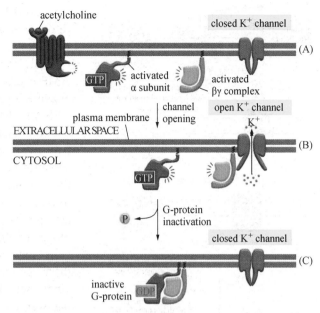

Figure 7-8 G-proteins-couple receptor activation to the opening of K^+ channels in the plasma membrane of heart muslce cells. (A) Binding the neurotransmitter acetylcholine to its G-protein-coupled receptor results in the dissociation of the G protein into an activated βγ complex and an activated α subunit. (B) The activated βγ complex binds to and opens a K^+ channel. (C) Inactivation of the α subunit by hydrolysis of bound GTP causes it to reassociate with the βγ complex to form an inactive G protein, allowing the K^+ channel to close (From: Alberts B, et al. Essential Cell Biology. 2004)

muscle cell membrane and opens the K^+ channels. K^+ ions flow out of the cell following K^+ electrochemical gradient causing a hyperpolarization of the heart muscle cell, which slows heartbeat down. When the α subunit inactivates itself by hydrolyzing its bound GTP and reassociates with the βγ complex, G protein is no longer active. Consequently, the K^+ channels reclose, and the signal shuts down (Figure. 7-8).

7.3.2 GPCR-initiated cyclic AMP pathway

Many extracellular signals acting via GPCRs go through the cyclic AMP pathway. The activated G protein α subunit ($G_{s\alpha}$) usually switches on adenylyl cyclase, causing a sudden increase in the concentration of cyclic AMP (cAMP) synthesized from ATP in the cell. Because this G protein stimulates **adenylyl cyclase**, it is also called G_s. A second enzyme that is continuously active inside the cell called cyclic AMP **phosphodiesterase** can rapidly convert cAMP back to AMP, thus shutting down the signaling process (Figure 7-9). Because cAMP can be broken down very quickly, the concentration of this second messenger can rise or fall ten-fold within seconds, thus responding to extracellular signals rapidly.

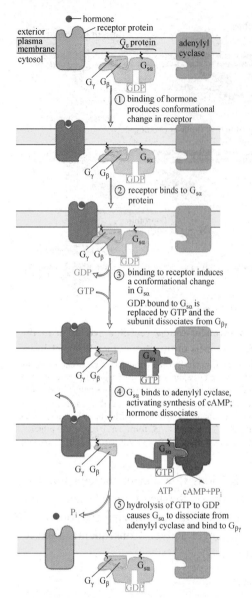

Figure 7-9 Activation of adenylyl cyclase following binding of an appropriate hormone (e.g., epinephrine, glucagon) to a Gs protein-coupled receptor. Following ligand binding to the receptor, the G_s protein relays the hormone signal to the effector protein adenylyl cyclase. G_s cycles between an inactive form with bound GDP and an active form with bound GTP. Dissociation of the active form yields the $G_{s\alpha}$, GTP complex, which directly activates adenylyl cyclase. Activation is short-lived because GTP is rapidly hydrolyzed. This terminates the hormone signal and leads to reassembly of the inactive $G\beta\gamma$. GDP form, returning the system to the resting state. Binding the activated receptor to $G_{s\alpha}$ promotes dissociation of GDP and its replacement with GTP (From: Lodish H, et al. Molecular Cell Biology. 2000)

cAMP is water-soluble, it can carry its signal throughout the cell, traveling from cell membrane (its synthetic site) to target proteins located in the cytosol, the nucleus, or other organelles. Cyclic AMP exerts its effect mainly through activating an enzyme **cyclic-AMP-dependent protein kinase (PKA)**. PKA is a normally inactive, as it is a heterotetramer made of two regulatory (R) and two catalytic (C) subunits and the regulatory subunits normally inhibit the catalytic activity of the enzyme. When binding to cAMP, PKA goes through a conformational change and releases its active catalytic subunits. The activated PKA then moves into the nucleus and phosphorylates specific gene regulatory proteins. Once phosphorylated, these proteins stimulate the transcription of a whole set of target genes. This type of signaling pathway controls many processes in cells, ranging from hormone synthesis in endocrine cells to the production of proteins involved in long-term memory in the brain. Figure 7-10 shows the process of such a pathway. Activated PKA can also catalyze the phosphorylation of particular serines or threonines on certain intracellular proteins in the cytosol, and changing their activity.

Some cAMP responses are very fast; such as the ones that happen in skeletal muscles, which only takes a few seconds to complete and does not depend on changes in gene transcription. Some cAMP do depend on changes in the transcription of specific genes and take minutes or even hours to develop.

Many cells signal through cAMP. However, different target cells respond very differently to the same extracellular signals that change intracellular cAMP concentrations. Although activation of PKA in a liver cell in response to epinephrine leads to the breakdown of glycogen, activation of the same enzyme in a kidney tubule cell in response to vasopressin causes an increase in the permeability of the membrane to water, and the secretion of thyroid hormone. The response of a given cell to cAMP is typically determined by the specific proteins phosphorylated by PKA. Different cell types have different sets of target proteins available to be phosphorylated; that is why the effects of cAMP vary with different target cells. Some of the known PKA target proteins are phosphorylase kinase α, protein phosphatase-1, pyruvate kinase, CREB, protein phosphatase inhibitor-1, etc.

7.3.3 GPCR-initiated inositol phospholipid pathway

Some extracellular signals exert their effects via activating a type of G protein that activates membrane-bound enzyme phospholipase C instead of adenylyl cyclase. The activated phospholipace C cleaves a lipid molecule that is a component of the cell membrane called **inositol phospholipid**. Inositol phospholipid is a phospholipid that has the sugar inositol attached to its head and is present in small quantities in the inner half of the plasma membrane lipid bilayer. Because inositol phospholipid is involved, this signaling pathway through the activation of phospholipase C is called the inositol phospholipid pathway (Figure 7-11). This pathway occurs in almost all eukaryotic cells and affects many different types of target proteins in those cells. Some examples of cell responses mediated by phospholipase C activation are glycogen breakdown, secretion of amylase, smooth muscle contraction and blood platelets aggregation, etc.

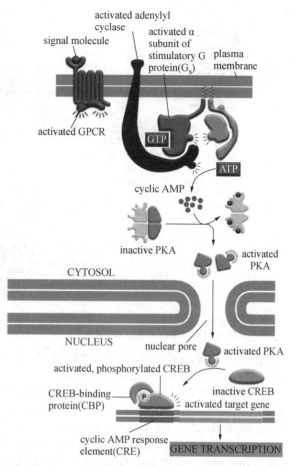

Figure 7-10 A rise in intracellular cyclic AMP can activate gene transcription. Binding a signal molecule to its G-protein-coupled receptor can lead to the activation of adenylyl cyclase and a rise in the concentration of intracellular cyclic AMP. In the cytosol, cyclic AMP activates PKA, which then moves into the nucleus, phosphorylates the gene regulatory protein CREB, and then stimulates a whole set of target gene transcription(From: Alberts B, et al. Molecular Biology of the Cell. 2008)

The activated phospholipase C chops the sugar-phosphate head off the inositol phospholipids, producing two small second messenger molecules — inositol 1, 4, 5-triphosphate (IP_3) and diacylglycerol (DAG) (Figure 7-11). Both molecules play a crucial part in signaling inside the cell. Therefore, phospholipase C activates IP_3 and DAG, two signaling cascades.

IP_3 is a small, hydrophilic sugar phosphate. It easily diffuses into the cytosol, where it binds to IP_3 receptor embedded in the endoplasmic reticulum membrane. This IP_3 receptor also functions as a tetrameric Ca^{2+} channel. Binding of IP_3 opens the Ca^{2+} channels so that Ca^{2+} stored inside the endoplasmic reticulum rushes out to the cytosol and triggers a sharp rise in the cytosolic concentration of free Ca^{2+} (Figure 7-11). Calcium ion is an intracellular secondary messenger that binds to various target proteins, triggering different responses. Some of IP_3 mediated cellular responses are muscle contraction, cyclic GMP

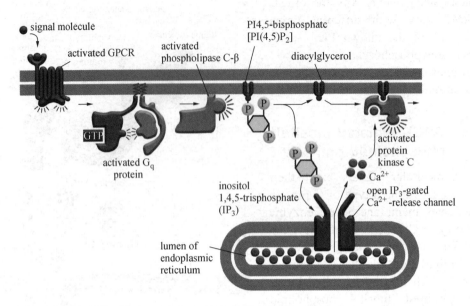

Figure 7-11 **Phospholipase C activates two signaling pathways** (From: Alberts B, et al. Molecular Biology of the Cell. 2008)

formation, actin polymerization, blood platelets shape change and aggregation, calcium mobilization, membrane depolarization, etc.

DAG is a lipid that remains embedded in the plasma membrane. With Ca^{2+}, DAG helps to recruit and activate an enzyme called protein kinase C (PKC), translocating PKC from the cytosol to the plasma membrane. Ca^{2+} is needed to bind to PKC for PKC to be activated (Figure 7-11). Once activated, PKC phosphorylates different sets of intracellular target proteins in different types of cells. PKC works in a similar way to PKA, although they have different target proteins. Some PKC-activated responses are serotonin release, histamine release, secretion of epinephrine, secretion of insulin, secretion of calcitonin, dopamine release, glycogen hydrolysis, etc.

7.4 Signaling through enzyme-coupled cell-surface receptors

Enzyme-coupled receptors usually have only one transmembrane segment, and the cytoplasmic domain of enzyme-coupled receptors acts as an enzyme or complex with another protein that acts as an enzyme. Some of these protein kinases and protein phosphatases are hydrophilic, while others are integral membrane proteins. Some of these enzymes have multiple proteins as substrates; others have only a single protein substrate. Many of the substrates of these enzymes are enzymes themselves, most often they are other protein kinases and phosphatases, or ion channels, transcription factors and various types of regulatory proteins.

7.4.1 Receptor tyrosine kinases

The largest class of enzyme-coupled receptors is made up of those with a cytoplasmic domain that acts as a tyrosine protein kinase, phosphorylating tyrosine side chains on selected proteins. These receptors are called **receptor tyrosine kinases** (**RTKs**). We will use receptor tyrosine kinase as an example to illustrate how external signals work through enzyme-coupled receptors(Figure 7-12).

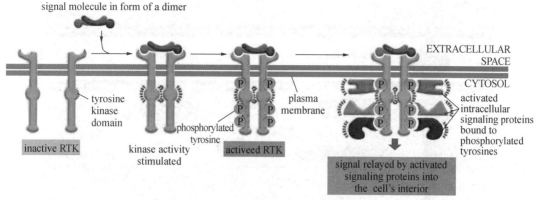

Figure 7-12 Activation of a receptor tyrosine kinase stimulates the assembly of an intracellular signaling complex (From: Alberts B, et al. Essential Cell Biology. 2004)

In many cases, an external signal first causes two enzyme-coupled receptors to come together in the membrane and to form a dimer. In figure 7-12 however, the signaling molecule itself is a dimer and thus can physically cross-link two receptor molecules. Within the dimer, contact with the two adjacent intracellular receptor tails activates their kinase function, causing each receptor to phosphorylate the other. In the case of receptor tyrosine kinase, the phosphorylations occur on specific tyrosine located on the cytosolic tail of the receptors. The phosphorylation then triggers the assembly of an elaborate intracellular signaling complex on the receptor tails. The newly phosphorylated tyrosines serve as binding sites for many kinds of intracellular signaling proteins, which become active upon binding. As long as the receptor is activated, the intracellular signaling complex transmits signals along several routes simultaneously to many destinations inside the cell, thus activating and coordinating the numerous biochemical changes that are required to trigger a complex response such as cell proliferation. The activated receptors are usually turned off by protein tyrosine phosphatases in the cell. These phosphatases remove the phosphates that were added to receptors previously in response to extracellular signal. In some cases, the activated receptors are endocytosed into the interior of the cell and then destroyed by digestion in lysosomes.

Different receptor tyrosine kinases recruit different sets of intracellular signaling molecules, producing different effects. The more widely used ones include a phospholipase that functions the same way as phospholipase C to activate the inositol phospholipid signaling pathway. They also include an enzyme called phosphatidyl-inositol 3-kinase (PI3-kinase), which phosphorylates inositol phospholipids in the plasma membrane making them docking sites for other intracellular signaling proteins. One of these is **protein kinase B (PKB)**, which phosphorylates target proteins on their serines and threonines and is essential in cell survival and growth. Some of these intracellular signaling molecules function solely as physical adaptors; they help to build a larger signaling complex by coupling receptors to other proteins, which in turn may bind to and activate yet other proteins to pass the signals along. One of the key players in these adaptor-assembled signaling complexes is **Ras**, a small protein that has a lipid tail bound to the cytoplasmic side of the plasma membrane.

Virtually all receptor tyrosine kinases activate Ras. The Ras protein is a member of a large family of small, single-subunit GTP-binding proteins called the **monomeric GTP-**

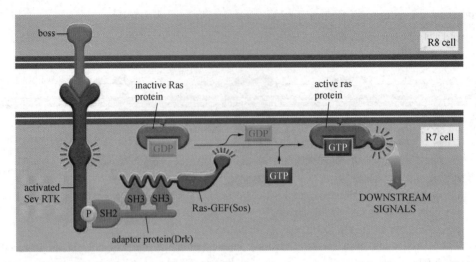

Figure 7-13 Sev RTK activates Ras in the fly eye. Sev (on the surface of the R7 precursor cell) is activated by the Boss protein (on the surface of R8). The adaptor protein Drk docks on a particular phosphotyrosine on the activated receptor. Drk recruits and stimulates a protein that functions as a Ras-activating protein Sos. Sos stimulates the inactive Ras protein to replace its bound GDP by GTP, which activates Ras to relay the signal downstream, inducing the R7 precursor cell to differentiate into a UV-sensing photoreceptor cell. Note that the Ras protein contains a covalently attached lipid group (red) that helps anchor the protein to the plasma membrane(From: Alberts B, et al. Molecular Biology of the Cell. 2008)

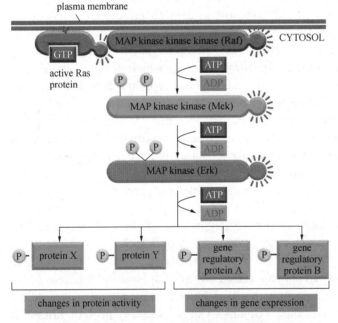

Figure 7-14 The steps of a generalized MAP kinase cascade. Activated Ras recruits the protein Raf to the membrane, where it is phosphorylated and thus activated. In the pathway depicted here, Raf (MAPKKK) phosphorylates and activates another kinase named MEK (MAPKK), which in turn phosphorylates and activates still another kinase termed ERK (MAPK). Erk in turn phosphorylates a variety of downstream proteins. The resulting changes in gene expression and protein activity cause complex changes in cell behaviour (From: Alberts B, et al. Molecular Biology of the Cell. 2008)

binding proteins. Ras is similar to subunit of a G protein in both its structure and function: it has GTP-bound active state and GDP-bound inactive state; it is activated when Ras interacts with an activating protein and exchanges its GDP for GTP; Ras switches itself off by hydrolyzing its GTP to GDP. Figure 7-13 shows how the Sev RTK activates Ras in the fly eye. The assembly of photoreceptor cells in a developing *Drosophila* ommatidium begins with R8 and ending with R7, which is the last photoreceptor cell to develop.

When activated, Ras triggers a phosphorylation domino cascade in which a series of protein kinases phosphorylate and activate one another in turn, relaying messages along the way from the plasma membrane all the way to the nucleus. This relay system is called a **MAP-kinase cascade**

(Figure 7-14). It takes its name from the final kinase in the cascade MAP-kinase (**mitogen-activated protein kinase**, **MAPK**). In this cascade, MAP-kinase is phosphorylated and activated by an enzyme called **MAP-kinase-kinase** (**MAPKK**), which itself is activated by an enzyme called **MAP-kinase-kinase-kinase** (**MAPKKK**), which is activated by Ras. MAPKKs are dual-specificity kinases, a term denoting that they can phosphorylate tyrosine as well as serine and threonine residues. All MAPKs have a tripeptide near their catalytic site with the sequence Thr-X-Tyr. MAPKK phosphorylates MAPK on both the threonine and tyrosine residue of this sequence, thereby activating the enzyme.

Once activated, the MAP-kinase relays the signal downstream by phosphorylating various proteins in the cell, including gene regulatory proteins and other protein kinases. The activated MAP-kinase **phosphorylates** certain **gene regulatory proteins** on serines and threonines, altering their ability to control gene transcription and thereby to change the pattern of gene expression. The final outcome is determined by what signals the cell receives and which genes are active in the cell.

7.4.2 Tyrosine-kinase-associated receptor

Not all enzyme-coupled receptors work through activating a sequence of protein kinases to pass messages to the nucleus. Some use a more direct way to control gene expression. **Cytokines** can bind to receptors that activate the latent gene regulatory proteins at the plasma membrane. Once activated, these regulatory proteins head straight for the nucleus

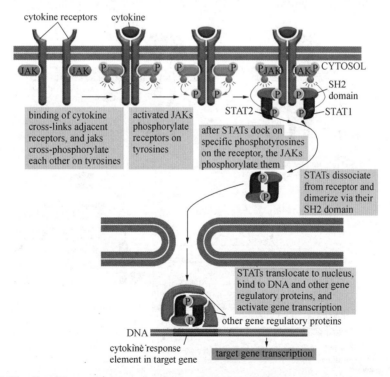

Figure 7-15 Cytokine receptors are associated with cytoplasmic tyrosine kinases. Binding a cytokine to its receptor causes JAKs to cross-phosphorylate and activate one another. The activated kinases then phosphorylate the receptor proteins on tyrosines. STATs present in the cytosol then attach to the phosphotyrosines on the receptor, and the JAKs phosphorylate and activate these proteins as well. The STATs then dissociate from the receptor proteins, dimerize, migrate to the nucleus, and activate the transcription of specific target genes(From: Alberts B, et al. Molecular Biology of the Cell. 2008)

where they stimulate the transcription of certain genes (Figure 7-15). For example, interferon, the cytokine that instructs cells to produce proteins that will make them more resistant to viral infection, uses such a pathway. Unlike the receptor tyrosine kinases, cytokine receptors have no intrinsic enzyme activity but are associated with **cytoplasmic tyrosine kinases** called JAKs (Janus kinases). Cytokine binding to **cytokine receptor** causes receptor dimerization and brings two JAKs into close proximity so that they transphosphorylate each other, thereby increasing the activity of their tyrosine kinase domain. JAKs then phosphorylate and activate **cytoplasmic gene regulatory proteins** called STATs (signal transducers and activators of transcription). STATs migrate to the nucleus where they stimulate transcription of specific **target genes**. Different cytokine receptors activate different STATs, thus have different effects. As any pathway that is turned on by phosphorylation, the cytokine signal is also shut off by removal of the phosphate groups from the activated signaling proteins with phosphatases.

7.4.3 Receptor serine/threonine kinases

There are other enzyme-coupled receptors — receptor serine/threonine kinases — use an even more direct signaling pathway. When activated, they directly phosphorylate and activate cytoplasmic gene regulatory proteins called SMADs. The hormones and local mediators that activate these receptors belong to the TGF-β (transforming growth factor-β) superfamily of extracellular proteins, which are essential in animal development, tissue repair and in immune regulation, as well as in many other processes.

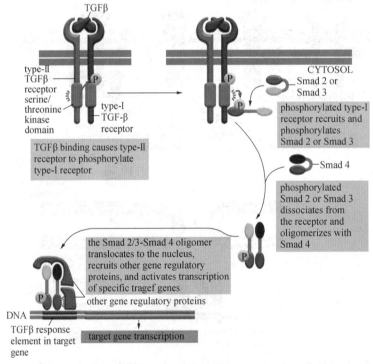

Figure 7-16 TGF-β receptors activate gene regulatory proteins directly at the plasma membrane. These receptor serine/threonine kinases phosphorylate themselves and then recruit and activate cytoplasmic SMADs. The SMADs then dissociate from the receptors and bind to other SMADs, and the complexes then migrate to the nucleus, where they stimulate transcription of specific target genes (From: Alberts B, et al. Molecular Biology of the Cell. 2008)

Members of the TGF-β superfamily act through receptor serine/threonine kinases that are single-pass transmembrane proteins with a serine/threonine kinase domain on the cytosolic side of the plasma membrane. There are two classes of receptor serine/threonine kinases, type I and type II. Typically, the ligand first binds to and activates a type-II receptor, which recruits, phosphorylates, and activates a type-I receptor, forming an active tetrameric receptor complex.

Once activated, the receptor complex uses a strategy for rapidly relaying the signal to the nucleus. The type-I receptor directly binds and phosphorylates a latent gene regulatory protein of the Smad family. Once one of these Smads has been phosphorylated, it dissociates from the receptor and binds to Smad 4, which can form a complex with the phosphorylated receptor-activated Smads. The Smad complex then moves into the nucleus, where it associates with other gene regulatory proteins, binds to specific sites in DNA, and activates a particular set of target genes (Figure 7-16).

7.5 Signal network system

Cells communicate with each other via signaling pathways. A cell receives information from its environment through the activation of various receptors. These receptors can bind only to specific ligands, thus neglecting a large variety of unrelated signaling molecules in the environment. A single cell may have many different receptors receiving and passing through many different signals into the cell simultaneously. Once transmitted into the cell, these signals are routed through selected pathways reaching their final target proteins to cause different effects in a cell such as causing the cell to divide, differentiate, change shape, relocate, turn on a particular metabolic pathway, send out a signal of its own, or even commit suicide. In previous sections, we reviewed some signaling pathways as if they acted linearly and independently. In real life, however, signaling pathways in the cell form a complex signal network system, where players in one pathway can participate in events of other pathways. Signals from different ligands binding to different unrelated receptors can converge to activate a common effector, such as Ras, leading to the same cellular responses (Figure 7-17).

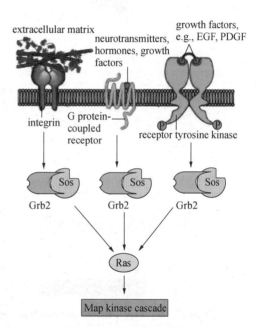

Figure 7-17 Signals transmitted from a G protein-coupled receptor, an integrin, and a receptor tyrosine kinase all converge on Ras and are then transmitted along the MAP kinase cascade (From: Karp G, et al. Cell and Molecular Biology — concepts and experiments. 2010)

Signals from the same ligand can diverge to activate different effectors, leading to diverse cellular responses. For example, insulin is a hormone that induces both immediate and long-term cellular responses. Within minutes, 10^{-9} to 10^{-10} M insulin may cause an increase in the rate of glucose uptake from the blood or fat cells into muscle cells, as well as modulate the activity of various enzymes of glucose metabolism. The long-term insulin effects include increasing expression of liver enzymes that synthesize **glycogen** and of adipocyte enzymes that synthesize **triacylglycerols**. Higher concentration of insulin (10^{-8} M) can also function as a growth factor for many cells, the effects of which are manifested in hours and require continuous exposure. Signals can be passed back and forth among different pathways; this is called **cross-talk** within the **signal network system**, and this cross-talk occurs in virtually all of the control systems of the cell. Figure 7-18 shows how five pathways that we discussed previously are interconnected with each other. It is through the complex signal network system that a cell integrates and responds to information from different sources accordingly.

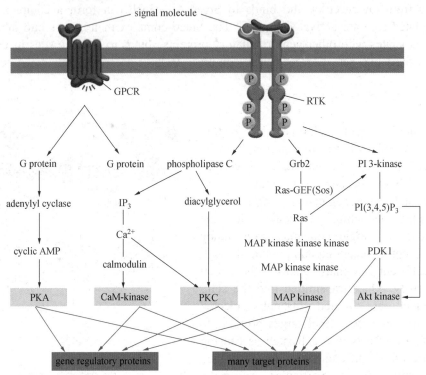

Figure 7-18 **Five parallel intracellular signaling pathways activated by G-protein-coupled receptors, receptor tyrosine kinases, or both.** In this schematic example, the five kinases (shaded yellow) at the end of each pathway phosphorylate target proteins (shaded red), some of which are phosphorylated by more than one of the kinases. The specific phospholipase C activated by the two types of receptors is different: G-protein-coupled receptors activate PLC-β, whereas receptor tyrosine kinases activate PLC-γ (not shown) (From: Alberts B, et al. Molecular Biology of the Cell. 2008)

Most cross-talks are mediated by the protein kinases that are present in each of the pathways. Protein kinases often phosphorylate -and hence regulate -components in other pathways. Hundreds of distinct types of protein kinases are thought to be present in a single mammalian cell. One example of this type of crosstalk involves cAMP. cAMP initiates the cascade leading to glucose mobilization. However, cAMP can also inhibit the growth of a variety of cells including fibroblasts and fat cells. This is achieved by cAMP activating PKA, the **cAMP-dependent kinase**, which can phosphorylate and inhibit Raf (a player in MAP kinase cascade), thus blocking signals transmitted through the MAP kinase cascade. These two pathways also intersect at another important signaling effector, the transcription factor CREB, which can be phosphorylated by PKA of cAMP-mediated pathways as well as by Rsk-2, a substrate of MAPK of the MAP kinase pathway. Both PKA and RSK-2 phosphorylate CREB on the same amino acid residue, Ser133, which endow the transcription factor with the same potential in both pathways (Figure 7-19).

In reality, the complexity of cell signaling is much greater than we are able to describe. We are still in the continuous quest for new links in the chains, new signaling molecules, new connections and even new pathways.

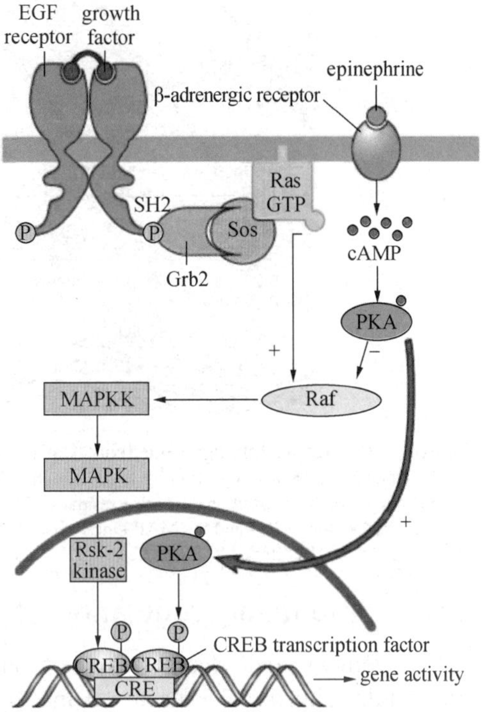

7.6 Cell signaling and the cytoskeleton

In most cells, cell adhesion and the organization of cytoskeleton directly affect cell functions. The receptors responsible for cell adhesion and cytoskeleton organization can also initiate intracellular signaling pathways that regulate other cell behavior such as gene expression. On the other hand, signaling molecules such as growth factors can induce cytoskeletal changes resulting in cell shape modification or cell movement. Thus components of cytoskeleton can act as both receptors and targets in cell signaling pathways.

7.6.1 Integrin signal transduction through FAK

Figure 7-19 An example of crosstalk between two major signaling pathways. Cyclic AMP acts in some cells, by means of the cAMP-dependent kinase PKA, to block the transmission of signals from Ras to Raf, which inhibits the activation of the MAP kinase cascade. In addition, both PKA and the kinases of the MAP kinase cascade phosphorylate the transcription factor CREB on the same serine residue, activating the transcription factor and allowing it to bind to specific sites on the DNA (From: Karp G, et al. Cell and Molecular Biology — concepts and experiments. 2010)

As briefly mentioned in chapter 6, the integrins are major adhesion proteins (receptors) that are responsible for the attachment of cells to the extracellular matrix. Integrins does so by interacting with components of the cytoskeleton to build a linkage between the extracellular matrix and adherent cells at the two types of cell-matrix junctions (focal adhesions and hemidesmosomes). Meanwhile, the integrins are also receptors that receive stimuli from extracellular matrix, activate intracellular signaling pathway, and subsequently control gene expression and other cell behavior.

Nonreceptor protein-tyrosin kinases, particularly **FAK** (**focal adhesion kinase**), play an important role in **integrin signal transduction**. Upon binding of integrin with extracellular matrix, FAK gets activated by autophosphorylating itself. In addition, members of the Src family of nonreceptor protein-tyrosin kinases bind to the autophosphorylation site of FAK and further phosphorylate some additional sites on FAK. Tyrosine-phosphorylated FAK provides binding sites for many downstream signaling molecules such as Grb2-Sos complex, which in turn activates Ras and MAP kinase cascades (Figure 7-20).

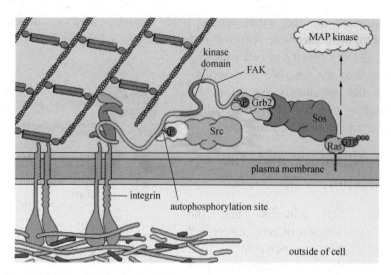

Figure 7-20   Model for signaling from the FAK protein-tyrosine kinase. Binding integrins to the extracellular matrix stimulates FAK activity, leading to its autophosphorylation. Src then binds to the FAK autophosphorylation site and phosphorylates FAK on additional tyrosine residues. These phosphotyrosines serve as binding sites for the Grb2-Sos complex, leading to activation of Ras and the MAP kinase cascade, as well as for additional downstream signaling molecules, including pI3-kinase(From: Geoffrey MC. The Cell — A Molecular Approach. 2000)

7.6.2   Rho family regulation of the actin cytoskeleton

Cell movement and changes in cell shape are two of common cell responses to its extracellular signals. As discussed in Chapter 6, these cell behaviors are governed by the actin cytoskeleton modeling and re-modeling underlying the plasma membrane. Therefore, assembly and disassembly of the actin filaments are crucial parts of cell response to extracellular signaling molecules.

Members of the **Rho family** of small GTP-binding proteins (including Rho, Rac and Cdc42) serve as universal regulators of the dynamic assembly and disassembly of the actin filaments, thus connecting extracellular signals to a variety of cell processes. Early studies of fibroblasts responding to growth factor stimulation first reveal the role of Rho family in different aspects of actin remodeling: Different members of the Rho family regulate the organization of actin to form filopodia (Cdc42), lamellipodia (Rac) and focal adhesion and stress fibers (Rho), respectively. Further studies show that the Rho family plays similar roles in regulating the actin cytoskeleton in all types of eukaryotic cells. Moreover, Rho family members are found to activate MAP kinase signaling pathways. So these small GTP-binding proteins regulate not only the cytoskeletal remodeling but also gene expression.

Several experiments have elucidated some of the pathways by which the Rho family members regulate cytoskeletal changes (Figure 7-21). A key target of Rho is a protein-serine/threonine kinase called Rho kinase. When Rho kinase is activated by Rho, it phosphorylates the light chain of myosin Ⅱ as well as inhibits myosin light chain phosphatase. The resulting increase in myosin Ⅱ light chain phosphorylation leads to the assembly of actin-myosin filaments, thus the formation of stress fibers and focal adhesions. On the contrary, Rac and Cdc42 stimulate the PAK (another protein-serine/threonine kinase), which in turn phosphorylates and inhibits myosin light chain kinase, causing a decrease in myosin Ⅱ phosphorylation. Inactivated myosin Ⅱ decreases its ability to interact with actin. As a result, cell surface protrusions occur. Both Rho and Rac also stimulate LIM-Kinase which regulates actin remodeling by phosphorylating cofilin, a key regulator of actin disassembly.

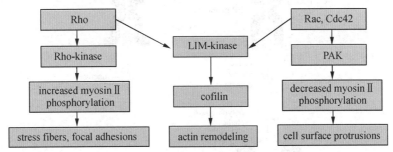

Figure 7-21 Targets of Rho family members (From: Geoffrey MC. The Cell — A Molecular Approach. 2000)

7.7 Cell communication in plants

Our knowledge of cell communication came mostly from animal cell studies. However, like animal cells, plant cells also communicate with each other constantly. Research on molecular mechanisms of cell signaling in *Arabidopsis* has shed some light on how plant cells communicate. Here we will discuss some current views on plant cell signaling processes.

7.7.1 Plant hormones

The **plant hormones** (also called growth regulators) are divided into five major classes: **auxins, gibberellins, cytokinins, abscisic acid, ethylene**. Plant hormones are all small molecules. They diffuse readily through cell walls and they can act locally or be transported to act on other parts of cells. Each hormone may have multiple effects. For example, two of the many functions of auxins include the induction of plant cell elongation and the regulation of cell division. However, each hormone has one main effect. For example: gibberellins affect stem elongation; cytokinins regulate cell division; abscisic acid makes cell senescence and leaf abscission; and ethylene influences fruit ripening.

7.7.2 Ethylene signaling pathway

Ethylene, a gas molecule, is an important plant hormone that has multiple effects on plant cells. These include the **promotion** of fruit ripening, leaf abscission and plant senescence. Ethylene also functions as a **stress signal** in response to flooding, wounding and infection.

Our understanding of the mechanism by which cells respond to ethylene comes from recent research on *Arabidopsis* as a model system. So far, several of the genes required for ethylene signaling pathway have been identified, including those encoding the ethylene receptors. Plants have a number of ethylene receptors which are all structurally similar: they are all dimeric transmembrane proteins and they all have an extracellular domain containing a copper atom that binds ethylene and an intracellular histidine-kinase-like domain. Ethylene signaling pathway is shown in figure 7-22.

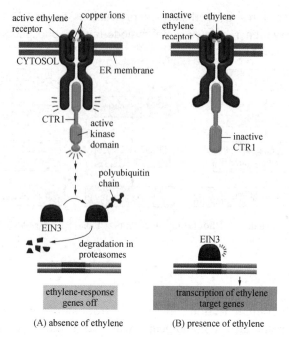

Figure 7-22 A current view of the ethylene two-component signaling pathway. (A) In the absence of ethylene, both the receptors and CTR1 are active, causing the ubiquitylation and destruction of the EIN3 protein, the gene regulatory protein in the nucleus that is responsible for the transcription of ethylene-responsive genes. (B) In the presence of ethylene, the binding of ethylene inactivates the receptors and disrupts the interaction between the receptors and CTR1. The EIN3 protein is not degraded and can therefore activate the transcription of ethylene-responsive genes (From: Alberts B, et al. Molecular Biology of the Cell. 2008)

7.7.3 Light receptor-rhytochromes

In plants, all **photoreceptors** (a.k.a. photoproteins) sense light through a covalently attached **light-absorbing chromophore** which changes its proteins conformation in response to light. The well-known plant photoproeins are the **phytochromes** which are present in all plants. The phytochromes are **dimeric cytoplasmic serine/threonine kinases** that respond to red and far-red light. Whereas red light generally activated the kinase activity of phytochromes; far-red light inhibited it. When activated by red light, the phytochrome autophosphorylates and then phosphorylates other proteins. In some cases the activated photoreceptor directly triggers signaling pathways in the cytoplasm that alter the cell behavior. In other cases, the activated phytochrome activates a gene regulatory protein in the cytoplasm which then moves into nucleus to regulate gene transcription. (Figure 7-23).

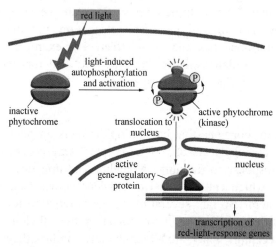

Figure 7-23 A current view of one way in which phytochromes mediate a light response in plant cells. When activated by light, the phytochrome, which is a dimer, phosphorylates itself and then moves into the nucleus, where it activates gene regulatory proteins to stimulate the transcription of specific genes. (From: Alberts B, et al. Molecular Biology of the Cell. 2008)

Summary

Cell communication is a universal and fundamental biological process that mediates interactions between a cell and its surroundings. Cell signaling requires not only extracellular signal molecules, but also a complementary set of receptor proteins in each cell that enable it to bind and respond to the signal molecules in a specific way. Some small hydrophilic signal molecules, including steroid or thyroid hormones and the dissolvable gas nitric oxide, signal by diffusing across the plasma membrane of target cell and activating intracellular receptor proteins or enzymes. However, most extracellular signal molecules are hydrophobic and can activate receptor proteins only on the surface of the target cells. These receptors act as signal transducers converting the extracellular signal to intracellular proteins (messengers) or other downstream target(s), which relay the signal by changing their level or activity, and ultimately effecting specific cell function(s). There are three main families of cell-surface receptors. Ion-channel-coupled receptors are transmitter-gated ion channels that open or close briefly in response to the binding of a signal molecule. G-protein-coupled receptors indirectly activate or inactivate plasma-membrane-bound enzymes via trimeric G proteins. Enzyme-coupled receptors either act directly as enzymes or act indirectly via their association with other enzymes. These enzymes are usually protein kinases that phosphorylate specific proteins in the target cell. Signaling pathways are highly interconnected into a complicated communication network system in which signals can diverge or converge, and they can go back and forth between two or more different signaling pathways.

In addition to the structural role, the integrins serve as receptors that activate intracellular signaling pathways, a nonreceptor protein-tyrosine kinase called FAK plays a key role in integrin signaling. A number of plant hormones, including ethylene, help coordinate plant development. Ethylene acts through receptor histidine kinases in a signaling pathway. Light has an important role in regulating plant development. These light response are mediated by a variety of light-sensitive photoproteins. For example, phytochromes are responsive to red light.

Cell signaling is not only important for the understanding of the functioning of a normal cell, but also is of vital importance to understand the growth and activity of an aberrant cell, or that of a cell that is combating adverse conditions.

Questions

1. Please define the following terms: signal molecule, cell signaling, signal transduction, A-kinase (PKA), C-kinase (PKC), G-protein, nitric oxide, phospholipase C, Ras, receptor tyrosine kinase, signaling cascade.
2. Describe the basic types of signal molecules and second messengers.
3. Describe the role that the inositol-lipid signaling pathway plays in the activation of protein kinase C.
4. What are the similarities and differences between the reactions that lead to the activation of G proteins and the reactions that lead to the activation of Ras?
5. ①Compare and contrast signaling by neurons which secrete neurotransmitters at synapses to signaling carried out by endocrine cells, which secrete hormones into the blood. ②Discuss the relative advantages of the two mechanisms.
6. Compare animal cells and plant cells intracellular signaling mechanisms.

（黄　燕　沈大棱）

Chapter 8

Nucleus and chromosomes

All organisms more complex than viruses consist of cells, aqueous compartments bounded by membranes, which under restricted conditions are capable of existing independently. Prokaryotes lack a defined nucleus and have a relatively simple internal organization. Under the electron microscope they appear relatively featureless. The genome of a prokaryote typically consists of a single small circular chromosome in which the DNA is not packaged in any obviously organized way. Eukaryotes have a much more complex intracellular organization with internal membranes, membrane-bound organelles including a nucleus, and a well-organized cytoskeleton. Typically there is a nucleus and cytoplasm, the latter comprising various organelles, membranes and an aqueous compartment known as the cytosol. The nucleus is the repository of the genetic material. Although some DNA is present within the mitochondria of animals, plants, and fungi and within the chloroplasts of plants, the vast majority (typically 99.5%) of DNA in most eukaryotes is found in the nucleus. It is usually the most prominent organelle in most plant and animal cells. In this chapter, first we begin by describing the structure and properties of the eukaryotic cell nucleus. We then consider how eukaryotic cell fold long DNA molecules into compact chromosomes. Finally, we discuss how about nucleolus and ribosome biogenesis.

8.1 The nucleus of a eukaryotic cell

Considering its importance in the storage and utilization of genetic information, the nucleus of a eukaryotic cell has a rather undistinguished morphology (Figure 8-1). The nucleus, the largest organelle in eukaryotic cells, is surrounded by two concentric membranes, each one a phospholipid bilayer containing many different types of proteins. The inner nuclear membrane defines the nucleus itself. In many cells, the outer nuclear membrane is continuous with the rough endoplasmic reticulum (RER), and the space between the inner and outer nuclear membranes is continuous with the lumen of the rough endoplasmic reticulum.

8.1.1 Organization of the eukaryotic nucleus

The contents of the nucleus are present as a viscous, amorphous mass of material enclosed by a complex **nuclear envelope** that forms a boundary between the nucleus and cytoplasm. Included within the nucleus of a typical interphase (i.e. nonmitotic) cell are: ① the chromosomes, which are present as highly extended nucleoprotein fibers, called **chromatin**; ②the **nuclear matrix**, which is a protein-containing fibrillar network; ③one or more **nucleoli**, which are irregular shaped electron-dense structures that function in the

Chapter 8 Nucleus and chromosomes

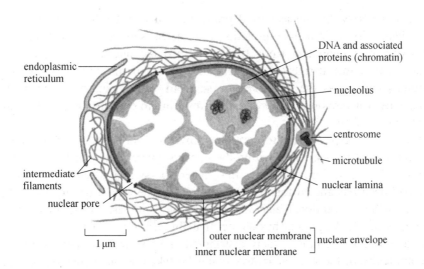

Figure 8-1 Diagrammatic view of a cross section of a typical cell nucleus. The nuclear envelope consists of two membranes, the outer one being continuous with the endoplasmic reticulum membrane. The space inside the endoplasmic reticulum (the ER lumen) is colored yellow; it is continuous with the space between the two nuclear membranes. The lipid bilayers of the inner and outer nuclear membranes are connected at each nuclear pore. Two networks of intermediate filaments (green) provide mechanical support for the nuclear envelope; the intermediate filaments inside the nucleus form a special supporting structure called the nuclear lamina. The nucleolus (gray) is the site of ribosomal RNA synthesis.

synthesis of rRNA and the assembly of ribosomes; ④the **nucleoplasm**, the fluid substance in which the solutes of the nucleus are dissolved (Figure 8-2).

8.1.2 The internal architecture of the eukaryotic nucleus

The light microscopy and early electron microscopy studies of eukaryotic cells revealed very few internal features within the nucleus. This apparent lack of structure led to the view that the nucleus has a relatively homogeneous architecture, a typical "black box" in common parlance. In recent years this interpretation has been overthrown and we now appreciate that the nucleus has a complex internal structure that is related to the variety of biochemical activities that it must carry out. Indeed, the inside of the nucleus is just as complex as the cytoplasm of the cell, the only difference being that, in contrast to the cytoplasm, the functional compartments within the nucleus are not individually enclosed by membranes, and so are not visible when the cell is observed using conventional light or electron microscopy techniques.

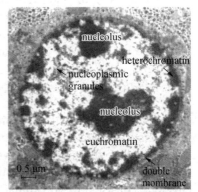

Figure 8-2 Transmission electron micrograph of a thin section of a rat liver cell nucleus. It shows a pair of nucleoli and scattered chromatin. Heterochromatin is evident around the entire inner surface of the nuclear envelope. Two prominent nucleoli are visible, and clumps of chromatin can be seen scattered throughout the nucleoplasm.

The revised picture of nuclear structure has emerged from two novel types of microscopy analysis. First, conventional electron microscopy has been supplemented by examination of mammalian cells that have been prepared in a special way. After dissolution of membranes by soaking in a mild non-ionic detergent such as one of the tween compounds, followed by treatment with a deoxyribonuclease to degrade the nuclear DNA, and salt extraction to remove the chemically basic histone proteins, the nuclear substructure has been revealed as a complex network of protein and RNA fibrils, termed the nuclear matrix, a proteinaceous scaffold-like network that permeates the cell, the matrix permeates the entire nucleus and includes regions defined as the **chromosome scaffold**, a component of the nuclear matrix which changes its structure during cell division, resulting in condensation of the chromosomes into their metaphase forms.

A second novel type of microscopy has involved the use of fluorescent labeling, designed specifically to reveal areas within the nucleus where particular biochemical activities are occurring. The nucleolus, which is the center for synthesis and processing of rRNA molecules, had been recognized for many years as it is the one structure within the nucleus that can be seen by conventional electron microscopy. Fluorescent labeling directed at the proteins involved in RNA splicing has shown that this activity is also localized into distinct regions, although these are more widely distributed and less well defined than the nucleoli.

The complexity of the nuclear matrix could be taken as an indication that the nucleus has a static internal environment, with limited movement of molecules from one site to another. Another new microscopy technique, called fluorescence recovery after photobleaching (FRAP), enables the movement of proteins within the nucleus to be visualized, and shows that this is not the case. The migration of nuclear proteins does not occur as rapidly as would be expected if their movement were totally unhindered, which is entirely expected in view of the large amounts of DNA and RNA in the nucleus, but it is still possible for a protein to traverse the entire diameter of a nucleus in a matter of minutes. Proteins involved in genome expression therefore have the freedom needed to move from one activity site to another, as dictated by the changing requirements of the cell. In particular, the linker histones continually detach and reattach to their binding sites on the genome. This discovery is important because it emphasizes that the DNA-protein complexes that make up chromatin are dynamic.

8.2 The nuclear envelope

The separation of a cell's genetic material from the surrounding cytoplasm may be single most important feature that distinguishes eukaryotes from prokaryotes, which makes the appearance of the **nuclear envelope** a landmark in biological evolution. The nuclear envelope consists of several distinct components (Figure 8-3). The core of the nuclear envelope consists of two cellular membranes arranged parallel to one another and separated by 10 to 50 nm. The membrane of the nuclear envelope, acting as a barrier that keeps ions, solutes, and macromolecules from passing between the nucleus and cytoplasm, maintain the nucleus as a distinct biochemical compartment. The two membranes are fused at sites forming circular pores that contain complex assemblies of proteins. The average mammalian cell contains thousands of nuclear pores, which actively transport selected molecules to and from the cytosol (Figure 8-3).

8.2.1 Structure of the nuclear envelope

The nuclear envelope has a complex structure, consisting of two nuclear membranes, an underlying **nuclear lamina**, and nuclear pore complexes (Figure 8-4). The nucleus is surrounded by a system of two concentric membranes, called the inner and outer nuclear membranes. The outer nuclear membrane is generally studded with ribosomes and is often seen to be continuous with the endoplasmic reticulum, so the space between the inner and outer nuclear membranes is directly connected with the lumen of the endoplasmic reticulum. In addition, the outer nuclear membrane is functionally similar to the membranes of the endoplasmic reticulum and has ribosomes bound to its cytoplasmic surface. In contrast, the inner nuclear membrane carries unique proteins that are specific to the nucleus (Figure 8-4).

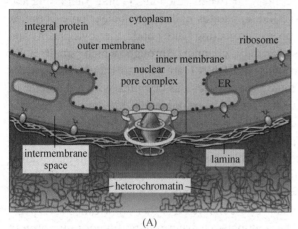

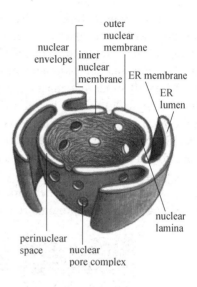

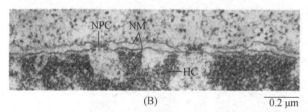

Figure 8-3 The nuclear envelope. (A) Schematic diagram showing the double membrane, nuclear pore complex, nuclear lamina, and the continuity of the outer membrane with the rough endoplasmic reticulum. (B) Electron micrograph of a section through a portion of the nuclear envelope of an onion root tip cell (From: Karp G. 2005).

Figure 8-4 The nuclear envelope. The double-membrane envelope is penetrated by nuclear pore complexes and is continuous with the endoplasmic reticulum. The ribosomes that are normally bound to the cytosolic surface of the ER membrane and outer nuclear membrane are not shown. The nuclear lamina is a fibrous meshwork underlying the inner membrane.

8.2.2 Functions of the nuclear envelope

The critical function of the nuclear membranes is to act as a barrier that separates the contents of the nucleus from the cytoplasm. Like other cell membranes, the nuclear membranes are phospholipid bilayers, which are permeable only to small nonpolar molecules. Other molecules are unable to diffuse through the phospholipid bilayer. The inner and outer nuclear membranes are joined at nuclear pore complexes, the sole channels through which small polar molecules and macromolecules are able to travel through the nuclear envelope. As discussed in the next section, the nuclear pore complex (NPC) is a complicated structure that is responsible for the selective traffic of proteins and RNA between the nucleus and the cytoplasm.

8.3 The nuclear pore complex

8.3.1 The structure of the nuclear pore complex

Nuclear pores are found at points of contact between the inner and outer nuclear membranes. They contain a complex, basketlike apparatus called the nuclear pore complex that appears to fill the pore like a stopper, projecting into both the cytoplasm and nucleoplasm (Figure 8-5A). NPCs are the only channels through which small polar molecules, ions, and macromolecules (proteins and RNA) are able to travel between the nucleus and the cytoplasm. The **NPC** is a huge, supramolecular complex with a diameter of about 120 nm and an estimated molecular masses of approximately 125 million Daltons about 15 to 30 times the mass of a ribosome. NPCs were the subject of early proteomic analysis designed to identify all of the proteins that make up these intricate structure. Despite their considerable size, NPCs contain only about 30 different proteins, called **nucleoporins**, which are largely conserved between yeast and vertebrates (Figure 8-5B). By controlling the traffic of molecules between the nucleus and cytoplasm, the NPC plays a fundamental role in the physiology of all eukaryotic cells. RNAs that are synthesized in the nucleus must be efficiently exported to the cytoplasm, where they function in protein synthesis. Conversely, proteins required for nuclear functions (e.g., transcription factors)

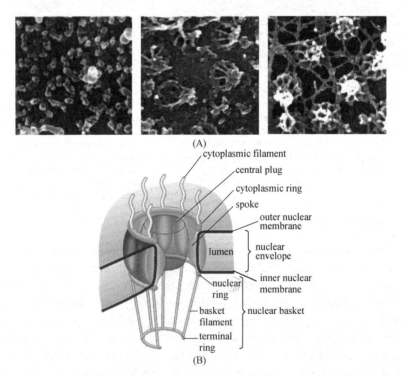

Figure 8-5 Model of the nuclear pore complex (NPC). (A) Nuclear envelopes of Xenopus oocytes visualized by field emission in-lens scanning electron microscopy. Left: Cytoplasmic face of nuclear pore complexes. Middle: Nucleoplasmic face of NPC, showing the "basket" structure. Right: Nucleoplasmic face of the nuclear envelope after removal of the nuclear membrane by mild detergent treatment. The nuclear lamin network, which inserts into the nuclear ring of the NPC, is exposed by this treatment. (B) Cut-away model of the NPC (From: Lodish H, et al. 2000).

must be transported into the nucleus from their sites of synthesis in the cytoplasm. In addition, many proteins shuttle continuously between the nucleus and the cytoplasm. The regulated traffic of proteins and RNA through the NPC thus determines the composition of the nucleus and plays a key role in gene expression.

Visualization of nuclear pore complexes by electron microscopy reveals a structure with eightfold symmetry organized around a large central channel, which is the route through which proteins and RNA cross the nuclear envelope. Detailed structural studies, including computer-based image analysis, have led to the development of three-dimensional models of the nuclear pore complex. These studies indicate that the nuclear pore complex consists of an assembly of eight spokes arranged around a central channel. The spokes are connected to rings at the nuclear and cytoplasmic surfaces, and the spoke-ring assembly is anchored within the nuclear envelope at sites of fusion between the inner and outer nuclear membranes. Protein filaments extend from both the cytoplasmic and nuclear rings, forming a distinct basketlike structure on the nuclear side. The central channel is approximately 40 nm in diameter, which is wide enough to accommodate the largest particles able to cross the nuclear envelope. It contains a structure called the central transporter, through which the active transport of macromolecules is thought to occur.

8.3.2 Selective transport of proteins to and from the nucleus

The basis for selective traffic across the nuclear envelope is best understood for proteins that are imported from the cytoplasm to the nucleus. Such proteins are responsible for all aspects of genome structure and function; they include histones, DNA polymerases, RNA polymerases, transcription factors, splicing factors, and many others. These proteins are targeted to the nucleus by specific amino acid sequences, called **nuclear localization signals** (**NLS**) that direct their transport through the NPC.

(1) Composition of the nuclear localization signals The first NLS to be mapped in detail was characterized by Robert Laskey and his coworkers at the Medical Research Council of England in 1982. They found that nucleoplasmin, one of the more abundant nuclear proteins of amphibian oocytes, contains a stretch of amino acids near its C-terminus that functions as a nuclear localization signal. This sequence enables a protein to pass through the nuclear pores and enter the nucleus. The best-studied, or "classical" NLSs, consist of one or two short stretches of positively charged amino acids. The T antigen encoded by the virus SV40, for example, contains an NLS identified as the seven-amino-acid sequence Pro-Lys-Lys-Lys-Arg-Lys-Val (Figure 8-6). The signal responsible for its nuclear localization was first identified by the finding that mutation of a single Lys residue prevents nuclear import, resulting instead in the accumulation of T antigen in the cytoplasm. Conversely, if this NLS

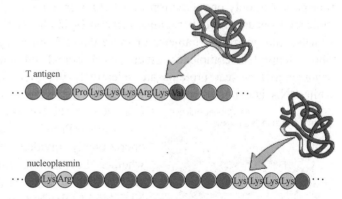

Figure 8-6 Nuclear localization signals (NLSs). The T antigen NLS is a single stretch of amino acids. In contrast, the NLS of nucleoplasmin is bipartite, consisting of a Lys-Arg sequence, followed by a Lys-Lys-Lys-Lys sequence located ten amino acids farther downstream.

is fused to a nonnuclear protein, such as serum albumin, and injected into the cytoplasm, the modified protein becomes concentrated in the nucleus. Not only was this sequence necessary for the nuclear transport of T antigen, but its addition to other, normally cytoplasmic, proteins was also sufficient to direct their accumulation in the nucleus.

NLSs have since been identified in many other proteins. Most of these sequences, like that of T antigen, are short stretches rich in basic amino acid residues (Lys and Arg). In many cases, however, the amino acids that form the NLS are close together but not immediately adjacent to each other. For example, the NLS of nucleoplasmin (a protein involved in chromatin assembly) consists of two parts: a Lys-Arg pair followed by four Lys located ten amino acids farther downstream (Figure 8-6). Both the Lys-Arg and Lys-Lys-Lys-Lys sequences are required for nuclear targeting, but the ten amino acids between these sequences can be mutated without affecting nuclear localization. Because this nuclear localization sequence is composed of two separated elements, it is referred to as bipartite. Similar bipartite motifs appear to function as the localization signals of many nuclear proteins; thus they may be more common than the simpler NLS of T antigen. In addition, some proteins, such as ribosomal proteins, contain distinct NLSs which are unrelated to the basic amino acid-rich NLSs of either nucleoplasmin or T antigen.

(2) Small molecular traffic through the NPC by passive diffusion   Depending on their size, molecules can travel through the NPC by one of two different mechanisms. Small molecules and some proteins with molecular mass less than about 50 kD pass freely across the nuclear envelope in either direction: cytoplasm to nucleus, and *vice versa*. These molecules diffuse passively through open aqueous channels, estimated to have diameters of approximately 9 nm, in the NPC. Most proteins and RNAs, however, are unable to pass through these open channels. Instead, these macromolecules pass through the NPC by an active process in which appropriate proteins and RNA are recognized and selectively transported in only one direction (trafficking in and out of the nucleus). The traffic of these molecules occurs through regulated channels in the NPC that, in response to appropriate signals, can open to a diameter of more than 25 nm-a size sufficient to accommodate large ribonucleoprotein complexes, such as ribosomal subunits. It is through these regulated channels that nuclear proteins are selectively imported from the cytoplasm to the nucleus while RNA is exported from the nucleus to the cytoplasm (Figure 8-7).

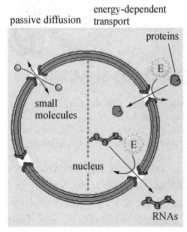

Figure 8-7   Molecular traffic through NPC

(3) Receptors for the NLS transport proteins to the nucleus   Protein import through the NPC can be operationally divided into two steps, distinguished by whether they require energy (Figure 8-8). In the first step, which does not require energy, proteins that contain NLS bind to the NPC but do not pass through the pore. In this initial step, NLS are recognized by a cytosolic receptor protein, and the receptor-substrate complex binds to the nuclear pore. The prototype receptor, called importin, consists of two subunits. One subunit (importin α) binds to the basic amino acid-rich NLS of proteins such as T antigen and nucleoplasmin. The second subunit (importin β) binds to the cytoplasmic filaments of the NPC, bringing the target protein to the nuclear pore. Other types of NLS, such as those of ribosomal proteins,

are recognized by distinct receptors which are related to importin β and function similarly to importin β during the transport of their target proteins into the nucleus.

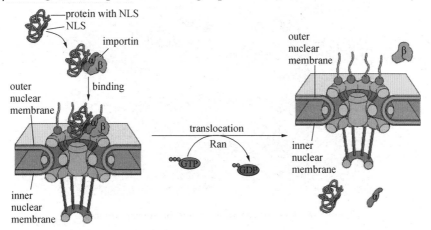

Figure 8-8 Protein import through the NPC

The second step in nuclear import, translocation through the NPC, is an energy-dependent process that requires GTP hydrolysis. A key player in the translocation process is a protein, Ran, a small GTP-binding protein that exists in two conformations, one when complexes with GTP and an alternative one when complexes with GDP (Figure 8-9). The conformation and activity of Ran is regulated by GTP binding and hydrolysis, like other GTP-binding proteins, such as several of the translation factors involved in protein synthesis. Enzymes that stimulate GTP binding to Ran are localized to the nuclear side of the nuclear envelope whereas enzymes that stimulate GTP hydrolysis are localized to the cytoplasmic side. Consequently, there is a gradient of Ran/GTP across the nuclear envelope, with a high concentration of Ran/GTP in the nucleus and a high concentration of Ran/GDP in the cytoplasm. This gradient of Ran/GTP is thought to determine the directionality of nuclear transport, and GTP hydrolysis by Ran appears to account for most (if not all) of the energy required for nuclear import. Importin β forms a complex with importin α and its associated target protein on the cytoplasmic side of the NPC, in the presence of a high concentration of Ran/GDP. This complex is then transported through the nuclear pore to the nucleus, where a high concentration of Ran/GTP is present. At the nuclear side of the pore, Ran/GTP binds to importin β, displacing importin α and the target protein. As a result, the target protein is released within the nucleus. The Ran/GTP-importin β complex is then exported to the cytosol, where the bound GTP is hydrolyzed to GDP, releasing importin β to participate in another cycle of nuclear import.

Some proteins remain within the nucleus following their import from the cytoplasm, but many others shuttle back and forth between the nucleus and the cytoplasm. Some of these proteins act as carriers in the transport of other molecules, such as RNA; others coordinate nuclear and cytoplasmic functions (e.g., by regulating the activities of transcription factors). Proteins are targeted for export from the nucleus by a Leu-rich sequences, called **nuclear export signals** (**NES**). Like NLS, the NES is recognized by receptors within the nucleus that direct protein transport through the NPC to the cytoplasm. Interestingly, the nuclear export receptors (called exportins) are related to importin β. Like importin β, the exportins bind to Ran, which is required for nuclear export as well as for nuclear import (Figure 8-9). Strikingly, however, Ran/GTP promotes the formation of

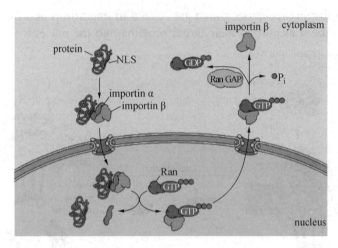

Figure 8-9 Role of the Ran protein in nuclear import

stable complexes between exportins and their target proteins, whereas it dissociates the complexes between importins and their targets. This effect of Ran/GTP binding on exportins dictates the movement of proteins containing nuclear export signals from the nucleus to the cytoplasm. Thus, exportins form stable complexes with their target proteins in association with Ran/GTP within the nucleus. Following transport to the cytosolic side of the nuclear envelope, GTP hydrolysis leads to dissociation of the target protein, which is released into the cytoplasm.

8.3.3 Transport of RNA between nucleus and cytoplasm

Whereas many proteins are selectively transported from the cytoplasm into the nucleus, most RNA molecules are exported from the nucleus to the cytoplasm. Since proteins are synthesized in the cytoplasm, the export of mRNA, rRNA, and tRNA is a critical step in gene expression in eukaryotic cells. Like protein import, the export of an RNA species through the NPC is an active, energy-dependent process that requires the Ran GTPase protein.

RNA molecules are transported across the nuclear envelope as RNA-protein complexes, which in some cases are large enough to visualize by electron microscopy. The substrates for transport are ribonucleoprotein complexes rather than naked RNA, and RNA molecules are targeted for transport from the nucleus by NES on the proteins bound to them. These proteins are recognized by exportins and transported from the nucleus to the cytoplasm. Pre-mRNAs and mRNAs are associated with a set of at least 20 proteins (forming heterogeneous nuclear ribonucleoproteins, hnRNPs) throughout their processing in the nucleus and eventual transport to the cytoplasm. At least two of these hnRNP proteins contain NES and are thought to function as carriers of mRNAs during their export to the cytoplasm. As discussed later, rRNA is assembled with ribosomal proteins in the nucleolus, and intact ribosomal subunits are then transported to the cytoplasm. Their export from the nucleus appears to be mediated by NES present on ribosomal proteins. For tRNA, the specific proteins that mediate nuclear export remain to be identified.

8.4 Chromatin and chromosomes

The presence of a nucleus and other membranous organelles characterizes eukaryotic cells. The nucleus houses the genetic material, DNA, which is complexed with an array of acidic and basic proteins into thin fibers. During nondivisional phases of the cell cycle, these fibers are uncoiled and dispersed into chromatin. During mitosis and meiosis, chromatin fibers coil and condense into structures called chromosomes. Chromosomes seem to appear out of nowhere at the beginning of mitosis and disappear once again when cell division has ended. The appearance and disappearance of chromosomes provided early cytologists with a challenging question: What is the nature of the chromosome in the nonmitotic cell? We are now able to provide a fairly comprehensive answer to this question.

Interphase chromatin is a tangled mass occupying a large part of the nuclear volume, in contrast with the highly organized and reproducible ultrastructure of mitotic chromosome. Chromatin structure is hierarchic, ranging from the two lowest levels of DNA packaging — the **nucleosome** and the 30 nm chromatin fiber — to the metaphase chromosomes, which represent the most compact form of chromatin in eukaryotes and occur only during nuclear division. After division, the chromosomes become less compact and cannot be distinguished as individual structures. The global structure of the interphase chromatin does not change visibly between divisions. No disruption is evident during the period of replication, when the amount of chromatin doubles. Chromatin is fibrillar, although the overall configuration of the fiber in space is hard to discern in detail. The fiber itself, however, is similar or identical to that of the mitotic chromosomes.

8.4.1 Heterochromatin and euchromatin

Although we know that the DNA of the eukaryotic chromosome consists of one continuous double-helical fiber along its entire length, we also know that the whole chromosome is not structurally uniform from end to end. In the early part of the twentieth century, it was observed that some parts of the chromosome remain condensed and stain deeply during interphase, while most parts are uncoiled and do not stain. In 1928, the terms euchromatin and heterochromatin were coined to describe the parts of chromosomes that are uncoiled and those that remain condensed, respectively.

Subsequent investigation revealed a number of characteristics that distinguish heterochromatin from euchromatin. These two types of chromatin can be seen in the nuclear section of Figure 8-2. In most regions, the chromatin fibers are much less densely packed than in the mitotic chromosome. This material is called euchromatin. It has a relatively dispersed appearance in the nucleus and occupies most of the nuclear region in Figure 8-2. Some regions of chromatin are very densely packed with fibers, displaying a condition comparable to that of the chromosome at mitosis. This material is called heterochromatin. It is typically found at centromeres, but occurs at other locations as well. It passes through the cell cycle with relatively little change in its degree of condensation. It forms a series of discrete clumps (Figure 8-2), but often the various heterochromatic regions, especially those associated with centromeres, aggregate into a densely staining chromocenter. Heterochromatin is divided into two classes, constitutive heterochromatin and facultative heterochromatin, depending on whether the chromatin is permanently or transiently compacted. Constitutive heterochromatin is a permanent feature of all cells and represents DNA that contains no genes and so can always be retained in a compact organization. This fraction includes centromeric and telomeric DNA as well as certain regions of some other chromosomes. For example, most of the human Y chromosome is made of constitutive heterochromatin. In contrast, facultative heterochromatin is not a permanent feature but is seen in some cells at some times. Facultative heterochromatin is thought to contain genes that are inactive in some cells or at some periods of the cell cycle. When these genes are inactive, their DNA regions are compacted into heterochromatin.

The extent of chromatin condensation varies during the life cycle of the cell. In interphase cells, the euchromatin is relatively decondensed and distributed throughout the nucleus. During this period of the cell cycle, genes are transcribed and the DNA is replicated in preparation for cell division. Most of the euchromatin in interphase nuclei appears to be in the form of 30 nm fibers, organized into large loops containing approximately 50 to 100 kb of DNA. About 10% of the euchromatin, containing the genes that are actively transcribed, is in a more decondensed state (the 10 nm conformation) that allows transcription. Chromatin structure is thus intimately linked to the control of gene expression in eukaryotes. In contrast to euchromatin, about 10% of interphase chromatin (heterochromatin) is in a very highly condensed state that resembles the chromatin of cells undergoing mitosis. Heterochromatin is transcriptionally inactive and contains highly repeated DNA sequences, such as those present at centromeres and telomeres.

8.4.2 Nucleosomes: the lowest level of chromosome organization

As we have seen, the genetic material of viruses and bacteria consists of strands of DNA or RNA that are nearly devoid of proteins. In eukaryotic chromatin, a substantial amount of protein is associated with the chromosomal DNA in all phases of the eukaryotic cell cycle. The associated proteins are divided into basic, positively charged histones and less positively charged nonhistones. Histones are present in enormous quantities (about 60 million molecules of several different types in each cell), and their total mass in chromosomes is about equal to that of the DNA itself. The condensed state of nucleic acid results from its binding to basic proteins. The positive charges of these proteins neutralize the negative charges of the nucleic acid. The structure of the nucleoprotein complex is determined by the interactions of the proteins with the DNA (or RNA).

Not only are the genomes of most eukaryotes much more complex than those of prokaryotes, but the DNA of eukaryotic cells is also organized differently from that of prokaryotic cells. The genomes of prokaryotes are contained in single chromosomes, which are usually circular DNA molecules. In contrast, the genomes of eukaryotes are composed of multiple chromosomes, each containing a linear molecule of DNA. Although the numbers and sizes of chromosomes vary considerably between different species, their basic structure is the same in all eukaryotes. The DNA of eukaryotic cells is tightly bound to small proteins that package the DNA in an orderly way in the cell nucleus to form eukaryotic chromatin.

The major proteins of chromatin are the histones — small proteins containing a high proportion of basic amino acids (Arg and Lys) that facilitate binding to the negatively charged DNA molecule. There are five major types of histones: called H1, H2A, H2B, H3, and H4—which are very similar among different species of eukaryotes. The histones are extremely abundant proteins in eukaryotic cells. In addition, chromatin contains an approximately equal mass of a wide variety of nonhistone chromosomal proteins. There are more than a thousand different types of these proteins, which are involved in a range of activities, including DNA replication and gene expression.

The general model for chromatin structure is based on the assumption that chromatin fibers, composed of DNA and protein, undergo extensive coiling and folding as they are condensed within the cell nucleus. X-ray diffraction studies confirm that histones play an important role in chromatin structure. Chromatin produces spaced diffraction rings, suggesting that repeating structural units occur along the chromatin axis. If the histone molecules are chemically removed from chromatin, the regularity of this diffraction pattern is disrupted. In the early 1970s, it was found that when chromatin was treated with nonspecific nucleases, most of the DNA was converted to fragments of approximately 200 bp in length. In contrast, a similar treatment of naked DNA produced a randomly sized population of fragments. This finding

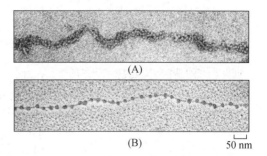

Figure 8-10   Electron of micrograph of nucleosomes. (A) chromatin isolated directly from an interphase nucleus appears in electron microscope as a thread 30 nm thick. (B) This electron micrograph shows a length of chromatin that has been experimentally unpacked, or decondensed, after isolation to show nucleosomes.

suggested that chromosomal DNA was protected from enzymatic attack, except at certain periodic sites along its length. It was presumed that the proteins associated with the DNA were providing the protection. In 1974, using the data from nuclease digestion and other types of information, Roger Kornberg then at Harvard University, proposed an entirely new structure for chromatin. He proposed that DNA and histones are organized into repeating subunits, called nucleosomes. When interphase nuclei are broken open very gently with nuclease and their contents examined under electron microscope, most of the chromatin is in the form of a fiber with a diameter of about 30 nm (Figure 8-10A). If this chromatin is subjected to treatments that cause it to unfold partially, it can then be seen under the electron microscope as a series of "beads on a string" (Figure 8-10B). The string is DNA, and each bead is a nucleosome core particle that consists of DNA wound around a core of proteins formed from histones.

The structure of nucleosomes was determined after first isolating them from the unfolded chromatin by digestion with particular enzymes, nuclease that break down DNA by cutting between the nucleotides. After digestion for a short period, the exposed DNA between the nucleosome core particles, the linker DNA, is degraded. An individual nucleosome core particle consists of a complex of eight histone proteins-two molecules each of histones H2A, H2B, H3, H4-and a double-stranded DNA of about 146 bp that winds around this **histone octamer** (Figure 8-11). The high-resolution structure of a nucleosome core particle was solved in 1997, revealing atomic detail the disc-shaped histone complex around which the DNA was tightly wrapped, making 1.65 turns in a left-handed coil.

Each nucleosome core particle is separated from the next by a region of linker DNA, which can vary in length from a few nucleotide pairs up to about 80 (The term "nucleosome" technically refers to a nucleosome core particle plus one of its adjacent DNA linkers, but is often used synonymously with "nucleosome core particle"). Formation of nucleosomes converts a DNA molecule into a chromatin thread approximately one-third of its initial length, and provides the first level of DNA packing.

All four of the histones that make up the core of the nucleosome, core are relatively small proteins with a high proportion of positively charged amino acids (Lys and Arg). The positive charges help the histones bind tightly to the negatively charged sugar-phosphate backbone of DNA. The interface between DNA and histone is extensive: 142 hydrogen

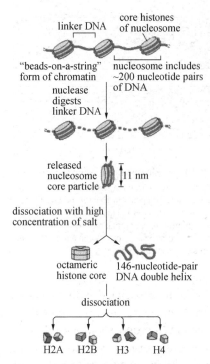

Figure 8-11 Nucleosomes contains DNA wrapped around a protein core of eight histone molecules. As indicated, the nucleosome core particle is released from chromatin by digestion of the linker DNA with a nuclease. After dissociation of the isolated nucleosome into its protein core and DNA, the length of the DNA that was wound around the core can be determined. Its length of 146 nucleotide pairs is sufficient to wrap almost twice around the histone core.

bonds are formed between DNA and the histone core in each nucleosome. Nearly half of these bonds form between the amino acid backbone of the histones and the phosphodiester backbone of the DNA. Numerous hydrophobic interactions and salt linkages also hold DNA and protein together in the nucleosome. For example, all the core histones are rich in Lys and Arg. These numerous interactions explain in part why DNA of virtually any sequence can bind to a histone core. Each of the core histones also has a long N-terminal amino acid "tail" (Figure 8-12), which extends out from the DNA-histone core. These histone tails are subject to several different types of covalent modification that control many aspects of chromatin structure.

As might be expected from their fundamental role in DNA packaging, the histones are among the most highly conserved eukaryotic proteins. For example, the amino acid sequence of histone H4 from a pea and a cow differ at only at 2 of the 102 positions. This strong evolutionary conservation suggests that the functions of histones involve nearly all of their amino acids, so that a change in any position is deleterious to the cell. This suggestion has been tested directly in yeast cells, in which it is possible to mutate a given histone gene *in vitro* and introduce it into the yeast genome in place of the normal gene. Recently, histones have been found in archaea-procaryotes that form a phylogenetic kingdom distinct from eukaryotes and eubacteria. This extreme evolutionary conservation reflects the vital structural role of histones in forming chromatin.

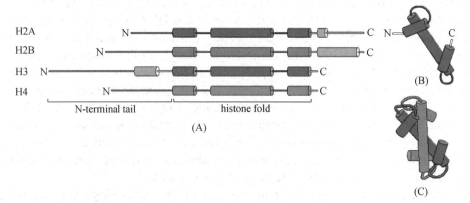

Figure 8-12 Overall structural organization of the core histones. (A) Each of the core histones contains an N-terminal tail, which is subject to several forms of covalent modification, and a histone fold region. (B) The structure of the histone fold, which is formed by all four of the core histones. (C) Histones H2A and H2B form a dimer through an interaction, histones H3 and H4 form a dimer through the same type of interaction.

Despite the high conservation of the core histones, many eukaryotic organisms also produce specialized variant core histones that differ in amino acid sequence from the main ones. For example, the sea urchin has five histone H2A variants, each of which is expressed at a different time during development. It is thought that nucleosomes that have incorporated these variant histones differ in stability from regular nucleosomes, and they may be particularly well suited for the high rates of DNA transcription and DNA replication that occur during these early stages of development.

8.4.3 Chromosomes have several levels of DNA packing

The packaging of DNA with histones yields a "beads on a string" form of chromatin. Although long strings of nucleosomes form on most chromosomal DNA, chromatin in the living cell rarely adopts the extended beads on a string seen in figure 8-10B. Instead, the nucleosomes are further packed upon one another to generate a more compact structure, the 30 nm fiber (Figure 8-10A). Packaging of nucleosomes into the 30-nm chromatin fiber depends on a fifth histone called histone H1, which is thought to pull the nucleosomes together into a regular repeating array. An overview of the various levels of chromatin organization, from the nucleosomal filament to a mitotic chromosome, is depicted in Figure 8-13.

As cells enter mitosis, their chromosomes become highly condensed so that they can be distributed to daughter cells. The loops of 30 nm chromatin fibers are thought to fold upon themselves further to form the compact metaphase chromosomes of mitotic cells, in which the DNA has been condensed nearly 10,000 fold

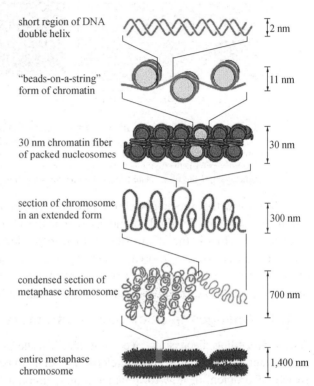

Figure 8-13 Higher levels of chromatin structure. Naked DNA molecules are wrapped around histones to form nucleosomes, which represent the lowest level of chromatin organization. Nucleosomes are organized into 30 nm fibers, which in turn are organized into looped domains. When cells prepare for mitosis, the loops become further compacted into mitotic chromosomes. Net result: Each DNA molecule has been packaged into a mitotic chromosome that is 10,000 fold shorter its extended length.

(Figure 8-13). Such condensed chromatin can no longer be used as a template for RNA synthesis, so transcription ceases during mitosis. Electron micrographs indicate that the DNA in metaphase chromosomes is organized into large loops attached to a protein scaffold, but we currently understand neither the detailed structure of this highly condensed chromatin nor the mechanism of chromatin condensation.

Several models have been proposed to explain how nucleosomes are packed in the 30 nm chromatin fiber; the one most consistent with the available data is a series of structural variations known collectively as the zigzag model (Figure 8-14). The 30 nm structure found in chromosomes is probably a fluid mosaic of the different zigzag variations.

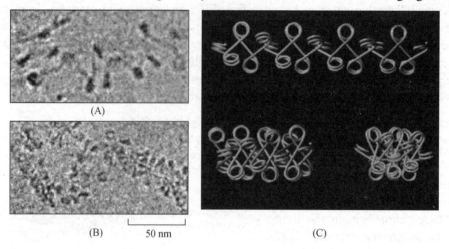

Figure 8-14 Chromatin fibers may be packed according to a zigzag model. (A and B) Electron microscopic evidence for the top and bottom-left model structures shown in (C). The structure of the 30 nm chromatin fiber may be a combination of these zigzag variations. An interconversion between these three variations may occur through an accordion-like expansion and contraction of the fiber. Note that the histone cores are omitted from the diagrams in (C).

We know that the 30 nm chromatin fiber can be compacted still further. During mitosis chromatin becomes so highly condensed that the chromosomes can be seen under the light microscope. How is the 30 nm fiber folded to produce mitotic chromosomes? The answer to this question is not yet known in detail, but it is thought that the 30 nm fiber is further organized into loops emanating from a central axis. Finally, this string of loops is thought to undergo at least one more level of packing to form the mitotic chromosome (Figure 8-13).

8.4.4 Changes in nucleosome structure allow access to DNA

As the preceding discussion, we know that histone proteins play an important structural role in packaging DNA into the nucleosome that make up chromatin. While solving the structural problem of how to organize a huge amount of DNA within the eukaryotic nucleus, a new problem was apparent: the chromatin fiber, when complexed with histones and folded into various levels of compaction, makes the DNA inaccessible to interaction with important nonhistone proteins. The variety of proteins that function in enzymatic and regulatory roles during the processes of replication and gene expression must interact directly with DNA. To accommodate these protein-DNA interactions, chromatin must be induced to change its structure, a process called **chromatin remodeling.** In the case of replication and gene expression, chromatin must relax its compact structure but be able to reverse the process during periods of inactivity.

Recently, it has been discovered that eukaryotic cells contain **chromatin remodeling complexes**, protein machines that use the energy of ATP hydrolysis to change the structure of nucleosomes temporarily so that DNA becomes less tightly bound to the histone core (Figure 8-15). The remodeled state may result from movement of the H2A-H2B dimers in the nucleosome core; the H3-H4 tetramer is particularly stable and would be difficult to rearrange.

Eukaryotic cells have several different chromatin remodeling complexes that differ subtly in their properties. Most are large protein complexes that can contain more than ten subunits. These complexes can make the underlying DNA more accessible to other proteins in the cell, especially those involved in DNA replication, and gene expression. During mitosis, at least some of the chromatin remodeling complexes are inactivated, which may help mitotic chromosomes maintain their tightly packed structure. Different remodeling complexes may have features specialized for each of these roles. It is thought that the primary role of some remodeling complexes is to allow access to nucleosomal DNA, whereas that of others is to re-form nucleosomes when access to DNA is no longer required.

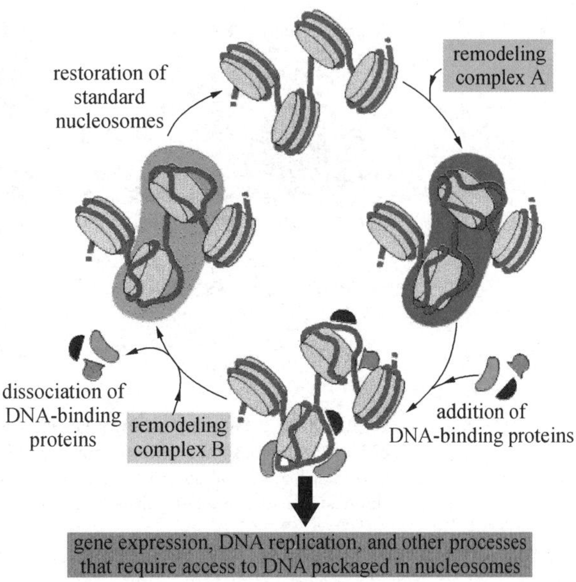

The remodeling of nucleosome structure has two important consequences. First, it permits ready access to nucleosomal DNA by other proteins in the cell, particularly those involved in gene expression, DNA replication, and repair. Even after the remodeling complex has dissociated, the nucleosome can remain in a "remodeled state" that contains DNA and the full complement of histones, but one in which the DNA-histone contacts have been loosened; only gradually does this remodeled state revert to that of a standard nucleosome. Second, remodeling complexes can catalyze changes in the positions of nucleosomes along DNA; some can even transfer a histone core from one DNA molecule to another.

Figure 8-15 Chromatin remodeling complexes alter nucleosome structure. According to this model, different chromatin remodeling complexes disrupt and re-form nucleosomes, although, in principle, the same complex might catalyze both reactions. The DNA-binding proteins could function in gene expression, DNA replication, or DNA repair.

8.4.5 Metaphase chromosomes

As we have seen, all eukaryotic cells in mitosis package their DNA tightly into chromosomes, i.e. metaphase chromosomes. They are so highly condensed that their morphology can be studied using the light microscope (Figure 8-16A). Several staining techniques yield characteristic patterns of alternating light and dark chromosome bands, which result from the preferential binding of stains or fluorescent dyes to AT-rich versus GC-rich DNA sequences. These bands are specific for each chromosome and appear to represent distinct chromosome regions. Genes can be localized to specific chromosome bands by *in situ* hybridization, indicating that the packaging of DNA into metaphase chromosomes is a highly ordered and reproducible process.

As the result of compaction, the chromosomes of a mitotic cell appear as distinct, rod-like structures. Examination of mitotic chromosomes under microscope reveals each of them to be composed of two distinct components. Each of these rod-shaped members is called a chromatid and is formed during replication in the previous interphase. The two sister chromatids of each mitotic chromosome are connected to each other securely at their centromeres and more loosely along their entire length (Figure 8-16B). The connections between chromatids consist of protein "glue" called cohesin that contains a number of

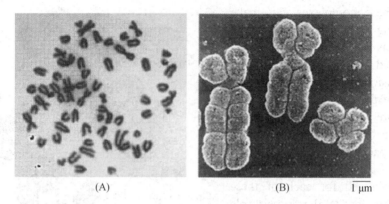

Figure 8-16 Human metaphase chromosomes. (A) A light micrograph of human chromosomes spread from a metaphase cell. (B) Scanning electron micrograph of several human metaphase chromosomes showing the paired identical chromatids associated along their length and joined tightly at the centromere.

highly conserved subunits. In vertebrates, most of the cohesin is released from the chromosomes during prophase, but some of the "glue" remains to hold the sister chromatids together. The sister chromatids are evident at this stage, and will give rise to the daughter chromosomes upon their separation during the anaphase stage of mitosis. Each chromatid consists of a fiber with a diameter of 30 nm and nubbly appearance. The DNA is five to ten times more condensed in chromosomes than in interphase chromatin.

As functioning organelles, eukaryotic chromosomes seem to require only three classes of DNA sequence element: centromeres, telomeres and origins of replication. This simple requirement has been verified by the successful construction of artificial chromosomes in yeast: large foreign DNA fragments behave as autonomous chromosomes when ligated to short sequences that specify a functional centromere, two telomeres and a replication origin.

(1) **Chromosomes as functioning organelles: the centromere** Each normal chromosome contains a condensed or constricted region called the **centromere**, which establishes the general appearance of each chromosome. It is seen under the microscope as the primary constriction, the region at which sister chromatids are joined. The centromere is essential for segregation during cell division. Chromosome fragments that lack a centromere (acentric fragments) do not become attached to the spindle, and so fail to be included in the nuclei of either of the daughter cells.

The centromere is a specialized region of the chromosome that plays a critical role in ensuring the correct distribution of duplicated chromosomes to daughter cells during mitosis. The cellular DNA is replicated during interphase, resulting in the formation of two copies of each chromosome prior to the beginning of mitosis. As the cell enters mitosis, chromatin condensation leads to the formation of metaphase chromosomes consisting of two identical sister chromatids. These sister chromatids are held together at the centromere, which is seen as a constricted chromosomal region. As mitosis proceeds, microtubules of the mitotic spindle attach to the centromere, and the two sister chromatids separate and move to opposite poles of the spindle. At the end of mitosis, nuclear membranes reform and the chromosomes decondense, resulting in the formation of daughter nuclei containing one copy of each parental chromosome.

The centromeres thus serve both as the sites of association of sister chromatids and as the attachment sites for microtubules of the mitotic spindle. They consist of specific DNA sequences to which a number of centromere-associated proteins bind, forming a specialized structure called the **kinetochore** (Figure 8-17). The binding of microtubules to kinetochore proteins mediates the attachment of chromosomes to the mitotic spindle. Proteins associated with the kinetochore then act as "molecular motors" that drive the movement of chromosomes along the spindle fibers, segregating the chromosomes to daughter nuclei.

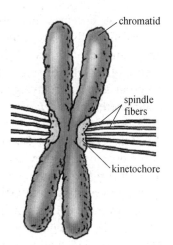

Figure 8-17 Centromere of a metaphase chromosome. Centromere is the region at which the two sister chromatids remain attached at metaphase. Specific proteins bind to centromeric DNA, forming the kinetochore, which is the site of spindle fiber attachment.

1) **The centromere sequences of yeast S. cerevisiae**: Centromeric DNA sequences have been defined best in yeast *S. cerevisiae*, where their function can be assayed by the segregation of plasmids at mitosis. Plasmids that contain functional centromeres segregate like chromosomes and are equally distributed to daughter cells following mitosis. In the absence of a functional centromere, however, the plasmid does not segregate properly, and many daughter cells fail to inherit plasmid DNA. Assays of this type have enabled determination of the sequences required for centromere function. Such experiments first showed that the centromere sequences of the well-studied *S. cerevisiae* are contained in approximately 120 bp consisting of three sequence elements: two short sequences of 9 and 11 base pairs separated by 80 to 90 bp of very AT-rich DNA (Figure 8-18).

TCACATGATGATATTTGATTTTATTATATTTTTAAAAAAAGTAAAAAATAAAAAGTAGTTTATTTTTAAAAAATAAAATTTAAAATATTTCACAAAATGATTTCCGAA
AGTGTACTACTATAAACTAAAATAATATAAAAATTTTTTTCATTTTTTATTTTTCATCAAATAAAAATTTTTTATTTTAAATTTTATAAAGTGTTTTACTAAAGGCTT
CDE-I   CDE-II 80~90 bp, >90% A+T   CDE-III

Figure 8-18 Three conserved regions can be identified by the sequence homologiesbetween yeast CEN elements

As summarized in figure 8-18, three types of sequence element can be distinguished in the centromeric DNA region (CEN) fragment:

Cell cycle-dependent element (CDE)-I is a sequence of 9 bp that is conserved with minor variations at the left boundary of all centromeres.

CDE-II is a >90% A-T-rich sequence of 80 to 90 bp found in all centromeres; its function could depend on its length rather than exact sequence. Its constitution is reminiscent of some short, tandemly repeated (satellite) DNAs. Its base composition may cause some characteristic distortions of the DNA double helical structure.

CDE-III is an 11 bp sequence highly conserved at the right boundary of all centromeres. Sequences on either side of the element are less well conserved, and may also be needed for centromeric function(CDE-III could be longer than 11 bp, if it turns out that the flanking sequences are essential).

Mutations in CDE-I or CDE-II reduce, but do not inactivate, centromere function, but point mutations in the central CCG of CDE-III completely inactivate the centromere.

The short centromere sequences defined in *S. cerevisiae*, however, do not appear to reflect the situation in other eukaryotes. More recent studies have defined the centromeres of the fission yeast *Schizosaccharomyces pombe* by a similar functional approach. Although *S. cerevisiae* and *S. pombe* are both yeasts, they appear to be as divergent from each other as either is from humans and are quite different in many aspects of their cell biology. These two yeast species thus provide complementary models for simple and easily studied eukaryotic cells. The centromeres of *S. pombe* span 40 to 100 kb of DNA; they are approximately a thousand times larger than those of *S. cerevisiae*. They consist of a central core of 4 to 7 kb of single-copy DNA flanked by repetitive sequences. Not only the central core but also the flanking repeated sequences are required for centromere function, so the centromeres of *S. pombe* appear to be considerably more complex than those of *S. cerevisiae*.

2) **The centromere sequences of *Drosophila melanogaster*:** Studies of a *Drosophila* chromosome have provided the first characterization of a centromere in higher eukaryotes. The *Drosophila* centromere spans 420 kb, most of which (more than 85%) consists of two highly repeated satellite DNAs with the sequences AATAT and AAGAG. The remainder of the centromere consists of interspersed transposable elements, which are also found at other sites in the *Drosophila* genome, in addition to a nonrepetitive region of AT-rich DNA. Deletion of the satellite sequences and transposable elements, as well as the nonrepetitive DNA, reduced the activity of the centromere in functional assays. Thus, both repetitive and nonrepetitive sequences appear to contribute to kinetochore formation and centromere function.

3) **Mammalian centromere:** Centromeres of humans and other mammals have not yet been defined by functional studies, but they have been identified by the binding of centromere-associated proteins. Mammalian centromeres are characterized by extensive regions of heterochromatin consisting of highly repetitive satellite DNA sequences. In humans and other primates the primary centromeric sequence is satellite DNA, which is a 171 bp sequence arranged in tandem repeats spanning up to millions of base pairs. The satellite DNA appears to play a role in centromere structure and function, since it has been found to bind centromere-associated proteins. However, the precise function of satellite DNA, as well as the potential roles of other sequences in mammalian centromeres, remains to be established. Consistent with their large size, mammalian centromeres form large kinetochores that bind 30 to 40 microtubules, whereas only single microtubules bind to the centromeres of *S. cerevisiae*.

(2) Chromosomes as functioning organelles: the telomeres   Another essential feature in all chromosomes is the **telomere**, which "seals" the chromosome ends. We know that the telomere must be a special structure, because chromosome ends generated by breakage are "sticky" and tend to react with other chromosomes, whereas natural ends are stable. They have several likely functions.

Maintaining the structural integrity of a chromosome. If a telomere is lost, the resulting chromosome end is unstable. It has a tendency either to fuse with the ends of other broken chromosomes, to be involved in recombination events or to be degraded. The loop structure of human telomeres means that natural chromosomes have no free DNA end.

Ensuring complete replication of the extreme ends of chromosomes. During DNA replication, synthesis of the lagging strand is discontinuous and requires the presence of some DNA ahead of the sequence which is to be copied to serve as the template for an RNA primer. However, at the extreme end of a linear molecule, there can never be such a template, and a different mechanism is required to solve the problem of replicating the ends of a linear DNA molecule. This problem has been solved by the evolution of a special mechanism, involving reverse transcriptase activity, to replicate telomeric DNA sequences. Maintenance of telomeres appears to be an important factor in determining the lifespan and reproductive capacity of cells, so studies of telomeres and telomerase have the promise of providing new insights into conditions such as aging and cancer.

Helping establish the three-dimensional architecture of the nucleus and/or chromosome pairing. Chromosome ends appear to be tethered to the nuclear membrane, suggesting that telomeres help position chromosomes.

Telomeres were initially recognized as distinct structures because broken chromosomes were highly unstable in eukaryotic cells, implying that specific sequences are required at normal chromosomal termini. This was subsequently demonstrated by experiments in which telomeres from the protozoan *Tetrahymena* were added to the ends of linear molecules of yeast plasmid DNA. The addition of these telomeric DNA sequences allowed these plasmids to replicate as linear chromosome-like molecules in yeasts, demonstrating directly that telomeres are required for the replication of linear DNA molecules.

The telomere DNA sequences of a variety of eukaryotes are similar, consisting of repeats of a simple-sequence DNA containing clusters of G residues on one strand (Figure 8-19). For example, human telomeres contain the sequence TTAGGG repeated from 500 to 5,000 times (Figure 8-20) and the telomere repeat in *Tetrahymena* is GGGGTT. All

CCCCAACCCCAACCCCAACCCCAACCCCAACCCCAA
GGGGTTGGGGTTGGGGTTGGGGTTGGGGTTGGGGTT

CCCCAACCCCAACCCCAA 5'
GGGGTTGGGGTTGGGGTTGGGGTTGGGGTTGGGGTT 3'

Figure 8-19　A typical telomere has a simple repeating structure with a G-T-rich strand that extends beyond the C-A-rich strand. The G-tail is generated by a limited degradation of the C-A-rich strand.

telomeric sequences can be written in the general form $C_n(A/T)_m$, where $n > 1$ and m is 1 to 4. These sequences are repeated hundreds or thousands of times, thus spanning up to several kilobases, and terminate with an overhang of single-stranded DNA. Recent results suggest that the repeated sequences of telomere DNA form loops at the ends of chromosomes, thereby protecting the chromosome termini from degradation.

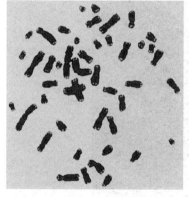

Figure 8-20　Telomeres *in situ* hybridization of a DNA probe containing the sequence TTAGGG, which localizes to the telomeres of human chromosome

(3) **Chromosomes as functioning organelles: origins of replication**　The DNA in most diploid cells normally replicates only once per cell cycle. The initiation of replication is controlled by *cis*-acting sequences that lie close to the points at which DNA synthesis is initiated. Probably these are sites at which *trans*-acting proteins bind. Eukaryotic origins of replication have been most comprehensively studied in yeast, where the presence of a putative replication origin can be tested by a genetic assay. To test the ability of a random fragment of yeast DNA to promote autonomous replication, it is incorporated into a bacterial plasmid together with a yeast gene that is essential for growth of yeast cells. This construct is used to transform mutant yeast that lacks the essential gene. The transformed cells can only form colonies if the plasmid can replicate in yeast cells. However, the bacterial replication origin in the plasmid does not function in yeast, the few plasmids that transform at high efficiency must possess a sequence within the inserted yeast fragment that confers the ability to replicate extrachromosomally at high efficiency -that is an autonomously replicating sequence (**ARS**) element.

ARS elements are thought to derive from authentic origins of replication and, in some cases, this has been confirmed by mapping a specific ARS element to a specific chromosomal location and demonstrating that DNA replication is indeed initiated at this location. ARS elements extend for only about 50 bp and consist of an AT-rich region which contains a conserved core consensus and some imperfect copies of this sequence. In addition, the ARS elements contain a binding site for a transcription factor and a multiprotein complex is known to bind to the origin.

Mammalian replication origins have been much less well defined because of the absence of a genetic assay. Some initiation sites have been studied, but such studies have not been able to identify a unique origin of replication. This has led to speculation that replication can be initiated at multiple sites over regions tens of kilobases long. Mammalian artificial chromosomes seem to work without specific ARS sequences being provided.

8.5 Nucleolus and ribosome biogenesis

The **nucleolus** is the most obvious structure seen in the nucleus of a eukaryotic cell when viewed in the light microscope. Consequently, it was so closely scrutinized by early cytologists that an 1898 review could list some 700 references. We now know that the nucleolus is the site of rRNA transcription and processing, and of ribosome assembly. Unlike other organelles in the cell, it is not bound by a membrane (Figure 8-21). Instead, it is a large aggregate of macromolecules, including the rRNA genes themselves, precursor rRNAs, mature rRNAs, rRNA-processing enzymes, snoRNPs, ribosomal protein subunits and partly assembled ribosomes. The close association of all these components presumably allows the assembly of ribosomes to occur rapidly and smoothly.

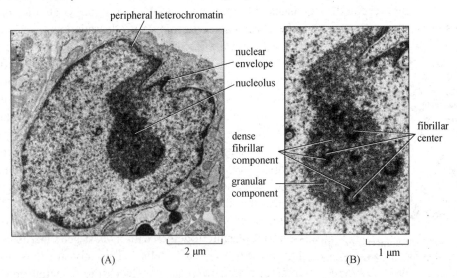

Figure 8-21 Electron micrograph of a thin section of a nucleolus in a human fibroblast, showing its three distinct zones. (A) View of entire nucleus. (B) High-power view of the nucleolus. It is believed that transcription of the rRNA genes takes place between the fibrillar center and the dense fibrillar component and that processing of the rRNAs and their assembly into ribosomes proceeds outward from the dense fibrillar component to the surrounding granular components.

As discussed before, cells require large numbers of ribosomes to meet their needs for protein synthesis. Actively growing mammalian cells, for example, contain 5 million to 10 million ribosomes that must be synthesized each time the cell divides. The nucleolus is a ribosome production factory, designed to fulfill the need for large-scale production of rRNA and assembly of the ribosomal subunits.

8.5.1 Organization of the nucleolus

Morphologically, nucleoli consist of three distinguishable regions: the fibrillar center, dense fibrillar component, and granular component (Figure 8-21). These different regions are thought to represent the sites of progressive stages of rRNA transcription, processing, and ribosome assembly. The rRNA genes are located in the fibrillar centers, with transcription occurring primarily at the boundary of the fibrillar centers and dense fibrillar component. Processing of the pre-rRNA is initiated in the dense fibrillar component and continues in the granular component, where the rRNA is assembled with ribosomal proteins to form nearly completed preribosomal subunits, ready for export to the cytoplasm.

Following each cell division, nucleoli form around the chromosomal regions that contain the 5.8S, 18S, and 28S rRNA genes, which are therefore called **nucleolar organizing regions**. The formation of nucleoli requires the transcription of 45S pre-rRNA, which appears to lead to the fusion of small prenucleolar bodies that contain processing factors and other components of the nucleolus. In most cells, the initially separate nucleoli then fuse to form a single nucleolus (Figure 8-22). The size of the nucleolus depends on the metabolic activity of the cell, with large nucleoli found in cells that are actively engaged in protein synthesis. This variation is due primarily to differences in the size of the granular component, reflecting the levels of ribosome synthesis.

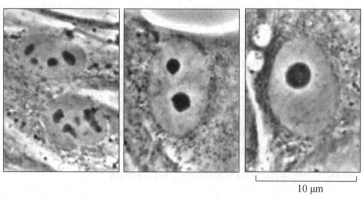

10 μm

Figure 8-22 Nucleolar fusion. These light micrographs of human fibroblasts grown in culture show various stages of nucleolar fusion. After mitosis, each of the ten human chromosomes that carry a cluster of rRNA genes begins to form a tiny nucleolus, but these rapidly coalesce as they grow to form the single large nucleolus typical of many interphase cells.

It is not yet understood how the nucleolus is held together and organized, but various types of RNA molecules play a central part in its chemistry and structure, suggesting that the nucleolus may have evolved from an ancient structure present in cells dominated by RNA catalysis. In present-day cells, the rRNA genes also have an important role in forming the nucleolus. In a diploid human cell, the rRNA genes are distributed into 10 clusters, each of which is located near the tip of one of the two copies of five different chromosomes. Each time a human cell undergoes mitosis, the chromosomes disperse and the nucleolus breaks up; after mitosis, the tips of the 10 chromosomes coalesce as the nucleolus reforms (Figure 8-22). The transcription of the rRNA genes by RNA polymerase I is necessary for this process.

As might be expected, the size of the nucleolus reflects the number of ribosomes that the cell is producing. Its size therefore varies greatly in different cells and can change in a single cell, occupying 25% of the total nuclear volume in cells that are making unusually large amounts of protein.

8.5.2　The arrangement of rRNA genes

The nucleolus is organized around the chromosomal regions that contain the genes for the 5.8S, 18S, and 28S rRNA. Eukaryotic ribosomes contain four types of RNA, designated the 5S, 5.8S, 18S, and 28S rRNA. The 5.8S, 18S, and 28S rRNA are transcribed as a single unit within the nucleolus by RNA polymerase Ⅰ, yielding a 45S ribosomal precursor RNA (Figure 8-23). The 45S pre-rRNA is processed to the 18S rRNA of the 40S ribosomal subunit and to the 5.8S and 28S rRNA of the 60S ribosomal subunit. Transcription of the 5S rRNA, which is also found in the 60S ribosomal subunit, takes place outside of the nucleolus and is catalyzed by RNA polymerase Ⅲ.

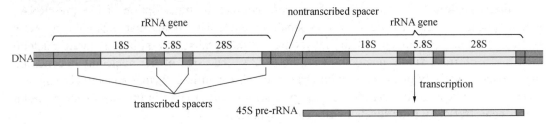

Figure 8-23　Ribosomal RNA genes. Each rRNA gene is a single transcription unit containing the 18S, 5.8S, and 28S rRNA and transcribed spacer sequences. The rRNA genes are organized in tandem arrays, separated by nontranscribed spacer DNA.

To meet the need for transcription of large numbers of rRNA molecules, all cells contain multiple copies of the rRNA genes. The human genome, for example, contains about 200 copies of the gene that encodes the 5.8S, 18S, and 28S rRNA and approximately 2,000 copies of the gene that encodes 5S rRNA. The genes for 5.8S, 18S, and 28S rRNA are clustered in tandem arrays on five different human chromosomes (chromosomes 13, 14, 15, 21, and 22); the 5S rRNA genes are present in a single tandem array on chromosome 1.

8.5.3　Transcription and processing of rRNA

Each nucleolar organizing region contains a cluster of tandemly repeated rRNA genes that are separated from each other by nontranscribed spacer DNA. These genes are very actively transcribed by RNA polymerase Ⅰ, allowing their transcription to be readily visualized by electron microscopy. In such electron micrographs, each of the tandemly arrayed rRNA genes is surrounded by densely packed growing RNA chains, forming a structure that looks like a Christmas tree. The high density of growing RNA chains reflects that of RNA polymerase molecules, which are present at a maximal density of approximately one polymerase per hundred base pairs of template DNA.

The primary transcript of the rRNA genes is the large 45S pre-rRNA, which contains the 18S, 5.8S, and 28S rRNA as well as transcribed spacer regions (Figure 8-24). External transcribed spacers are present at both the 5' and 3' ends of the pre-rRNA, and two internal transcribed spacers lie between the 18S, 5.8S, and 28S rRNA sequences. The initial processing step is a cleavage within the external transcribed spacer near the 5' end of the pre-rRNA, which takes place during the early stages of transcription. This cleavage requires the U3 small nucleolar RNP that remains attached to the 5' end of the pre-rRNA, forming the characteristic knobs. Once transcription is complete, the external transcribed spacer at the 3' end of the molecule is removed. In human cells, this step is followed by a cleavage at the 5' end of the 5.8S region, yielding separate precursors to the 18S and 5.8S + 28S rRNA. Additional cleavages then result in formation of the mature rRNA.

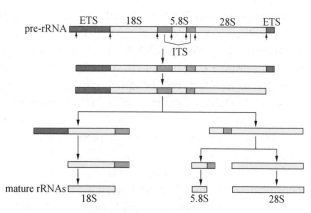

Figure 8-24 Processing of pre-rRNA. The 45S pre-rRNA transcript contains external transcribed spacers (ETS) at both ends and internal transcribed spacers (ITS) between the sequences of 18S, 5.8S, and 28S rRNA. The pre-rRNA is processed via a series of cleavages (illustrated for human pre-rRNA) to yield the mature rRNA species.

The processing of pre-rRNA requires the action of both proteins and RNA that are localized to the nucleolus. Nucleoli contain a large number (about 200) of small nucleolar RNA (snoRNA) that functions in pre-rRNA processing. Like the spliceosomal small nuclear RNA (snoRNA), the snoRNA are complexed with proteins, forming snoRNPs. Individual snoRNPs consist of single snoRNA associated with eight to ten proteins. The snoRNPs then assemble on the pre-rRNA to form processing complexes in a manner analogous to the formation of spliceosomes on pre-mRNA.

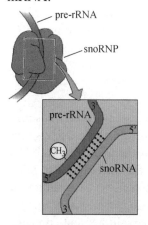

Figure 8-25 Role of snoRNA in base modification of pre-rRNA. The snoRNAs contain short sequences complementary to rRNA. Base pairing between snoRNAs and pre-rRNA targets the enzymes that catalyze base modification (e. g., methylation) to the appropriate sites on pre-rRNA.

Some snoRNA are responsible for the cleavages of pre-rRNA into 18S, 5.8S, and 28S products. For example, the most abundant nucleolar snoRNA is U3, which is present in about 200,000 copies per cell. As already noted, U3 is required for the initial cleavage of pre-rRNA within the 5' external transcribed spacer sequences. Similarly, U8 snoRNA is responsible for cleavage of pre-rRNA to 5.8S and 28S rRNAs, and U22 snoRNA is responsible for cleavage of pre-rRNA to 18S rRNA.

The majority of snoRNA, however, function to direct the specific base modifications of pre-rRNA, including the methylation of specific ribose residues and the formation of pseudouridines (Figure 8-25). Most of the snoRNA contain short sequences of approximately 15 nucleotides that are complementary to 18S or 28S rRNA. Importantly, these regions of complementarity include the sites of base modification in the rRNA. By base pairing with specific regions of the pre-rRNA, the snoRNA act as guide RNAs that target the enzymes responsible for ribose methylation or pseudouridylation to the correct site on the pre-rRNA molecule.

8.5.4 Ribosome assembly

The formation of ribosomes involves the assembly of the

ribosomal precursor RNA with both ribosomal proteins and 5S rRNA (Figure 8-26). The genes that encode ribosomal proteins are transcribed outside of the nucleolus by RNA polymerase II, yielding mRNA that is translated on cytoplasmic ribosome. The ribosomal proteins are then transported from the cytoplasm to the nucleolus, where they are assembled with rRNA to form preribosomal particles. Although the genes for 5S rRNA are also transcribed outside of the nucleolus, in this case by RNA polymerase III, 5S rRNA similarly is assembled into preribosomal particles within the nucleolus.

The association of ribosomal proteins with rRNA begins while the pre-rRNA is still being synthesized, and more than half of the ribosomal proteins are complexed with the pre-rRNA prior to its cleavage. The remaining ribosomal proteins and the 5S rRNA are incorporated into preribosomal particles as cleavage of the pre-rRNA proceeds. The smaller ribosomal subunit, which contains only the 18S rRNA, matures more rapidly than the larger subunit, which contains 28S, 5.8S, and 5S rRNA. Consequently, most of the preribosomal particles in the nucleolus represent precursors to the large subunit. The final stages of ribosome maturation follow the export of preribosomal particles to the cytoplasm, forming the active 40S and 60S subunits of eukaryotic ribosomes.

In addition to its important role in ribosome biogenesis, the nucleolus is also the site where other RNAs are produced and other RNA-protein complexes are assembled (Figure 8-26). For example, the U6 snRNP, which, as we have seen, functions in pre-mRNA splicing, is composed of one RNA molecule and at least seven proteins. The U6 snRNA is chemically modified by snoRNA in the nucleolus before its final assembly there into the U6 snRNP. Other important RNA protein complexes, including telomerase and the signal recognition particle, are also believed to be assembled at the nucleolus. Finally, the tRNAs that carry the amino acids for protein synthesis are processed there as well. Thus, the nucleolus can be thought of as a large factory at which many different noncoding RNAs are processed and assembled with proteins to form a large variety of ribonucleoprotein complexes.

8.5.5 Nucleolar dynamics

The nucleolus is a key organelle that coordinates the synthesis and assembly of ribosomal subunits and forms in the nucleus around the repeated ribosomal gene clusters. Because the production of ribosomes is a major metabolic activity, the function of the nucleolus is tightly linked to cell growth and proliferation, and recent data suggest that the nucleolus also plays an important role in cell-cycle regulation, senescence and stress responses. A quantitative analysis of the proteome of human nucleoli has been performed, using mass-spectrometry-based organelle proteomics and stable isotope labeling. *In vivo* fluorescent imaging techniques are directly compared to endogenous protein changes measured by proteomics. It is characterized that the flux of 489 endogenous nucleolar proteins in response to three different metabolic inhibitors that each affects nucleolar morphology. Proteins that are stably associated, such as RNA polymerase I subunits and small nuclear **ribonucleoprotein** particle complexes, exit from or accumulate in the nucleolus with similar kinetics, whereas protein components of the large and small ribosomal subunits leave the nucleolus with markedly different kinetics. The data establish a quantitative proteomic approach for the temporal characterization of protein flux through cellular organelles and demonstrate that the nucleolar proteome changes significantly over time in response to changes in cellular growth conditions.

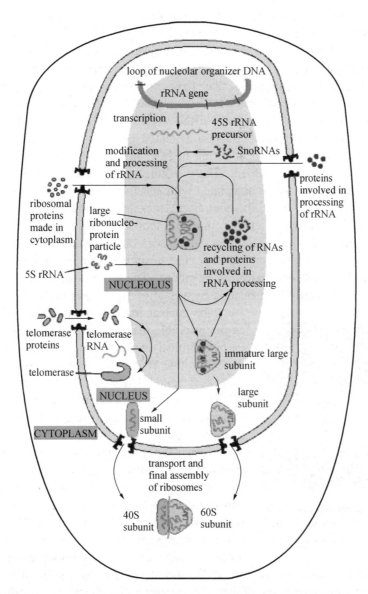

Figure 8-26 Function of the nucleolus in ribosome and other ribonucleoprotein synthesis.
The 45S precursor rRNA is packaged in a large ribonucleoprotein particle containing many ribosomal proteins imported from the cytoplasm. While this particle remains in the nucleolus, selected pieces are added and others discarded as it is processed into immature large and small ribosomal subunits. The two ribosomal subunits are thought to attain their final functional form only as each is individually transported through the nuclear pores into the cytoplasm. Other ribonucleoprotein complexes, including telomerase shown here, are also assembled in the nucleolus.

8.6 The nuclear matrix

When isolated nuclei are treated with a mild nonionic detergent and high salt (e. g., 2 mol/L NaCl), which remove lipids and nearly all of the histone and nonhistone proteins of the chromatin, the DNA is seen as a halo surrounding a residual nuclear core. This highly organized view of the nucleus implies that there is an underlying nuclear substructure. It has been known for many years that when mammalian cells are treated with DNase I, a fibrillar network of protein and RNA remains in the region of the nucleus. This protein network has been called the **nuclear matrix**, or **nuclear skeleton**, a proteinaceous scaffold-like network that permeates the cell. This protein network is composed of actin and numerous other protein components that have not been fully characterized, including components of the chromosomal scaffold that rearranges and condenses to form metaphase chromosomes during mitosis. However, snRNPs remain associated with the nuclear matrix prepared from detergent-extracted, DNase I-treated cells. Moreover, when the nuclear matrix is prepared with a low concentration of salt, pre-mRNA associated with the matrix undergo splicing when ATP is added. These results suggest that the RNA-processing foci observed microscopically may be associated with specific regions of the nuclear matrix.

Full relaxation of the supercoils requires one nick/85 kb, identifying the average length of "closed" DNA. This region could comprise a loop or domain similar in nature to those identified in the bacterial genome. Loops can be seen directly when the majority of proteins are extracted from mitotic chromosomes. The resulting complex consists of the DNA associated with 8% of the original protein content. The protein-depleted chromosomes take the form of a central scaffold surrounded by a halo of DNA (Figure 8-27).

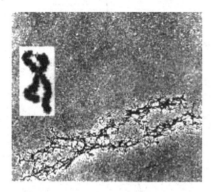

Figure 8-27 Histone-depleted chromosomes consist of a protein scaffold to which loops of DNA are anchored

The nuclear matrix, or scaffold, has been defined as the insoluble material left in the nucleus after a series of biochemical extraction steps. Some of the proteins that constitute it can be shown to bind specific DNA sequences called SARs (scaffold attachment regions) or MARs (matrix attachment regions), as the same sequences appear to attach to the protein substructure in both metaphase and interphase cells. These DNA sequences have been postulated to form the base of chromosomal loops (Figure 8-28), or to attach chromosomes to the nuclear envelope and other structures in the nucleus. It is not understood how the folded 30 nm fiber is anchored to the chromosome axis, but evidence suggests that the base of chromosomal loops is rich in DNA topoisomerases, which are enzymes that allow DNA to swivel when anchored. By means of such chromosomal attachment sites, the matrix might help to organize chromosomes, localize genes, and regulate gene expression and DNA replication. For instance, if cells are incubated with fluorescently or radioactively labeled RNA or DNA precursors for a brief period, nearly all of the newly synthesized nucleic acid is found to be associated with the fibrils of the nuclear matrix. It still remains uncertain, however, whether the matrix isolated by cell biologists represents a structure that is present in intact cells.

(1) **The metaphase scaffold consists of a set of specific proteins**   The metaphase scaffold consists of a dense network of fibers. Threads of DNA emanate from the scaffold, apparently as loops of average length 10~30 μm (30~90 kb). The DNA can be digested without affecting the integrity of the scaffold, which consists of a set of specific proteins. This suggests a form of organization in which loops of DNA of 60 kb are anchored in a central proteinaceous scaffold.

The appearance of the scaffold resembles a mitotic pair of sister chromatids. The sister

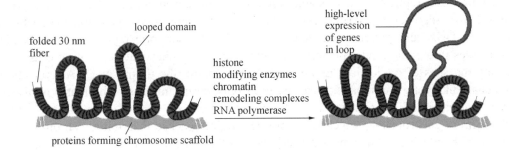

Figure 8-28 A model for the structure of an interphase chromosome. A section of an interphase chromosome is shown folded into a series of looped domains, each containing 20,000 ~ 100,000 bp condensed into a 30 nm fiber. Individual loops can decondense, perhaps in part through an accordion-like expansion of the 30 nm fiber, when the cell requires direct access to the DNA packaged in these loops.

scaffolds usually are tightly connected, but sometimes are separate, joined only by a few fibers. Could this be the structure responsible for maintaining the shape of the mitotic chromosomes? Could it be generated by bringing together the protein components that usually secure the bases of loops in interphase chromatin?

(2) No consensus sequences in MAR fragments   A surprising feature is the lack of conservation of sequence in MAR fragments. They are usually 70% A·T-rich, but otherwise lack any consensus sequences. However, other interesting sequences often are in the DNA stretch containing the MAR. *Cis*-acting sites that regulate transcription are common. And a recognition site for topoisomerase II is usually present in the MAR. It is therefore possible that an MAR serves more than one function, providing a site for attachment to the matrix, but also containing other sites at which topological changes in DNA are effected.

(3) The relationship between the chromosome scaffold and the nuclear matrix   What is the relationship between the chromosome scaffold of dividing cells and the matrix of interphase cells? Are the same DNA sequences attached to both structures? In several cases, the same DNA fragments that are found with the nuclear matrix *in vivo* can be retrieved from the metaphase scaffold. And fragments that contain MAR sequences can bind to a metaphase scaffold. It therefore seems likely that DNA contains a single type of attachment site, which in interphase cells is connected to the nuclear matrix, and in mitotic cells is connected to the chromosome scaffold.

The nuclear matrix and chromosome scaffold consist of different proteins, although there are some common components. Topoisomerase II is a prominent component of the chromosome scaffold, and is a constituent of the nuclear matrix, suggesting that the control of topology is important in both cases.

Summary

The presence of a nucleus and other membranous organelles characterizes eukaryotic cells. The nucleus is the largest structure and the most prominent organelle in the eukaryotic cell. It consists of nuclear envelope (including nuclear pore complex), chromatin, nucleolus and nuclear matrix. The nuclear envelope and the nuclear pore complex may be single most important feature that distinguishes eukaryotes from prokaryotes. The nucleus contains most of the genetic material of the cell. The genetic material is complexed with protein and is organized into a number of linear structures called chromosomes. Three functional elements are required for replication and stable inheritance of chromosome. The centromere is essential for segregation during cell division. Telomeres are specialized structures, comprising DNA and protein, which cap the ends of eukaryotic chromosomes and play critical roles in chromosome replication and maintenance. Eukaryotic replication origins are not only the sites of replication initiation, but also control the timing of DNA replication. Individual eukaryotic chromosomes can be seen only during mitosis. Chromatin is divided into euchromatin and heterochromatin. During interphase, the general mass of chromatin is in the form of euchromatin, which is slightly less tightly packed than mitotic chromosomes. Regions of heterochromatin remain densely packed throughout interphase. Chromosomes in eukaryotic cells consist of DNA tightly bound to a roughly equal mass of specialized proteins. These proteins fold the DNA into a more compact form so that it can fit into a cell nucleus. Nucleoli are extremely dense structures in the nuclei and are highly active in rRNA synthesis. The nuclear matrix consists of DNA, nucleoproteins, and structural proteins.

Questions

1. Please explain the following terms: constitutive heterochromatin and facultative heterochromatin, euchromatin and heterochromatin, fluorescence recovery after photobleaching (FRAP), matrix-associated regions (MAR) or scaffold attachment region (SAR), nuclear envelope, nuclear localization signals (NLS), nuclear pore complex (NPC), nucleolus, nucleosome, nucleosome remodeling.
2. What role do the following cellular components play in the storage, expression, or transmission of genetic information: (a) chromatin, (b) nucleolus, (c) ribosome, (d) centromere?
3. Multiple-choice questions
   (1) Concerning the nucleus:
      (a) The diameter is generally about 5 μm
      (b) Euchromatin is not actively expressed
      (c) Nucleoli are very active in the synthesis of messenger RNA (mRNA)
      (d) All cells in the human body have nuclei
   (2) Within the nucleus:
      (a) Chromatin consists of DNA and RNA only
      (b) Histone proteins are negatively charged
      (c) The nucleosome core protein consists of eight histone subunits
      (d) H1 histone is not found in the core protein
4. Short-answer questions: What is a nucleosome and what are the proteins present in this structure?
5. Define the following terms and their relationships to one another:
   (a) Interphase chromosome; (b) Mitotic chromosome; (c) Chromatin; (d) Heterochromatin; (e) Histones; (f) Nucleosome.
6. If a human nucleus is 10 μm in diameter, and it must hold as much as 2 m of DNA, which is complexed into nucleosomes that during full extension are 11 nm in diameter, what percentage of the volume of the nucleus is occupied by the genetic material?

(曲志才)

Chapter 9

Cell cycle and cell division

Most eukaryotic cells proceed through an ordered series of events, constituting the **cell cycle**, during which their chromosomes are duplicated and division to each of two daughter cells. Regulation of the cell cycle is critical for the normal development of multicellular organisms. Loss of control ultimately leads to cancer, an all-too-familiar disease that kills one in every six people in the developed world. In the late 1980s, it became clear that the molecular processes regulating the main events in the cell cycle, i. e. chromosome replication and **cell division**, are fundamentally similar in all eukaryotic cells. Because of this similarity, research with diverse organisms, each with its own particular experimental advantages, has contributed to a growing understanding of how these events are coordinated and controlled. In this chapter we focus on the various events of the cell division and how they are regulated in eukaryotes.

9.1 An overview of the cell cycle

The division cycle of most cells consists of four coordinated processes: cell growth, DNA replication, distribution of the duplicated chromosomes to daughter cells, and cell division (Figure 9-1). In bacteria, these four events occur throughout most of the cell cycle. In eukaryotes, however, the cell cycle is more complex and consists of four discrete phases. Although cell growth is a continuous process, DNA is synthesized during only one phase of the cell cycle, and the replicated chromosomes are then distributed to daughter nuclei by a complex series of events preceding cell

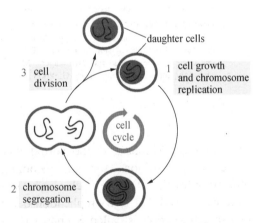

Figure 9-1 Four processes of cell cycle in eukaryotic cell (From: Alberts B, et al. 2008)

division. Progression between these stages of the cell cycle is controlled by a conserved regulatory apparatus, which not only coordinates the different events of the cell cycle, but also links the cell cycle with extracellular signals that control cell proliferation.

A cell reproduces by its contents and then divides into two. In unicellular organisms, such as bacteria and yeasts, each cell division produces a complete new organism, whereas a new functioning organism is produced in multicellular cells. A cell reproduces by carrying out an orderly sequence of events in which it duplicates its contents and then divides into two. This cycle of duplication and division is the essential mechanism by which all living things reproduce. The cell cycle can be regarded as the life cycle of an individual cell, which can be divided into two major phases based on cellular activities readily visible with a light microscope, **interphase** and **M phase** (Figure 9-2). Interphase, the initial stage of the cycle, is a time when the cell grows engages in diverse metabolic activities. M phase includes: the process of mitosis, during which duplicated chromosomes are separated into two nuclei and finally the cell splits up into two.

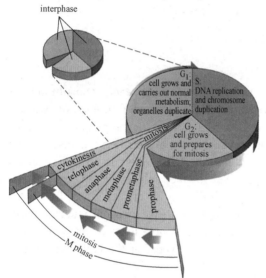
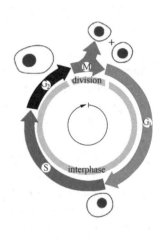

Figure 9-2 An overview of the eukaryotic cell cycle. This diagram of the cell cycle indicates the stages through which a cell passes from one division to the next. The cell cycle can be divided into two stages: M phase and interphase (From: Karp G. 2005)

Figure 9-3 The eukaryotic cell cycle is comprised of four successive phases. Nuclear division and then cytoplasmic division occur in M phase. Interphase is divided into three phases: G_1, S and G_2 phases. The cell grows continuously during interphase but stops growing during M phase. (From: Alberts B, et al. 2002)

Although M phase is the period when the contents of a cell are actually divided, numerous preparations for an upcoming mitosis occur during interphase, including replication of the DNA of each chromosome. Studies in the early 1950s on asynchronous cultures showed that DNA replication occurs during a defined period of the cell cycle. And this specific period is called S phase (DNA synthesis phase). Investigations of this nature show two periods during interphase when no DNA synthesis occurs, one before and one after S phase, which are designated G_1 (gap I) and G_2 (gap II), respectively. During both of these intervals, as well as during S, intensive metabolic activity, cell growth, and cell differentiation. By the end of G_2, the volume of the cell has roughly doubled, DNA has been replicated, and mitosis (M phase) is initiated. Following mitosis, continuously dividing cells then repeat this cycle (G_1, S, G_2, M) over and over (Figure 9-3).

Analysis of a large variety of cells has revealed great variability in the duration of the cell cycle. While M phase usually lasts only an hour or so in mammalian cells, interphase may extend for days, weeks and even longer, which depend on the cell type and the environmental conditions. For example, cell cycles can range from as short as 30 min in a cleaving frog embryo, whose cell cycles lack both G_1 and G_2 phases, and the time of the cycle occupies several months in slowly growing tissues, such as the mammalian liver. G_1 is the most variable among four stages of cell cycle. While the total time of S, G_2 and M phases is relatively fixed, with a few notable exceptions. For example, cell that has stopped dividing, whether temporarily or permanently, whether in the body or in culture, are present in a stage preceding the initiation of DNA synthesis. Cells that are arrested in this state are described as be in the G_0 (G zero) stage. Once the G_0 phase cells are activated by suitable extracellular conditions and signals, they will return to cell cycle and continue through mitosis.

9.2 Regulation of the cell cycle

The events of the cell cycle must occur in a particular sequence, and this sequence must be preserved even if one of the steps takes longer than usual. To coordinate these activities, eukaryotic cells have evolved a complex network of regulatory proteins, known as the cell-cycle control system. All of the nuclear DNA, for example, must be replicated before the nucleus begins to divide. If DNA synthesis is slowed down or stalled, mitosis and cell division will also be delayed. The same events also occur at other stages of the cell cycle. It is the cell-cycle control system that ensures correct progression through the cell cycle by regulating the cell-cycle machinery.

For many years, it was not even clear whether there was a separate control system, or whether the processes of DNA synthesis, mitosis, and cytokinesis somehow controlled themselves. A major breakthrough came in the late 1980s with the identification of the key proteins of the control system, along with the realization that they are distinct from the proteins that perform the processes of DNA replication, chromosome segregation, and so on. In this case, a central controller triggers each process in a set sequence (Figure 9-4).

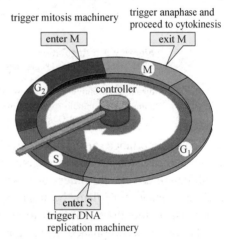

Figure 9-4 The control of the cell cycle.
The essential processes of the cell cycle — such as DNA replication, mitosis, and cytokinesis are triggered by a cell-cycle control system. By analogy with a washing machine, the cell-cycle control system is shown here as a central arm — the controller — that rotates clockwise, triggering essential processes when it reaches specific points on the outer dial (From: Alberts B, et al. 2002)

9.2.1 Cell cycle control by cell growth and extracellular signals

The progression of cells through the division cycle is regulated by **extracellular signals** from the environment, as well as by internal signals that monitor and coordinate the various processes that take place during different cell cycle phases. An example of cell cycle regulation by extracellular signals is provided by the effect of growth factors on animal cell proliferation. In addition, different cellular processes, such as cell growth, DNA replication, and mitosis, all must be coordinated during cell cycle progression. This is accomplished by a series of control points that regulate progression through various phases of the cell cycle.

A major cell cycle regulatory point in many types of cells occurs late in G_1 and controls progression from G_1 to S. This regulatory point was first defined by studies of budding yeast *S. cerevisiae*, where it is known as **START** (Figure 9-5). Once cells have passed START, they are committed to entering S phase and undergoing one cell division cycle. However, passage through START is a highly regulated event in the yeast cell cycle, where it is controlled by external signals, such as the availability of nutrients and cell size.

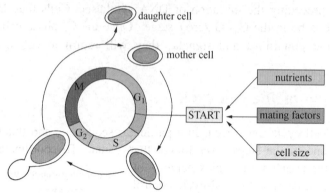

Figure 9-5 Regulation of the cell cycle of budding yeast. Buds form just after START and continue growing until they separate from the mother cell after mitosis. The daughter cell formed from the bud is smaller than the mother cell and therefore requires more time to grow during the G_1 phase of the next cell cycle. Although G_1 and S phases occur normally, the mitotic spindle begins to form during S phase, so the cell cycle of budding yeast lacks a distinct G_2 phase (From: Alberts B, et al. 2002)

The importance of START regulation is particularly evident in budding yeasts, in which cell division produces progeny cells of very different sizes: a large mother cell and a small daughter cell. In order for yeast cells to maintain a constant size, the small daughter cell must grow more than the large mother cell does before they divide again. Thus, cell size must be monitored in order to coordinate cell growth with other cell cycle events. This regulation is accomplished by a control mechanism that requires each cell to reach a minimum size before it can pass START. Consequently, the small daughter cell spends a longer time in G_1 and grows more than the mother cell.

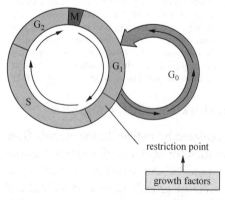

Figure 9-6 Regulation of animal cell cycles by growth factors. The availability of growth factors controls the animal cell cycle at the restriction point. If growth factors are not available during G_1, the cells enter a quiescent stage of the cycle called G_0 (From: Alberts B, et al. 2002)

The proliferation of most animal cells is similarly regulated in the G_1 phase. In particular, a decision point in late G_1, called the **restriction point** in animal cells, functions analogously to START in yeasts (Figure 9-6). In contrast to yeasts, however, the passage of animal cells through the cell cycle is regulated primarily by the extracellular growth factors that signal cell proliferation, rather than by the availability of nutrients. In the presence of the appropriate growth factors, cells pass the restriction point and enter S phase. Once it has passed through the restriction point, the cell is committed to proceed through S phase and the rest of the cell cycle, even in the absence of further growth factor stimulation. On the other hand, if appropriate growth factors are not available in G_1, progression through the cell cycle stops at the restriction point. Such arrested cells then enter a quiescent stage of the cell cycle called G_0, in which they can remain for long periods of time without proliferating. G_0 cells are metabolically active, although they cease growth and have reduced rates of protein synthesis. As already noted, many cells in animals remain in G_0 unless called on to go back to G_1 phase by appropriate growth factors or other extracellular signals. For example, skin fibroblasts are arrested in G_0 until they are stimulated to divide as required to repair damage resulting from a wound. The proliferation of these cells is triggered by platelet-derived growth factor, which is released from blood platelets during clotting and signals the proliferation of fibroblasts in the vicinity of the injured tissue.

Although the proliferation of most cells is regulated primarily in G_1, some cell cycles are instead controlled principally in G_2. For example, vertebrate oocytes can remain arrested in G_2 for long periods of time (several decades in humans) until their progression to M phase is triggered by hormonal stimulation. Extracellular signals can thus control cell proliferation by regulating progression from the G_2 to M as well as the G_1 to S phases of the cell cycle.

9.2.2 Cell cycle checkpoints

The controls discussed in the previous section regulate cell cycle progression in response to cell size and extracellular signals, such as nutrients and growth factors. In addition, the events that take place during different stages of the cell cycle must be coordinated with one another so that they occur in the appropriate order. For example, it is critically important that the cell not begin mitosis until replication of the genome has been completed. The alternative would be a catastrophic cell division, in which the daughter cells failed to inherit complete copies of the genetic material. In most cells, this coordination between different phases of the cell cycle is dependent on a system of **checkpoints** and feedback controls that prevent entry into the next phase of the cell cycle until the events of the preceding phase have been completed.

At least three major checkpoints exist within the cell cycle, where the cell is monitored or "checked" before it can proceed to the next stage of the cycle to ensure that incomplete or damaged chromosomes are not replicated and passed on to daughter cells (Figure 9-7).

The first is G_1/S checkpoint, which monitors the cell size achieved following the previous mitosis, and whether the DNA has been damaged. If the cell has not achieved an adequate size, or if the DNA has been damaged, further progress through the cycle is arrested until these conditions are "corrected", so to speak. If both conditions are initially "normal", then the checkpoint is traversed and the cell proceeds to the S phase of the cycle.

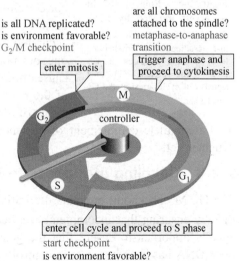

Figure 9-7 Checkpoints in the cell-cycle control system. Information about the completion of cell-cycle events, as well as signals from the environment, can cause the control system to arrest the cycle at specific checkpoints. The most prominent checkpoints occur at locations marked with yellow boxes (From: Alberts B, et al. 2008)

An interesting related finding involves the protein product of the *p53* gene in humans and its involvement during scrutiny at the G_1/S checkpoint. This protein functions during the regulation of apoptosis, the genetic process whereby programmed cell death occurs. In mammalian cells, arrest at the G_1/S checkpoint is mediated by the action of the *p53* gene protein, which is rapidly induced in response to damaged DNA (Figure 9-8). Interestingly, the gene encoding *p53* is frequently mutated in human cancers. Loss of *p53* function as a result of these mutations prevents G_1 arrest in response to DNA damage, so the damaged DNA is replicated and passed on to daughter cells instead of being repaired.

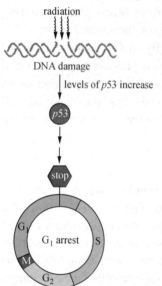

Figure 9-8 Role of *p53* in G_1 arrest induced by DNA damage. DNA damage, such as that resulting from irradiation, leads to rapid increases in *p53* levels. The protein P53 then signals cell cycle arrest at the G_1 checkpoint.

The second checkpoint is the G_2/M checkpoint, where physiological conditions in the cell are monitored prior to entering mitosis. It prevents the initiation of mitosis until DNA replication is completed. This checkpoint also checks whether centrosome has finished duplication. This G_2/M checkpoint senses unreplicated DNA and centrosomes, which generates a signal that leads to cell cycle arrest. Operation of the G_2/M checkpoint therefore prevents the initiation of M phase before completion of S phase, so cells remain in G_2 until the genome has been completely replicated. Progression through the cell cycle is also arrested at the G_2/M checkpoint in response to DNA damage after replication, such as that resulting from irradiation. This arrest allows time for the damage to be repaired, rather than being passed on to daughter cells.

The final checkpoint occurs during mitosis and is called the M checkpoint. It maintains the integrity of the genome occurs toward the end of mitosis. This checkpoint monitors the correct alignment of chromosomes on the mitotic spindle, thus ensuring that a complete set of chromosomes is distributed accurately to the daughter cells. For example, the failure of one or more chromosomes to align properly on the spindle causes mitosis to arrest at metaphase, prior to the segregation of the newly replicated chromosomes to daughter nuclei. As a result of this checkpoint, the chromosomes do not separate until a complete complement of chromosomes has been organized for distribution to each daughter cell.

9.2.3 Coupling of S phase to M phase

The G_2/M checkpoint prevents the initiation of mitosis prior to the completion of S phase, thereby ensuring that incompletely replicated DNA is not distributed to daughter cells. It is equally important to ensure that the genome is replicated only once per cell cycle. Thus, once DNA has been replicated, control mechanisms must exist to prevent initiation of a new S phase prior to mitosis. These controls prevent cells in G_2 from reentering S phase and block the initiation of another round of DNA replication until after mitosis, at which point the cell has entered the G_1 phase of the next cell cycle.

Initial insights into this dependence of S phase on M phase came from cell fusion experiments carried out by Potu Rao and Robert Johnson of the University of Colorado in 1970 (Figure 9-9). These investigators isolated cells in different phases of the cycle and then fused these cells to each other to form cell hybrids. When interphase phases cells fused with M phase cells, different shapes of chromosomes condensation have found in the fused cells, which called premature chromosome condensation (PCC). It thus appeared that there was a kind of factor which can induce chromosome to condense. This factor was named as maturation promoting factor (mitosis promoting factor).

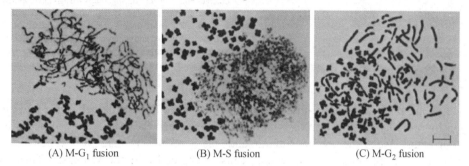

(A) M-G_1 fusion (B) M-S fusion (C) M-G_2 fusion

Figure 9-9 Cell fusion experiments demonstrating the maturation promoting factor in M phase. When G_1 cells were fused with M phase cells, the G_1 nucleus immediately began to condense as single line chromosomes (A) In contrast, when S cells were fused with M phase cells, the chromosomes look like powder (B) The fusion of G_2 phase cells with M phase cells, the G_2 nucleus condensed to double-line chromosomes. All these premature chromosomes condensations were induced by maturation promoting factor in M phase cells.

9.2.4 The role of cyclins

While the cell fusion experiments revealed the existence of some factors that regulated the whole cell cycle, they provided no information about the biochemical properties of these factors. Insights into the nature of the agents that trigger DNA replication and promote a cell into **mitosis** (or **meiosis**) were first gained in a series of experiments on the oocytes and early embryos of frogs and invertebrates. Results of these experiments showed that entry of a cell into M phase is initiated by a protein kinase called MPF. MPF was originally called the maturation-promoting factor. It consists of two subunits: ① a subunit with kinase activity that transfers phosphate group from ATP to specific Ser and Thr residues of specific protein substrates; ② a regulatory subunit called cyclin (Figure 9-10A). The term "cyclin" was coined because the concentration of this regulatory protein rises and falls in a predictable pattern with each cell cycle progresses (Figure 9-11).

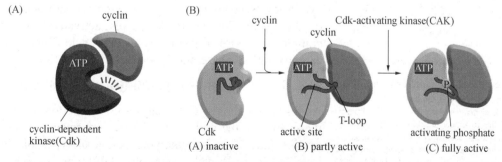

Figure 9-10 The structural basis of the cyclin-dependent protein kinases. (A) Cdk is a complex formed by a regulated subunit called cyclin and a catalytic subunits having kinase activity. Without cyclin, Cdk is inactive. (B) The active site of Cdk2 is blocked by a region of the protein called the T-loop without cyclin bound. When cyclin binds to the Cdk, the T-loop move out of the active site, causing the partial activation of the Cdk2. After the phosphorylation of Cdk2 by CAK at a threonine residue in the T-loop, the enzyme change its shape and the fully activation was aroused (From: Alberts B, et al. 2008)

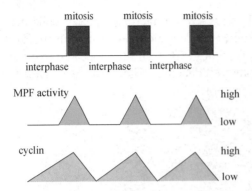

Figure 9-11 **Fluctuation of cyclin and MPF levels during the cell cycle.** This drawing depicts the cyclical changes that occur during early frog development when mitotic divisions occur synchronously in all cells of the embryo. The top tracing shows the alternation between periods of mitosis and interphase, the middle tracing shows the cyclical changes in the concentrations of cyclins that control the relative activity of the MPF kinase (From: Karp G. 2005)

Cyclins are proteins that accumulate continuously throughout the cell cycle and are then destroyed by proteolysis during mitosis. For example, when the cyclin concentration is low, the protein kinase lacks the cyclin subunit and, as a result, is inactive. When the cyclin concentration rises, the kinase is activated, causing the cell to enter M phase. These results suggested that: ①progression of cells into mitosis depends on an enzyme whose sole activity is to phosphorylate other proteins; ②the activity of this enzyme is controlled by a subunit whose concentration varies from one stage of the cell cycle to another. The cyclin that helps drive cells into M phase is called M-cyclin, including cyclin A, cyclin B etc. Synthesis of M-cyclin starts immediately after cell division continues steadily throughout interphase and carries out their regulatory functions in M phase. Whereas, the cyclin that becomes active toward the end of G_1 phase and is responsible for driving the cell into S phase is called G_1-cyclin, for example, cyclin C and cyclin D.

9.2.5 Cyclin-dependent kinase

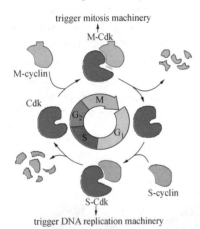

Figure 9-12 **A simplified view of the core of the cell cycle control system.** Cdk associates successively with different cyclins to trigger the different events of the cycle. Cdk activity is usually terminated by cyclin degradation. For simplicity, only the cyclins that act in S phase (S-cyclin) and M phase (M-cyclin) are shown, and they interact with a single Cdk; as indicated, the resulting cyclin-Cdk complexes are referred to as S-Cdk and M-Cdk, respectively (From: Alberts B, et al. 2002)

Over the past two decades, a large number of laboratories have focus on MPF-like enzymes, called **cyclin-dependent kinases (Cdks)** which are composed of cyclins and cell-cycle kinases. It has been found that Cdks are not only involved in M phase, but are the key agents that orchestrate activities throughout the cell cycle. Cyclins have no enzymatic activity themselves, but they have to bind to the cell-cycle kinases before the kinases can become enzymatically active. The cyclical changes in cyclin levels result in the cyclin assembly and activation of the cyclin-Cdk complexes; this activation of these complexes in turn triggers various cell-cycle events, such as entry into S phase and M phase (Figure 9-12).

9.2.6 Progression regulatory of cell cycle

As discussed above, Cdks plays a core function in timing the events of the cell cycle. There are several types of cyclins and, in most eukaryotes, several types of Cdks involve in cell-cycle control. Different cyclin-Cdk complexes trigger different steps of the cell cycle. Research on the genetic control of the cell cycle in yeast began in the 1970s in two laboratories, Leland Hartwell at the University of Washington and Paul Nurse at the

University of Oxford. Both laboratories discovered a gene that, when mutated, would cause the cycle of cells at elevated temperature to stop at certain points. In fission yeast, the first found product of this gene was called *cdc2* or Cdk1. Its sequence was eventually found to be homologous to that of the catalytic subunit of MPF. The *cdc* is an abbreviation for "cell division cycle". Subsequent studies on yeast and many different mammalian cells have supported the concept that the progression of a eukaryotic cell through its cell cycle is regulated at distinct stages. The cell cycle control system achieves each step by means of molecular brakes that can stop the cycle at various checkpoints.

The first transition point, i. e. START in fission yeast, occurs before the end of G_1. Passage through START requires the activation of *cdc2* by one or more G_1-cyclins, whose levels rise during late G_1 (Figure 9-13). Activation of *cdc2* by these cyclins leads to the initiation of replication at sites where prereplication complexes had previously assembled. The transition from G_2 to mitosis requires activation of *cdc2* in fission yeast. Cdks containing an M-cyclin phosphorylate substrates that are required for the cell to enter mitosis. The concentration of each type of cyclin rises gradually, and then falls sharply, at a specific time in the cell cycle as a result of degradation by the ubiquitin pathway. The increase in concentration helps to activate the appropriate Cdk partner, while the rapid falls returns the Cdk to its inactive state. Exit from mitosis and entry into G_1 depends on a rapid decrease in Cdk activity that results from a plunge in concentration of the M-cyclins (Figure 9-13).

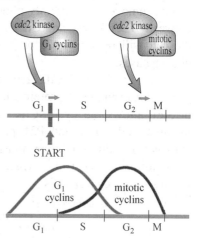

Figure 9-13 **A model for cell cycle regulation in fission yeast.** The cell cycle is controlled primarily at two points, START and the G_2 to M transition. In fission yeast, cyclins can be divided into two groups, G_1 cyclins and M cyclins. Passage of a cell through these two critical junctures requires the activation of the same *cdc2* kinase by a different kind of cyclin(From: Karp G. 2005)

Cyclin-dependent kinases are often described as the "engines" that drive the cell cycle through its various stages. The activities of these enzymes are regulated by a variety of "brakes" and "accelerators" that operate in combination with one another. These include:

(1) *Cyclin binding*　A rise and fall of cyclin levels plays an important part in regulating Cdk activity during the cell cycle. When a cyclin is present in the cell, it binds to the catalytic subunit of the Cdk partner. The binding causes a major change in the conformation of the catalytic subunit and the movement of a flexible loop of the Cdk polypeptide chain away from the opening to the enzyme's active site, allowing the Cdk to phosphorylate its protein substrates (Figure 9-10B).

(2) *Cdk phosphorylation state*　Activity of Cdk is also regulated by phosphorylation and dephosphorylation. The cyclin-Cdk complex to be active it has to be phosphorylated at one or more sites by a specific protein kinase, and be phosphorylated at other sites by a specific protein phosphatase. This change mainly includes following steps. In step 1, one of the kinases, called CAK (Cdk-activiting kinase), phosphorylates a critical Thr residue (Thr 161 in Figure 9-14). Phosphorylation of this residue is necessary, but not sufficient, for the Cdk to be active. A second protein kinase shown in step 1, called Wee1, phosphorylates a Thr residue and a Tyr residue in the enzyme (Thr14 and Tyr 15 in Figure 9-14A). If these two residues are phosphorylated, the enzyme is inactive, regardless of the phosphorylation state of any other residue. In other words, the effect of Wee1 overrides the effect of CAK,

keeping the Cdk in an inactive state. Line 2 of figure 9-14B shows the phenotype of the cells with a mutant *wee1* gene. These mutants cannot maintain the Cdk in an inactive state to divide at an early stage in the cell cycle producing smaller cells, hence the name "wee." In normal cells, Wee1 keeps the Cdk inactive until the end of G_2. Then, at the end of G_2, the inhibitories of phosphate at Thr14 and Tyr15 are removed by the third enzyme, a phosphatase named Cdc25 (step 3, Figure 9-14). Removal of this phosphate switches the stored cyclin-Cdk molecules into the active state, driving the yeast cell into mitosis. Line 3 of figure 9-14B shows the phenotype of cells with a mutant *Cdc25* gene. These mutants cannot remove the inhibitory phosphate from the Cdk and cannot enter mitosis. The balance between Wee1 kinase and Cdc25 phosphatase activities, which normally determines whether the cell will remain in G_2 or progress into mitosis, is regulated by still other kinases and phosphatases.

(3) *Cdk inhibitors* Cdk activity can be blocked by a variety of inhibitors. In budding yeast, for example, a protein called Sic1 acts as a Cdk inhibitor during G_1. The degradation of Sic1 allows the cyclin-Cdk that is present in the cell to initiate DNA replication. In mammal cells, P27 and P21 proteins can suppress G_1/S-Cdk and S-Cdk activities when cells get into differentiation and DNA damage respectively. The protein P16 is also a kind of inhibitor for G_1-Cdk, which can prevents cells over-dividing. So P16 is frequently inactivated in cancer cells.

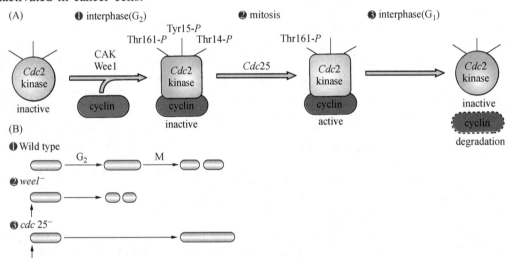

Figure 9-14 **Progression through the fission yeast cell cycle requires the phosphorylation and dephosphorylation of critical Cdc2 residues** (From: Karp G. 2005)

(4) *Controlled proteolysis* From figures 9-11 and 9-13 we can see that the variety of the cyclin concentration leads to change in the activity of Cdk. Cells regulate the concentration of cyclins and other key cell cycle proteins by adjusting both the rate of synthesis and destruction at different points of the cell cycle. Degradation is accomplished by means of the ubiquitin-proteasome pathway. Regulation of the cell cycle requires two important classes of multisubunit complexes, SCF and APC complexes. These complexes recognize proteins to be degraded and link these proteins to a polyubiquitin chain, which ensures their destruction in a proteasome. The SCF complex is active from late G_1 through early mitosis and mediates the destruction of G_1 cyclins, Cdk inhibitors, *et al* (Figure 9-15A). These proteins become targets for an SCF after they are phosphorylated by the protein kinases. The subunits called

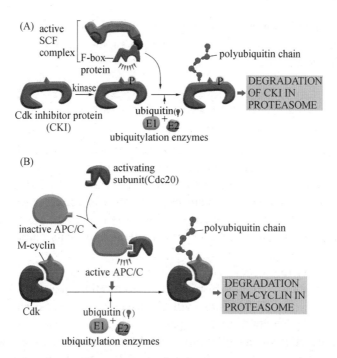

Figure 9-15 Control of proteolysis by SCF and APC complex during the cell cycle
(From: Alberts B, et al. 2008)

F-box protein help SCF recognize the target proteins and then SCF catalyzes ubiquitylation of the targets. For example, SCF ubiquitylates a Cdk inhibitor protein (CKI) in late G_1 stage to help control DNA replication and cells entering into S phage. APC is an abbreviation for "anaphase promoting complex". The activity of APC acts in mitosis and degrades a number of key mitotic proteins, including the mitotic cyclins. When M-Cdk accumulates in M phage, it will stimulate Cdc20 activity. The Cdc20 protein, a subunit of APC complex triggers initial activation of APC to degrade M-Cdk kinases. Destruction of the mitotic cyclins allows a cell to exit mitosis and enter a new cell cycle (Figure 9-15B).

As noted above, the proteins and processes that control the cell cycle are remarkably conserved among eukaryotes. As in yeast, successive waves of synthesis and degradation of different cyclin play a key role in driving mammalian cells from one stage to the next. The mammalian cells have several different versions of this protein kinase. The pairing between individual cyclin and Cdks is highly specific, and only certain combinations are found (Figure 9-16). For example, the activity of a cyclin E-Cdk2 complex triggers the cell into S phase, whereas activity of a cyclin B1-Cdk1 complex drives the cell into mitosis. Cdks do not always stimulate activities, but can also inhibit inappropriate events. For example, cyclin B1-Cdk1 activity

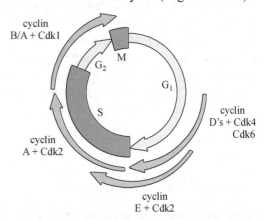

Figure 9-16 Combinations between various cyclins and cyclin-dependent kinases at different stages in the mammalian cell cycle (From: Karp G. 2005)

during G_2 prevents a cell from re-replicating DNA that has already been replicated earlier in the cell cycle. This helps ensuring that each region of the genome is replicated once and only once per cell cycle.

9.3 Cell division

According to the third tenet of the cell theory, new cells originate only from other living cells. The process by which this occurs is called **cell division**. For a multicellular organism, such as a human or an oak tree, countless divisions of a single-cell zygote produce an organism of astonishing cellular complexity and organization. Cell division does not stop with the formation of the mature organism but continues in certain tissues throughout life. Millions of cells residing within the marrow of your bones or the lining of your intestinal tract are undergoing division at this very moment. This enormous output of cells is needed to replace cells that have aged or died.

Although cell division occurs in all organisms, it takes place very differently in prokaryotes and eukaryotes. We will restrict discussion to the eukaryotic version. Here, two distinct types of eukaryotic cell division will be discussed. Mitosis leads to production of cells that are genetically identical to their parent, whereas meiosis leads to production of cells with half the genetic content of the parent. Mitosis serves as the basis for producing new cells, meiosis as the basis for producing sexually reproducing organisms. Together, these two types of cell division form the links in the chain between parents and their offspring and, in a broader sense, between living species and the earliest eukaryotic life forms present on Earth.

9.3.1 Mitosis

Mitosis is a process of nuclear division in which replicated DNA molecules of each chromosome are faithfully partitioned into two nuclei. The cell division that follows is called **cytokinesis**, the separation of the two cells by division of the plasma membrane. The two daughter cells resulting mitosis and cytokinesis possess a genetic content identical to each other and to the mother cell from which they arose. Therefore mitosis maintains a constant amount of genetic material from cell generation to cell generation. Mitosis can take place in either haploid or diploid cells. Although mitosis proceeds as a continuous sequence of events, it is generally divided into six stages: prophase, prometaphase, metaphase, anaphase, telophase and cytokinesis, each characterized by a particular series of events (Figure 9-17). We must keep in mind that the division of mitosis into arbitrary phases is done only for the sake of discussion and experimentation.

(1) The replicated chromosomal material condenses to form compact chromosomes, consisting of two associated sister chromatids
(2) Cytoskeleton is disassembled and mitotic spindle is assembled
(3) Golgi complex and ER fragment
(4) Nuclear envelope disperses

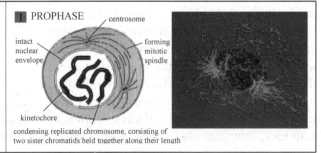

PROPHASE
centrosome
intact nuclear envelope
forming mitotic spindle
kinetochore
condensing replicated chromosome, consisting of two sister chromatids held together along their length

Chapter 9 Cell cycle and cell division

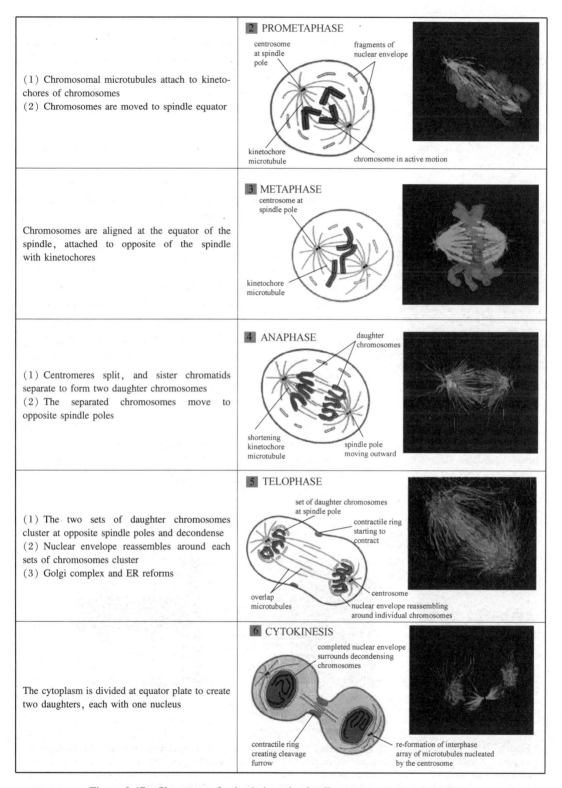

Figure 9-17 Six stages of mitosis in animal cells (From: Albert B, et al. 2008)

(1) **Prophase** During the first stage of mitosis, **prophase**, the duplicated chromosomes

are prepared for segregation, and the mitotic machinery is assembled.

1) Formation of the mitotic chromosome: The extended state of interphase replicated begin to condense and to be converted into much shorter, thicker structures by a remarkable process of **chromosome compaction** (or **chromosome condensation**) , and they are easily visible in the light microscope as chromosomes (Figure 9-17).

As the result of compaction, the chromosomes of mitotic cell appear as distinct, rod-like structures. Each duplicated chromosome is seen as a pair of sister chromatids joined by the duplicated but unseparated centromere (Figure 9-18). Sister chromatids are a result of replication in the previous interphase.

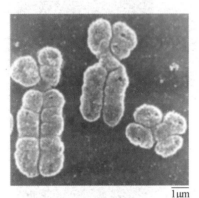

Figure 9-18 Each mitotic chromosome is comprised of a pair of sister chromatids connected to one another by the protein complex cohesin (From: Karp G. 2005)

2) Centromeres and kinetochores: The most notable landmark on a mitotic chromosome is an indentation or primary constriction, which marks the position of the **centromere** (Figure 9-18). The centromere is the residence of highly repeated DNA sequences (satellite DNA) that serve as the binding sites for specific proteins. Examination of sections through a mitotic chromosome reveals the presence of a proteinaceous, button-like structure, called **kinetochore**, at the outer surface of the centromere of each sister-chromatid (Figure 9-19). The kinetochore assembles on the centromere during prophase, and they form the centromere-kinetochore complex. As will be apparent shortly, the kinetochore functions as: ①the site of attachment of the chromosome to the dynamic microtubules of the mitotic spindle (Figure 9-19A, B); ②the residence of several motor proteins involved in chromosome motility (Figure 9-19C); ③a key component in the signaling pathway of an important checkpoint.

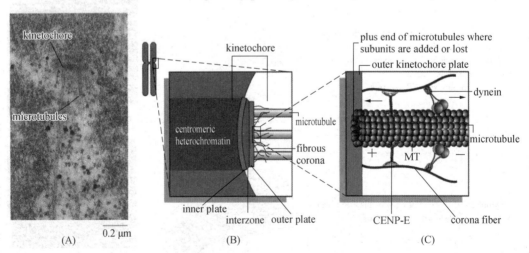

Figure 9-19 The kinetochore. (A) Electron micrograph of a section through the kinetochore of a mammalian metaphase chromosome, showing its three-layered structure. (B) The kinetochore contains an electron-dense inner and outer plate separated by a lightly staining interzone. The inner plate is thought to consist of a specialized layer of chromatin that is attached to the centromeric heterochromatin of the chromosome, whereas the outer plate contains proteins that are thought to the plus end of microtubules. (C) A schematic model of the disposition of motor proteins at the outer surface of the kinetochore. These motors play a role in tethering the microtubule to the kinetochore. (From: Karp G. 2005)

3) **Formation of the mitotic spindle**: As discussed earlier, we knew how microtubule assembly in animal cells is initiated by a special microtubule-organizing structure, the centrosome. As a cell progresses past G_2 and into mitosis, the microtubules of the cytoskeleton undergo sweeping disassembly in preparation for their reassembly as components of a complex, microtubule-containing "machine" called the **mitotic spindle**. The rapid disassembly of the interphase cytoskeleton is thought to be accomplished by the inactivation of proteins that stabilize microtubules (e.g., **microtubule-associated proteins**, or **MAPs**) and the activation of proteins that destabilize these polymers.

As DNA replication begins in the nucleus at the onset of S phase, each centriole duplicates its centrosome to produce two daughter centrosomes, which initially remain together at one side of the nucleus. At the beginning of prophase, the two daughter centrosomes separate (Figure 9-20). They organize their own array of microtubules and begin to move to opposite poles of the cell and gradually form the framework of the mitotic spindle. The two centrosomes that give rise to these microtubules are called **spindle poles**. The interacting microtubules are called **interpolar microtubules**, the microtubules with free plus ends are called **astral microtubules**, and the microtubules which catch chromosomes are called **kinetochore microtubules**.

4) **Dissolution of nuclear envelope and the partitioning of cytoplasmic organelles**: The nuclear envelope disappears at the end of prophase. Meanwhile, this signals the beginning of the next stage of cell cycle called prometaphase. The disassembly of the nuclear lamina, which constitutes the inner layer of the nuclear envelope, is promoted by phosphoralation of the lamin molecules by the mitotic Cdk kinase (M-Cdk). The phosphoralation phenomenon leads to disassembly of the nuclear lamina and the nuclear envelope is fragmented into a population of small vesicles (nuclear lamina A) or soluble monomer proteins (nuclear lamina B) that disperse throughout the mitotic cell. During the process of mitosis, the Golgi membranes also

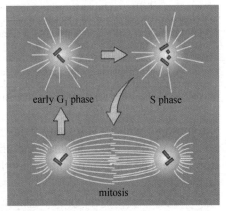

Figure 9-20 **The centrosome cycle of an animal cell** (From: Karp G. 2005)

become fragmented to form a distinct population of small vesicles that are divided between daughter cells. It is generally held that the endoplasmic reticulum undergoes fragmentation during prophase to give rise to large numbers of vesicles that become dispersed throughout the cytoplasm.

(2) **Prometaphase** Upon nuclear envelope breakdown, the cell enters the second phase of mitosis, **prometaphase**, during which mitotic spindle assembly is formed and the chromosomes attach to microtubules in the spindle via their kinetochores and move into the center of the cell (Figure 9-17). At the beginning of prometaphase, compacted chromosomes are scattered throughout the space that was the nuclear region. As the microtubules of the spindle penetrate into the central region of the cell, the free (plus) ends of the microtubules are seen to grow and shrink in a highly dynamic fashion, searching for a chromosome. Once attached, the chromosomes start to align along the metaphase plate in the center of the spindle.

The spindle microtubules end up attached to the chromosomes through specialized

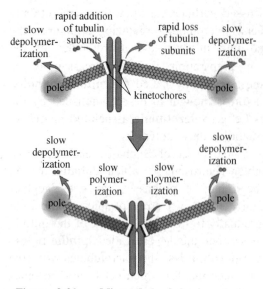

Figure 9-21 Microtubule behavior during formation of the metaphase plate. Initially, the chromosome is connected to microtubules from opposite poles that are very different in length. As prometaphase continues, this imbalance is corrected as the result of the shortening of microtubules from one pole, due to the rapid loss of tubulin subunits at the kinetochore, and the lengthening of microtubules from the opposite pole, due to the rapid addition of tubulin subunits at the kinetochore (From: Karp G. 2005)

protein complexs called kinetochores, which assemble on each centromere of chromosome. Once the nuclear envelope has broken down, the microtubule encounters a chromosome and captures it. As kinetochores on sister chromatids face in opposite directions, they tend to attach to microtubules from opposite poles of the spindle. So, the two sister chromatids of each mitotic chromosome ultimately become connected by their kinetochores to microtubules that extend to form opposite poles. With the chromosomes moving toward the center of the mitotic spindle, the longer microtubules are shortened, while the shorter microtubules are elongated. Shortening and elongation of microtubules carry out primarily by loss or gain of subunits at the plus end of the microtubule (Figure 9-21). Remarkably, this dynamic activity occurs while the plus end of each microtubule remains attached to a kinetochore. The motor proteins located at the kinetochore (Figure 9-19C) may play a key role in tethering the microtubule to the chromosome while it loses and gains subunits.

Eventually, each chromosome aligns at the center of the spindle, so that microtubules from each pole are equivalent in length. It marks the end of prometaphase and the beginning of metaphase.

(3) Metaphase Metaphase is characterized by the lining up of the chromosomes along the equator of the spindle, midway between the spindle poles (Figure 9-22). The plane of alignment of the chromosomes at metaphase is called the **metaphase plate**. The paired kinetochore microtubules on each chromosome attach to opposite poles of the spindle. The mitotic spindle of the metaphase cell contains a highly organized array of microtubules that is ideally suited for separating the duplicated chromatids positioned at the center of the cell.

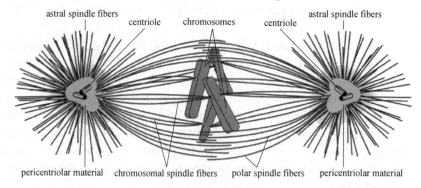

Figure 9-22 The mitotic spindle of an animal cell during metaphase. Each spindle pole contains a pair of centrioles surrounded by amorphous pericentriolar material at which the microtubules are nucleated. All of the spindle microtubules have their minus ends at centrosome (From: Karp G. 2005).

In animal cells, the microtubules of the metaphase spindle can be functionally divided into three types (Figure 9-23).

1) **Astral microtubules**: That radiate outward from the **centrosome** into the region outside the body of the spindle. They contact the cell cortex and help position the spindle apparatus in the cell.

2) **Chromosomal (or kinetochore) microtubules**: That stretch from the centrosome to the kinetochores of the chromosomes. During metaphase, the chromosomal microtubules exert a pulling force on the kinetochore. As a result, the chromosomes are maintained in the equatorial plane by a "tug-of-war" between balanced pulling forces exerted by chromosomal spindle fibers from opposite poles. During anaphase, chromosomal microtubules are required for the movement of the chromosomes toward the poles.

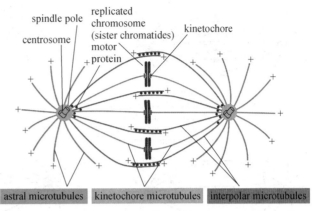

Figure 9-23   Three classes of microtubules make up the mitotic spindle. Schematic drawing of a spindle with chromosomes attached, showing the three types of spindle microtubules: astral microtubules, kinetochore microtubules, and inter polar microtubules (From: Alberts B, et al. 2008)

3) **polar microtubules**: That extend from the centrosome past the chromosomes. Inter polar microtubules from one centrosome overlap with their counterparts from the opposite centrosome. The inter polar microtubules form a structural basket that maintains the mechanical integrity of the spindle.

(4) Anaphase   The paired chromatids synchronously separate to form two daughter chromosomes at the onset of anaphase, and the chromatids begin their poleward migration (Figure 9-17). The movement of each chromosome toward a pole results from the shortening of the chromosomal microtubules attached to its kinetochore. As the chromosome moves during anaphase, its centromere is seen at its leading edge with the arms of the chromosome trailing behind. Anaphase can be divided into two continuous stages, anaphase A and anaphase B. In anaphase A, the kinetochore microtubules are shortened by depolymerization, and the attached chromosomes move poleward. In anaphase B, after the sister chromatids have separated, the polar microtubules are gradually elongated, and the two spindle poles move farther apart. At the same time, the astral microtubules at each spindle pole pull the poles away from each other (Figure 9-24).

In anaphase A, the driving force for the movements of each chromatid is provided mainly by the action of microtubule motor proteins. The loss of tubulin subunits at the kinetochore depends on a catastrophin that is bound to both the microtubule and the konetochore. This process uses the energy of ATP hydrolysis to remove tubulin subunits from the microtubule.

In anaphase B, the driving forces for pushing the spindle poles and the sister chromatids farther apart are thought to be from two sets of motor proteins-members of the **kinesin** and **dynein** families. The dyneins which interact with kinetochore microtubules and astral microtubules walk towards the centrosome to help the sister chromatids separate each other and fix the centrosomes at the two poles of the cell. The elongation of the interpolar

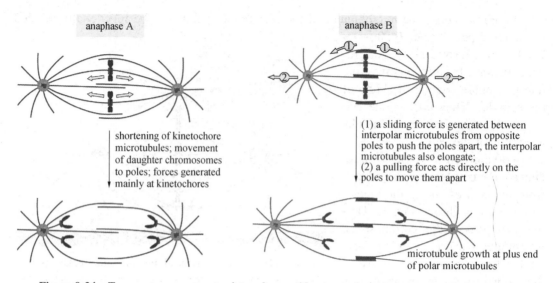

Figure 9-24 Two processes separate sister chromatids at anaphase (From: Alberts B, et al. 2008)

microtubules during anaphase B is accompanied by the net addition of tubulin subunits to the plus ends of the polar microtubules. Thus, the kinesins connecting two interpolar spindles move to the plus ends of the microtubules and push the poles apart each other (Figure 9-25).

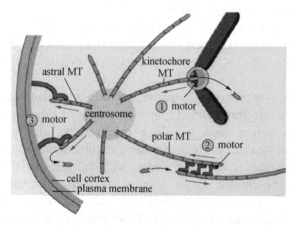

Figure 9-25 Microtubules and motors in the spindle. In anaphase A, ①motors (dyneins) move to the minus ends of kinetochore microtubules and help the sister chromatids separate apart. In anaphase B, ② motors (kinesins) walk to the plus ends of interpolar microtubules and push the ploes to be far away each other. ③motors (dyneins) which is on the astral MT will fix two centrosomes at the poles of the cells.

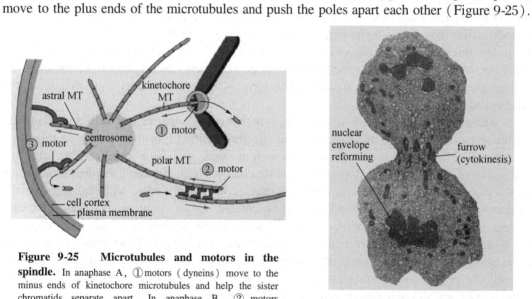

Figure 9-26 Telophase. Elcectron micrograph of a section through an ovarian granulose cell in telophase (From: Karp G. 2005)

(5) Telophase   During telophase, chromatids arrive at opposite poles of cell, and new membranes form around the daughter nuclei. The nuclear envelope reforms and the chromosomes become disperse and are no longer visible under the light microscope. The division of the cytoplasm begins with the assembly of the contractile ring (Figures 9-17, 9-26). Meanwhile, the spindle fibers disperse and the kinetochore microtubules begin to disintegrate at this stage.

(6) Cytokinesis

1) Cytokinesis in animal cells: The final stage in the process of cell division is known as **cytokinesis**, which usually begins in anaphase but is not completed until after the two daughter nuclei have formed. During cytokinesis, the cytoplasm divided by a process termed cleavage, driven by the tightening of a **contractile ring** composed of actin and myosin protein subunits. The first visible sign of cytokinesis is when the cell begins to pucker in, a process called **furrowing**. As the ring of cytoskeletal proteins contracts, a cleavage furrow is formed perpendicular to the mitotic spindle and gradually splits the cytoplasm and its contents into two daughter cells. At the beginning of cytokinesis, a narrow band forms around the cell surface. As time progresses, the indentation deepens to form a furrow completely encircling the cell. The plane of the furrow lies in the same plane previously occupied by the chromosome of the metaphase plate, so that the two sets of chromosomes are ultimately partitioned into different cells. The furrow continues to deepen until opposing surfaces make contact with one another and fuse in the center of the cell, thus splitting the cell in two (Figure 9-27).

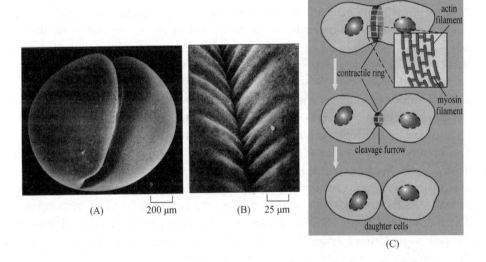

Figure 9-27 Formation and operation of the contractile ring during cytokinesis. (A) The cleavage furrow of a cleaving frog egg in the low-magnification scanning electron. (B) The surface of the furrow at higher magnification (From: Alberts B, et al. 2008). (C) Actin filaments become assembled in a ring at the cell equator. Contraction of the ring, which requires the action of myosin, causes the formation of a furrow that splits the cell into two (From: Karp G. 2005)

The contractile ring that produces cell cleavage is composed of an organized cytoskeletal network that includes actin and bipolar myosin-II filaments working together in a sliding action that mimics muscle contraction (Figure 9-28). In early telophase, after sister chromatids separate, actin and myosin filaments which release when cell enters mitosis begin to polymerize rapidly on the membrane to form the contractile ring. Once assembled, the contractile ring is capable of exerting a force strong enough to tend a fine glass needle inserted into the cell prior to cytokinesis. The force-generating mechanism operating during cytokinesis is thought to be similar to the actinomyosin-based contraction of muscle cells. Whereas the sliding of actin filaments of a muscle cell brings about the shortening of the muscle fiber, sliding of the filaments of the contractile ring pulls the cortex and attached plasma membrane toward the center of the cell. As a result, the contractile ring constricts the equatorial region of the cell, which much likes pulling on a purse string narrows the opening of a purse.

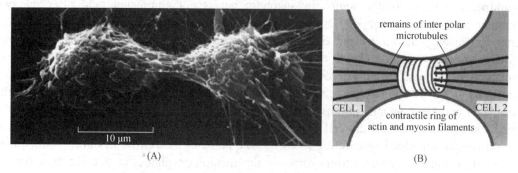

Figure 9-28 **The contractile ring divides the cell into two.** (A) Scanning electron micrograph of an animal cell in culture in the last stages of dividing. (B) Schematic diagram of the midregion of a similar cell showing the contractile ring beneath the plasma membrane and the remains of the two sets of interpolar microtubules.

2) **Cytokinesis in plant cells**: In higher plant cells, cytokinesis is regulated by the cell wall and occurs by a different mechanism. Unlike animal cells, plant cells must construct an extracellular wall inside a living cell. Wall formation starts in the center of the cell and grows outward to meet the existing lateral walls. The formation of a new wall begins with the construction of a simple precursor, which is called the **cell plate**.

The new cell wall starts to assemble in the cytoplasm between the two sets of segregated chromosomes in late anaphase. The assembly process is guided by a structure called the **phragmoplast**, which is formed by the remains of the interpolar microtubules at the equator of the old mitotic spindle. After formation of the phragmoplast, small Golgi-derived secretory vesicles move into the region, probably transported along the microtubules, and become aligned along a plane between the daughter nuclei. These vesicles are filled with polysaccharide and glycoproteins which are pivotal elements of the new cell wall. As more vesicles go to the equator, the cell plate expands until it bumps into the cell membrane and fuses with it (Figure 9-29). Thus, the plant cell divides into two with new cell wall between them. Later, cellulose microfibrils are laid down within the matrix to complete the construction of the new cell wall.

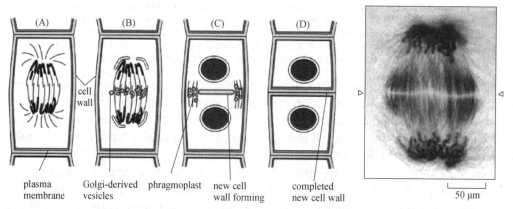

Figure 9-29 **Cytokinesis in a plant cell is guided by phragmoplast.** The membrane-enclosed vesicles, derived from the Golgi apparatus and filled with cell wall material, fuse to form the growing new cell wall, which grows outward to reach the plasma membrane and original cell wall. The plasma membrane and the membrane surrounding the new cell wall fuse, completely separating the two daughter cells.

9.3.2 Meiosis

Meiosis is a special kind of cell division, which only occurs at some stage of the sexual cell reproduction. The production of offspring by sexual reproduction includes the union of two cells, each with a complete haploid set of chromosomes. The doubling of the chromosomes at fertilization is compensated by an equivalent reduction in chromosome number at meiosis. Meiosis ensures production of a haploid phase in the life cycle, and fertilization ensures a diploid phase. Without meiosis, the chromosome number would double with each generation, and sexual reproduction would not be possible. Meiosis involves a single round of DNA replication that duplicates the chromosomes, followed by two successive cell divisions, called meiosis I (the first meiotic division) and meiosis II (the second meiotic division), without further DNA replication. Meiosis I, which results in reduction in the number of chromosomes, can be factitiously divided into five stages: prophase I, prometaphase I, metaphase I, anaphase I and telophase I. Meiosis II is very similar to a mitotic division. As is the case in any cell division, the interphase of meiosis includes G_1 phase, S phase and G_2 phase. In order to distinguish from mitosis, the interphase of meiosis is called premeiotic interphase. The premeiotic S phase has two marked characteristics, one is that it takes several times longer than the premeiotic S phase; the other is that partial DNA (from 99.7% to 99.9%) is replicated.

(1) **The first meiotic division**   A series of complex events occurs during the long prophase of meiotic division I: duplicated homologous chromosomes pair, genetic recombination is initiated between nonsister chromatids, and each pair of duplicated homologs assembles into an elaborate structure called the **synaptonemal complex** (SC). In some organisms, genetic recombination begins before the synaptonemal complex assembles and is required for the complex to form; in others, the complex can form in the absence of recombination. In all organisms, however, the recombination process is completed while the DNA is held in the synaptonemal complex, which serves to space out the crossover events along each chromosome.

1) **Prophase I**: Prophase I is traditionally divided into five sequential stages: **leptotene**, **zygotene**, **pachytene**, **diplotene**, and **diakinesis** — defined by the morphological changes associated with the assembly (synapsis) and disassembly (desynapsis) of the synaptonemal complex (Figure 9-30).

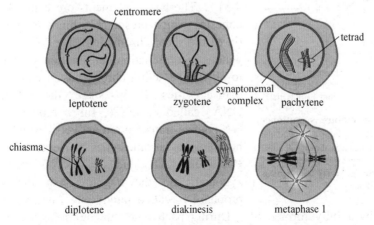

Figure 9-30   Stages of prophase I (From: Karp G. 2005)

Leptotene: The first stage of prophase I is **leptotene** (leptonema), during which the chromosomes condense and become gradually visible in the light microscope. Although the chromosomes have replicated at an earlier stage, there is no indication that each chromosome is actually composed of a pair of identical chromatids. In the electron microscope, however, the chromosomes are revealed to be composed of paired chromatids. The telomeres of maize leptotene chromosomes are distributed throughout the nucleus. Then, near the end of leptotene, there is a dramatic reorganization so that the telomeres become localized at the inner surface of the nuclear envelope at one side of the nucleus. Clustering of telomeres at one end of the nuclear envelope occurs in a wide variety of eukaryotic cells and causes the chromosomes to resemble a bouquet of flowers. And the clustering of telomeres at the nuclear envelope is thought to facilitate the alignment of homologues in preparation for synapsis.

Zygotene: The second stage of prophase I, which is called zygotene (pairing stage), is marked by the visible association of homologues with one another. This process of association is called synapsis and the complex formed by a pair of synapsed homologous chromosomes is called a bivalent or tetrad. The former term reflects the fact that the complex contains two homologues, whereas the latter term calls attention to the presence of four chromatids. It had been assumed for years that interaction between homologous chromosomes first begins as chromosomes initiate synapsis. Studies on yeast cells demonstrated that homologous regions of DNA from homologous chromosomes are already in contact with one another during leptotene. Chromosome compaction and synapsis during zygotene simply make this arrangement visible under the microscope. The first step in genetic recombination is the occurrence of double-stranded breaks in aligned DNA molecules. Studies in both yeast and mice suggest that the DNA breaks occur in leptotene, well before the chromosomes are visibly paired.

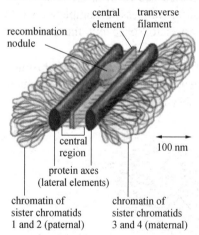

Figure 9-31 A mature synaptonemal complex. Only a short section of the long ladder-like complex is shown. A similar synaptonemal complex is present in organisms as diverse as yeasts and humans (From: Karp G. 2005)

Electron micrographs indicate that chromosome synapsis is accompanied by the formation of a complex structure called the synaptonemal complex. The SC is a ladder-like structure with transverse protein filaments connecting the two lateral elements. The chromatin of each homologue is organized into loops that extend from one of the lateral elements of the SC (Figure 9-31). The lateral elements are composed primarily of cohesin, which presumably binds together the chromatin of the sister chromatids. The SC is not required for genetic recombination but functions mainly as a scaffold to allow interacting chromatids to complete their crossover activities. In this stage, the rest 0.1% ~ 0.3% DNA, called zygDNA, finish replicating. The zygDNA happen transcription actively, which probably make homologous chromosomes pair each other.

Pachytene: The end of synapsis marks the end of zygotene and the beginning of the next stage of prophase I, called **pachytene** (pachynema), which is characterized by a fully formed SC. During pachytene, the homologues are held closely together along their length by the SC. The DNA of sister chromatids is extended into parallel loops (Figure 9-31). Under the electron microscope, a number of electron-dense bodies about 100 nm in diameter are seen within the center of the SC. These structures have been named recombination nodules because they correspond to the sites where crossing over is taking place. Recombination nodules contain the enzymatic machinery that facilitates genetic recombination, which is completed by the end of pachytene.

Diplotene: Diplotene follows the pachytene, in which each chromosome continue shortening and thickening and the two chromosomes in each bivalent begin to repel each other and a split occurs between the chromosomes (Figure 9-32). The separation is not accomplished, but the homologous chromosomes stick together at certain points, the **chiasmata** (singular chiasma). Chiasmata are formed by covalent junctions between a chromatid from one homologue and nonsister chromatid from extent of genetic recombination, which play a crucial part in holding the compact homologs together. The chiasmata are made more visible by a tendency for the homologues to separate from one another at the diplotene stage.

Diakinesis: During the final stage of meiotic prophase I, called **diakinesis**, the meiotic spindle is assembled, and the chromosomes are prepared for separation. The chromosomes condense and become more compact. Diakinesis ends with the disappearance of the nucleolus, the breakdown of the nuclear envelope, and the movement of the tetrads to the metaphase plate. In vertebrate oocytes, these events are triggered by an increase in the level of the protein kinase activity of MPF.

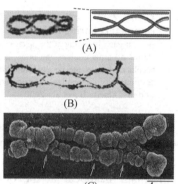

In some organism, just before or during diakinesis the chiasmata move to the ends of the chromosome arms. While in most eukaryotic species, chiasmata can still be seen in homologous chromosomes aligned at the metaphase plate of meiosis I. In humans and other vertebrates, every pair of homologues typically contains at least one chiasma, and the longer chromosomes tend to have two or three of them. It is thought that some mechanism exists to ensure that even the smallest chromosomes form a chiasma. If a chiasma does not occur between a pair of homologous chromosomes, the chromosomes of that bivalent tend to separate from one another after dissolution of the SC.

Figure 9-32 Visible evidence of crossing over. Pairs of diplotene bivalents from the grasshopper showing the chiasmata formed between chromatids of each homologous chromosome. The accompanying inset indicates the crossover that has presumably occurred within the bivalent. The chromatids of each diplotene chromosome are closely apposed except at the chiasmata (From: Karp G. 2005)

This premature separation of homologues often results in the formation of nuclei with an abnormal number of chromosomes.

2) Metaphase I and anaphase I: In metaphase I, the bivalent become aligned in the center and their own homologous chromosomes are connected to the chromosomal fibers from opposite poles (Figure 9-33). In contrast, the kinetochores of sister chromatids are connected as a unit to microtubules from the same spindle pole. The orientation of the maternal and paternal chromosomes of each bivalent on the metaphase I plate is random; the maternal member of a particular bivalent has an equal likelihood of facing either pole. During anaphase I, the homologous chromosomes separate and move to the opposite pole, while the sister chromatids remain together. Thus, anaphase I is the cytological event that corresponds to Mendel's law of independent assortment. As a result of independent assortment, organisms are capable of generating a nearly unlimited variety of gametes.

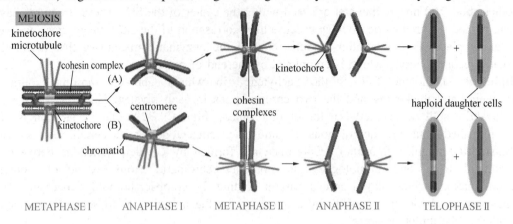

Figure 9-33 Separation of homologous chromosomes during meiosis Ⅰ and separation of chromotids during meiosis Ⅱ. (A)In meiosis Ⅰ, the two sets of kinetochore microtubules are located side-by-side on each homolog at the sister centromeres. The proteolytic destruction of the cohesion complexes along the sister chromatid arms unglues the arms and destroy the crossovers, which allow the duplicated homologs to separate at anaphase Ⅰ, yet the residual cohesin complexes at the centromeres keep the sisters together. (B) The proteolytic destruction of the residual cohesin complexes at the centromeres allows the sister chromatids to separate at anaphase Ⅱ (From: Alberts B, et al. 2008)

Separation of homologous chromosomes at anaphase Ⅰ require the dissolution of the chiasmata that hold the bivalent together. The chiasmata are maintained by cohesin between sister chromatids in regions that flank these sites recombination (Figure 9-33A). The chiasmata disappear at the onset of anaphase Ⅰ, as the arms of the chromatids of each bivalent lose cohesin. Loss of cohesin between the arms is accomplished by proteolytic cleavage of the cohesin molecules in those regions of the chromosome. In contrast, cohesion between the joined centromeres of sister chromatids remains strong, because the cohesin situated there is protected from proteolytic attack. As a result, sister chromatids remain firmly attached to one another as they move together toward a spindle pole together during anaphase Ⅰ.

3) **Telophase Ⅰ**: Telophase Ⅰ of meiosis is very similar to the telophase of mitosis. During telophase Ⅰ, the homologous chromosomes pairs complete their migration to the two poles as a result of the action of the spindle. A haploid set of chromosomes is at each pole, with each chromosome still having two chromatids. The nuclear envelope may or may not reform around each chromosome set, the spindle disappears, and cytokinese follows. Meiotic division Ⅰ is far more complex and requires much more time than either mitosis or meiotic division Ⅱ. Even the preparatory DNA replication during meiotic division Ⅰ tends to take much longer than an ordinary S phase, and cells can then spend days, months, or even years in prophase Ⅰ, which depend on the species and on the gamete being formed.

(2) The second meiotic division When meiotic division Ⅰ endcs, nuclear membranes reform around the two daughter nuclei and the short-lived interphase of meiotic division Ⅱ begins. During this period, the chromosomes may decondense somewhat, but usually they soon recondense and prophase Ⅱ begins (because there is no DNA synthesis during this interval, in some organisms the chromosomes seem to pass almost directly from one division phase into the next). Prophase Ⅱ is brief: If the nuclear envelope had reformed in telophase Ⅰ, it is broken down again. The chromosomes become recompacted and line up

at the metaphase plate. Unlike metaphase Ⅰ, the kinetochores of sister chromatids of metaphase Ⅱ face opposite poles and become attached to opposing sets of chromosomal spindle fibers (Figure 9-33B). Anaphase Ⅱ begins with the synchronous splitting of the centromeres, which had held the sister chromatids together, allowing them to move toward opposite poles of the cell. Meiosis Ⅱ ends with telophase Ⅱ, in which the chromosomes are once again enclosed by a nuclear envelope. The products of meiosis are haploid cells with a 1C amount of nuclear DNA.

In female animals, to end meiosis, the germ cell only produces one mature egg (oocyte), the rest three haploid cells called polar body will disintegrate. However, in male animals, after meiosis, the germ cell will produce four haploid products called sperm.

Summary

Cell division is one of the important directions of cell. The division cycle of most cells consists of four coordinated processes: cell growth, DNA replication, distribution of the duplicated chromosomes, and cell division. The cell cycle can be regarded as the life cycle of an individual cell. It can be divided into two major phases based on cellular activities readily visible with a light microscope, M phase and interphase. Interphase is the interval between divisions during which the cell undergoes its functions and prepares for mitosis. The events of the cell cycle must also occur in a particular sequence, and this sequence must be preserved even if one of the steps takes longer than usual. To coordinate these activities, eukaryotic cells have evolved a complex network of regulatory proteins, known as the cell-cycle control system. The cell cycle is tightly regulated at three checkpoints: G_1/S, G_2/M and M phases. Cyclins control the cell cycle by interacting with cyclin-dependent kinases (CDKs) complexes. These in turn are activated and stimulate cell cycle progression by phosphorylation of specific target in the cell.

Mitosis is a process of nuclear division in which replicated DNA molecules of each chromosome are faithfully partitioned into two nuclei. Mitosis is conventionally divided into six stages: prophase, prometaphase, metaphase, anaphase, telophase and cytokines each characterized by a particular series of events.

Meiosis is a special kind of cell division, which only occurs at some stage of the sexual cell reproduction. Meiosis ensures production of a haploid phase in the life cycle, and fertilization ensures a diploid phase. Without meiosis, the chromosome number would double with each generation, and sexual reproduction would not be possible. Meiosis involves a single round of DNA replication that duplicates the chromosomes, followed by two successive cell divisions, called meiosis Ⅰ and meiosis Ⅱ. Meiosis Ⅰ, which results in reduction in the number of chromosomes, can be factitiously divided into five stages, prophase Ⅰ, prometaphase Ⅰ, metaphase Ⅰ, anaphase Ⅰ, and telophase Ⅰ. Prophase Ⅰ is traditionally divided into five sequential stages, leptotene, zygotene, pachytene, diplotene, and diakinesis defined by the morphological changes associated with the assembly (synapsis) and disassembly (desynapsis) of the synaptonemal complex. Meiosis Ⅱ is very similar to a mitotic division.

Questions

1. Give short definitions of the following terms: bivalent and tetrad, Cdk-activating kinase (CAK), cell cycle and cell cycle control, cell cycle checkpoint, centromere and kinetochore, chiasma (chiasmata) and crossing over, cyclin and cyclin-dependent kinase (Cdk), homologous chromosome, maturation-promoting factor (MPF), SCF

and APC complexes, microtubule-associated protein (MAP), mitosis and meiosis, phragmoplast, restriction point, sister chromatid, START, synapsis and synaptonemal complex (SC).
2. Short-answer questions
   (1) Describe the chief events of mitosis and show how these differ from meiosis.
   (2) Describe the phases of the cell cycle and the events that characterize each phase.
   (3) What "checkpoints" occur in the cell cycle? What is the role of each?
3. Essay questions:
   (1) Describe the function mechanism of cyclin in the regulation of the cell cycle?
   (2) What types of force-generating mechanisms might be responsible for chromosome movement during anaphase?
   (3) Contrast the events that take place during mitosis and meiosis, and their roles in the lives of a plant or animal?
   (4) What is the main formation of the spindle, and their function respectively?
   (5) What is the cell cycle? What are the stages of the cell cycle? How does the cell cycle vary among different types of cells?

(任 华 李秀兰)

Chapter 10

Cell differentiation

10.1 Introduction

Cell differentiation is the process by which cells acquire the specialized functions that allow them to play their roles in the tissues and organs of the adult animal. This process occurs through changes in gene expression. It is from the proteins expressed in a cell that a cell acquires its properties, and it is through the regulation of gene expression that different cells express different proteins and thus become different from each other.

10.1.1 The biology of cell differentiation

Molecular biology of cell differentiation in development with special emphasis on mammalian systems, which focuses on biological systems, *in vivo* and in cell culture, and examines molecular mechanisms of gene expression during differentiation.

(1) **Cell determination and differentiation** With few exceptions, all of the cells in a multicellular organism contain the same DNA. This is not surprising, given that organisms develop from mitotic divisions of one original cell, called a **zygote**. However, it is clear that there are many different cell types in the bodies of multicellular organisms. How do these cells "mature" to take on specific roles for an organism? The processes by which cells mature, commonly referred to as cell determination and differentiation.

During early embryogenesis, cells divide and gradually become committed to specific patterns of gene activity through a process called cell **determination**. Specific genes are associated with the determination event. Because the daughter cells of each "determined" cell have the same limited potential as their parent cell, determination is considered heritable. Determination is permanent under normal conditions but it is possible to reverse the process experimentally.

```
totipotent cells
      ↓
  determination
      ↓
pruripotent/multipotent cells
      ↓
  differentiation
      ↓
 specialized cells
```

In fact, cell determination is the process by which portions of the genome are selected for expression in different embryonic cells. This involves developmental decisions that gradually restrict cell fate. Cells can progress from totipotent to pluripotent to determined. Regulated gene expression underlies cell differentiation while cell differentiation is stable but can occasionally be reversed. Currently, epigenetic mechanisms account for the stability of determination and differentiation.

(2) **Cell differentiation plays important roles in development** Cell differentiation also describes the progressive restriction of the developmental potential and increasing specialization of function which takes place during the development of the embryo and leads to the formation of specialized cells, tissues, and organs.

Cell differentiation leads the continuous loss of physiological and cytological characters of young cells, resulting in getting the characters of adult cells. The unspecialized cells become modified and specialized for the performance of specific functions. Differentiation results from the controlled activation and de-activation of genes. When young cells develop as a result of cell division, they become differentiated to suit the function they were made for. This is done by the genetic information present in the cell, which will have genes "switched on" and "switched off" to determine the function of the cell.

(3) **Cell differentiation involves in wide range of mechanisms** Cell differentiation covers molecular biology of differentiation; embryonic patterning; carcinogenesis and cancer; mechanisms involved in cell growth and division especially relating to cancer; establishment of cellular polarity; cell and tissue interactions *in vivo* and *in vitro*; signal transduction pathways in development and differentiation; transgenic and targeted mutagenesis models of differentiation; organogenesis and plant cell differentiation.

(4) **Cell differentiation occurs through changes in gene expression** It is from the proteins expressed in a cell that a cell acquires its properties, and it is through the regulation of gene expression that different cells express different proteins and thus become different from each other. A wide variety of proteins are capable of affecting a cell's differentiated state. Many of the key proteins are **transcription factors**, but intercellular signaling plays a critical role in influencing the presence and activity of the transcription factors in a cell. On occasion, even the presence of a cytoskeletal protein can globally alter a cell's properties. Ultimately, before differentiation can be considered to be complete, the new state of gene expression must be stabilized. The mechanisms by which chromatin proteins stabilize the differentiated state, even through rounds of DNA replication and cell division, are largely unknown, and represent an area of great interest in current research.

10.1.2 The majority of organisms consist of many types of cells

Living organisms are made up of cells that are usually too small to see with the naked eye. Many cells, for example most animal cells, plant and bacterial cells, have a fixed shape. However, large organisms contain many cells, and the human body contains billions of cells of many different types.

(1) **Multicellular organisms have many specialized cells** The single-celled procaryotic and eucaryotic cells lacked the specialization and organization we see in many plants and animals today. Single-celled organisms need to perform every life function within the confines of a single cell. Multicellular organisms, like ourselves, can have cells specialized for a particular function. Thus, a liver cell doesn't have to "worry" about all the chemical reactions and interactions with other cells that a retinal cell in the back of your eye has to worry about.

(2) **This specialization allows for more efficiency** Progressive restriction of the developmental potential and increasing specialization of function which takes place during the development of the embryo and leads to the formation of specialized cells, tissues, and organs. Some cells provide protection; some give structural support or assist in locomotion; others offer a means of transporting nutrients. All cells develop and function as part of the organized system — the organism — they make up.

10.2 Cells with potency of differentiation

Multicellular organisms are formed from a single **totipotent** stem cell — a single fertilized egg or zygote. As this cell and its progeny undergo cell divisions, the potential of the cells becomes restricted, and they specialize to generate cells of a certain lineage. In the majority of organisms, there are more than one type of cell. Indeed, about 200 different types of cells — many highly specialized — make up the tissues and organs of the human body.

10.2.1 Cells with different potency of differentiation

According to the potency of differentiation, cells could be classified as: **totipotent cells**; **pluripotent cells**; **multipotent cells** (Recently); terminal differentiated cell, or specialized cell.

Except of totipotent cells, the pluripotent cells and multipotent cells are true stem cells, which are further divided as embryonic stem cells and adult stem cells, therefore they are also named as pluripotent stem cells and multipotent stem cells. multipotent stem cells finally produce terminal and specialized cells, such as red blood cells, white blood cells (Figure 10-1).

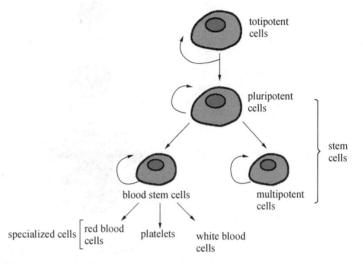

Figure 10-1 Totipotent cells, pluripotent cells, multipotent cells

10.2.2 Human body's development

In order to throughly understand totipotent, pluripotent and multipotent cells, it is important to view them in the context of human development (Figure 10-2).

At the time of fertilization, the one zygote cell produced is capable of forming an entire organism. These cells are classified as totipotent, which means that the potential of the cell is unlimited. For a short time, zygote division creates identical totipotent cells. Any of these cells formed during the first division after fertilization could be placed in a woman's uterus and develop into a fetus.

By the fourth day, the cells begin to form a **blastocyte**, or bundle of cells. The outer layer of the blastocyte forms the placenta and other necessary tissues in the uterus required for the fetus to develop. The inner cluster of cells will continue to develop into nearly all of the tissues of the human body. Although those inner cells will form virtually every type of tissue in the body, they cannot give rise to the placenta or other supporting tissues for the uterus. Thus, they are unable to form an organism on their own if placed in a woman's uterus, and are therefore referred to as pluripotent. As the pluripotent cells continue to specialize, they become cells that only lead to the development of specific tissues. Some will lead to bone marrow, while others will lead to blood or skin. The stem cells that carry this extra specialization are considered multipotent.

It is well known that multipotent stem cells play a vital role in fetal development, however, multipotent cells can still be found during the course of a person's adult life. Although usually found in very small quantities, adult stem cells play a critical role in sustaining life. For example, red blood cells are continuously replaced, and the production of new red blood cells is initiated by multipotent blood stem cells. Virtually any body function that requires growth involves stem cells.

The difference between totipotent, pluripotent, and multipotent cells is essential in understanding cell differentiation as well as stem cells.

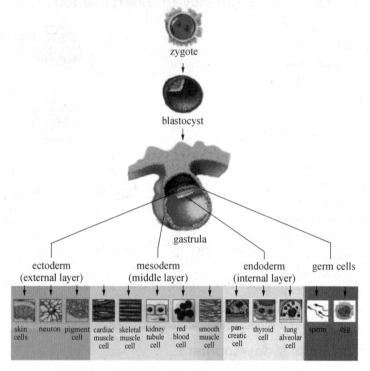

Figure 10-2 Differentiation of human tissues

10.2.3 Totipotent cells

A cell that has the capacity to independently develop into a complete organism is called totipotent. The only totipotent cells are the fertilized egg and the first 4 or so cells produced by its cleavage (as shown by the ability of mammals to produce identical twins, triplets, etc.). An example is a fertilized egg in the uterus. A totipotent cell by itself can give rise to an entire embryo.

Totipotent cells are the most determined and most specialized cells in all of development because they are the least differentiated. Progressively the developing cells lose, in fact, the ability to use this information. No information is lost only the ability to use it is lost.

As described above, totipotent cells are not true stem cells, just like fertilized egg is not listed as stem cell.

10.2.4 pluripotent cells

Pluripotent cells have the potential to make any differentiated cell in the body, but cannot contribute to making the extraembryonic membranes (which are derived from the trophoblast). A pluripotent cell cannot by itself give rise to an embryo, but it can give rise to all the different cells found in the organs of the body.

In contrast to totipotent cells, pluripotent cells have a more restricted capacity to differentiate. By introducing pluripotent cells in different tissues they may be induced to act like the surrounding cells. However, a single pluripotent cell can not develop into an independent organism. Pluripotent cells can be re-transformed into totipotent cells by means of transplantation of nuclei (see below). Pluripotent stem cells have the biggest importance for scientists in stem cell research.

10.2.5 multipotent cells

Multipotent cells can only differentiate into a limited number of types. They found in adult animals; perhaps most organs in the body (e.g., brain, liver, bone marrow, and skin.) contain them where they can replace dead or damaged cells. A multipotent cell can give rise to more than one type of cell but not all the different cells in all the organs. For example, hematopoietic stem cell can only produces mature blood cells.

Multipotent cells are also sometimes called adult stem cells, such as umbilical cord stem cells. While they are abundant in the umbilical cord blood, they are also present in the organs of newborns, children, and adults. Some populations of multipotent stem cells are more active than others, such as those in the skin and bone marrow.

10.2.6 terminal differentiated cells

Scientists believed that all cells carry out basic functions, such as respiration, growth, division, synthesis. Most cells have specialized capabilities, and traditionally, cells in their final, differentiated state usually have very unique characteristics, reflected in their morphology and function. These cells are terminal differentiated cell. They may be characterized by: ① No proliferative capacity whatsoever; ② Never seem to divide; ③Irreplaceable; ④Have a long life span and live in protected environments. Scientists now find exceptions that some terminal differentiated cells could undergo: non-constantly renewing; return to active cycling (cell cycle/division) in response to critical cell depletion; renewal by simple duplication of existing differentiated cells division gives rise to daughter cells of the same type.

The evidence that cell production in vertebrates is controlled by intrinsic processes is most compelling for non-neuronal tissue. For example, proliferation in the liver, which is normally characterized by very low cell turnover, can be dramatically stimulated in the partial hepatectomy model. After surgical removal of 2/3 of the liver, the remaining hepatocytes exit G_0 and synchronously enter the cell cycle in order to restore liver mass.

10.3 Stem cells

Stem cells are present both in the embryo and in the adult. Depending on where the stem cells originate, they have different properties. According to this classification scheme, the three kinds of stem cells are: **Embryonic stem cells**, embryonic germ cells, and **adult stem cells**.

Stem cells are important for living organisms for many reasons. In the 3-to 5-day-old embryo, called a blastocyst, stem cells in developing tissues give rise to the multiple specialized cell types that make up the heart, lung, skin, and other tissues. In some adult tissues, such as bone marrow, muscle, and brain, discrete populations of adult stem cells generate replacements for cells that are lost through normal wear and tear, injury, or disease.

10.3.1 The Unique Properties of All Stem Cells

Stem cells have two important characteristics that distinguish them from other types of cells. First, they are unspecialized cells that renew themselves for long periods through cell division. The second is that under certain physiological or experimental conditions, they can be induced to become cells with special functions such as the beating cells of the heart muscle or the insulin-producing cells of the pancreas.

Stem cells differ from other kinds of cells in the body. All stem cells—regardless of their source—have three general properties: stem cells are capable of dividing and renewing themselves for long periods; stem cells unspecialized; stem cells give rise to specialized cell types.

(1) Stem cells must self-renew   One of the most important issues in stem cell biology is to understand the mechanisms that regulate self-renewal. Self-renewal is crucial to stem cell function, because it is required by many types of stem cells to persist for the lifetime of the animal. Moreover, whereas stem cells from different organs may vary in their developmental potential, all stem cells must self-renew and regulate the relative balance between self-renewal and differentiation. Understanding the regulation of normal stem cell self-renewal is also fundamental to understanding the regulation of cancer cell proliferation, because cancer can be considered to be a disease of unregulated self-renewal.

(2) Stem cells are unspecialized   One of the fundamental properties of a stem cell is that it does not have any tissue-specific structures that allow it to perform specialized functions. A stem cell cannot work with its neighbors to pump blood through the body (like a heart muscle cell); it cannot carry molecules of oxygen through the bloodstream (like a red blood cell); and it cannot fire electrochemical signals to other cells that allow the body to move or speak (like a nerve cell). However, unspecialized stem cells can give rise to specialized cells, including heart muscle cells, blood cells, or nerve cells.

(3) Stem cells are capable of dividing and renewing themselves for long periods
Unlike muscle cells, blood cells, or nerve cells—which do not normally replicate themselves—stem cells may replicate many times. When cells replicate themselves many times over it is called proliferation. A starting population of stem cells that proliferates for many months in the laboratory can yield millions of cells. If the resulting cells continue to be unspecialized, like the parent stem cells, the cells are said to be capable of long-term self-renewal.

(4) Stem cells can produce specialized cells   When unspecialized stem cells give rise to specialized cells, the process is called differentiation. Scientists are just beginning to understand the signals inside and outside cells that trigger stem cell differentiation. The internal signals are controlled by a cell's genes, which are interspersed across long strands of DNA, and carry coded instructions for all the structures and functions of a cell. The external signals for cell differentiation include chemicals secreted by other cells, physical contact with neighboring cells, and certain molecules in the microenvironment.

10.3.2 Embryonic stem cells

Embryonic stem cells (ES cells) are from the inner cell mass, which is part of the early (5~6 day old) embryo called the blastocyst. Once removed, the cells of the inner cell mass can be cultured into embryonic stem cells.

In 1998 we saw the publication of two papers describing the growth *in vitro* of human embryonic stem (ES) cells derived either from the inner cell mass (ICM) of the early blastocyst or the primitive gonadal regions of early aborted fetuses. Work on murine ES cells over many years had already established the amazing flexibility of ES cells, essentially able to differentiate into almost all cells that arise from the three germ layers. The realization of such pluripotentiality has, of course, resulted in the field of stem cell research going into overdrive (Figure 10-2).

(1) **The generation of embryonic stem cells** Embryonic stem cells, as their name suggests, are derived from embryos. The embryos from which human embryonic stem cells are derived are typically four or five days old and are a hollow microscopic ball of cells called the blastocyst. The blastocyst is constructed with three structures: the trophoblast, which is the layer of cells that surrounds the blastocyst; the blastocoel, which is the hollow cavity inside the blastocyst; and the inner cell mass, which is a group of approximately 30 cells at one end of the blastocoel.

(2) **ES differentiation** The capacity of embryonic stem cells for virtually unlimited self-renewal and differentiation capacity has opened up the prospect of widespread applications in biomedical research and regenerative medicine. For the latter, the cells provide hope that it will be possible to overcome the problems of donor tissue shortage and also, by making the cells immunocompatible with the recipient, implant rejection.

Stem cells begin their transformation into the different types of cells that make up the organisms during a phase in the development process called cell differentiation. In vertebrates, differentiation begins during a stage called gastrulation, when distinct tissue layers first form. Like most other developmental processes, differentiation is controlled by genes, the genetic instructions encoded in the DNA of every cell. Genes instruct each cell to build the proteins that allow it to create the structures, and ultimately perform the functions, specific to its type of cell.

(3) **Embryonic stem cells grown *in vitro*** Growing cells in the laboratory is known as cell culture. Human embryonic stem cells are isolated by transferring the inner cell mass into a plastic laboratory culture dish that contains a nutrient broth known as culture medium. The cells divide and spread over the surface of the dish. The inner surface of the culture dish is typically coated with mouse embryonic skin cells that have been treated so they will not divide. This coating layer of cells is called a feeder layer.

Over the course of several days, the cells of the inner cell mass proliferate and begin to cover the culture dish. When this occurs, they are removed gently and plated into several fresh culture dishes. The process of replicating the cells is called subculturing, and repeated many times and for many months. Each cycle of subculturing the cells is referred to as a passage. After six months or more, the original 30 cells of the inner cell mass yield millions of embryonic stem cells.

(4) **Induced differentiation of embryonic stem cells** As long as the embryonic stem cells in culture are grown under certain conditions, they can remain undifferentiated (unspecialized). To generate cultures of specific types of differentiated cells—heart muscle cells, blood cells, or nerve cells, the specific conditions should be applied to control the differentiation of embryonic stem cells. These methods are: change the chemical

composition of the culture medium; alter the surface of the culture dish; modify the cells by inserting specific genes.

10.3.3 Adult stem cells

An adult stem cell is an undifferentiated cell found among differentiated cells in a matured tissue or organ, can renew itself, and can differentiate to yield the major specialized cell types of the tissue or organ (Figure 10-3).

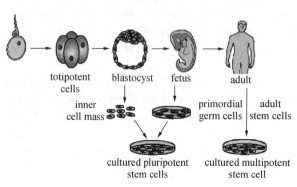

Figure 10-3 **This picture illustrates three sources of stem cells.** Embryonic stem cells can be isolated from the blastocyst, embryonic germ cells from the fetus, and adult stem cells from the tissues of adults (From: www.nih.gov)

(1) **The history of adult stem cell researches** The history of research on adult stem cells began about 50 years ago. In the 1960s, researchers discovered that the bone marrow contains at least two kinds of stem cells. One population, called hematopoietic stem cells, forms all the types of blood cells in the body. A second population, called bone marrow stromal cells, was discovered a few years later. Stromal cells are a mixed cell population that generates bone, cartilage, fat, and fibrous connective tissue.

Also in the 1960s, scientists who were studying rats discovered two regions of the brain that contained dividing cells, which become nerve cells. Despite these reports, most scientists believed that new nerve cells could not be generated in the adult brain. It was not until the 1990s that scientists agreed that the adult brain does contain stem cells that are able to generate the brain's three major cell types—astrocytes and oligodendrocytes, which are non-neuronal cells, and neurons, or nerve cells. Therefore, it was found that adult stem cells in many more tissues than they once thought possible.

(2) **Location of the adult stem cells** Adult stem cells have been identified in many organs and tissues. One important point to understand about adult stem cells is that there are a very small number of stem cells in each tissue. Stem cells are thought to reside in a specific area of each tissue where they may remain quiescent (non-dividing) for many years until they are activated by disease or tissue injury. The adult tissues reported to contain stem cells include brain, gut epithelium, bone marrow, peripheral blood, blood vessels, skeletal muscle, skin, pancreas, renal, connective tissue, liver and so on.

(3) **The importance of adult stem cell** The primary roles of adult stem cells in a living organism are to maintain and repair the tissue in which they are found. Some scientists now use the term somatic stem cell instead of adult stem cell. Unlike embryonic stem cells, which are defined by their origin (the inner cell mass of the blastocyst), the origin of adult stem cells in mature tissues is unknown.

Adult stem cells originate from mature adults. These can also be referred to as multipotent stem cells, as the number of cell types which they can differentiate into is limited. Adult stem cells serve as a fresh source of cells in living organisms. They replace cells that need to be replaced on a regular basis in a living organism, such as blood (which has a 120 day lifespan) and other connective tissues.

Scientists in many laboratories are trying to find ways to grow adult stem cells in cell culture and manipulate them to generate specific cell types so they can be used to treat injury or disease. Some examples of potential treatments include replacing the dopamine-producing cells in the brains of Parkinson's patients, developing insulin-producing cells for type I diabetes and repairing damaged heart muscle following a heart attack with cardiac muscle cells.

(4) **The plasticity and transdifferentiatin of adult stem cell**   As reported, adult stem cells occur in many tissues and that they enter normal differentiation pathways to form the specialized cell types of the tissue in which they reside. Adult stem cells may also exhibit the ability to form specialized cell types of other tissues, which is known as transdifferentiation or plasticity.

In the adult, organ formation and regeneration was thought to occur through the action of organ- or tissue-restricted stem cells (i.e., haematopoietic stem cells giving rise to all the cells of the blood, neural stem cells making neurons, astrocytes, and oligodendrocytes). However, it is now believed that stem cells from one organ system, for example the haematopoietic compartment can develop into the differentiated cells within another organ system, such as the liver, brain or kidney. Thus, certain adult stem cells may turn out to be as malleable as ES cells and so also be useful in regenerative medicine.

The following list offers examples of adult stem cell plasticity that have been reported during the past few years.
- Hematopoietic stem cells may differentiate into: three major types of brain cells (neurons, oligodendrocytes, and astrocytes), skeletal muscle cells, cardiac muscle cells, and liver cells.
- Bone marrow stromal cells may differentiate into: cardiac muscle cells and skeletal muscle cells.
- Brain stem cells may differentiate into: blood cells and skeletal muscle cells.
- Muscle precursors can give rise to hematopoietic cells, osteogenic and adipogenic differentiation potential that is normally retrieved in mesoderm-derived stromal cells.

Current research is aimed at determining the mechanisms that underlie adult stem cell plasticity. If such mechanisms can be identified and controlled, existing stem cells from a healthy tissue might be induced to repopulate and repair a diseased tissue.

10.4   Controls of cell differentiation

The progressive restriction of differentiation potential from pluripotent embryonic stem cells, via multipotent progenitor cells to terminally differentiated, mature somatic cells, involves step-wise changes in transcription patterns that are tightly controlled by the coordinated action of key transcription factors and changes in epigenetic modifications.

10.4.1   Differentiation involves gene expression

Every cell from different tissue or organ has the same set of genes in a same individual. A heart cell nucleus contains skin cell genes, as well as the genes that instruct stomach cells how to absorb nutrients. This suggests that in order for cells to differentiate — to become different from one another — certain genes must somehow be activated, while others remain inactive. That is that the differential gene expression leads to cell differentiation, because all adult cells have the genes necessary to become any type of cells of individual.

(1) **Differentiation is controlled by genes**   The formation and function of tissues requires the acquisition of cellular phenotype during differentiation. Differentiation involves expression of genes that specify cellular fate and down-regulation of genes that hold cells in

an original state. Once a tissue forms, changes in gene expression are also important as cells within a tissue adapt to environmental changes through a pattern of new gene expression. These cellular adaptations allow cells within a tissue to survive and maintain function. Thus, the control of gene expression is important for the formation, function, and maintenance of cells.

(2) **Different cells dependent on different proteins**   Indeed, about 200 different types of cells — many highly specialized — make up the tissues and organs of the human body. In the human body, virtually every cell contains the same genetic material, the same DNA. Although humans have many different types of cell, for example liver cells or nerve cells, each of them contain all the information needed to make all human proteins. Even if a cell does not actually make a protein, it has the genetic information to do so. The different cell types of an organism are distinguishable both by the amounts and types of proteins they produced. Some proteins may be unique to liver cells, some to nerve cells etc. At the same time, most proteins are the same from one cell type to another; for example, human actins in one cell type is much the same as in another.

10.4.2 Transcription factors associated with cell differentiation

DNA binding transcription factors (TFs) play critical roles in initiating global changes in gene activity, such as activation or repression. The binding of TFs to target genes leads to the recruitment of co-factors which include chromatin remodeling and modification complexes. The regulation of gene target networks by specific TFs at different stages constitutes the transcriptional programs that direct cellular differentiation.

A wide variety of proteins are capable of affecting a cell's differentiated state. Many of the key proteins are transcription factors, but intercellular signaling plays a critical role in influencing the presence and activity of the transcription factors in a cell. Before differentiation can be considered to be complete, the new state of gene expression must be stabilized.

(1) **Pluripotent transcription factors**   ES cells have the ability to self-renew or differentiate into cells of all three germ layers (endoderm, ectoderm, and mesoderm). In ES cells, OCT4, SOX2, and NANOG form a core transcriptional network influencing the stem cell self-renewal machinery. Several hundred target genes co-occupied by OCT4, SOX2, and NANOG can be classified into two groups of downstream genes exerting opposing functions. One group includes actively transcribed genes associated with proliferation and transcription factors necessary to maintain the ES cell state. The other group includes transcriptionally silent genes encoding developmental regulators that are only activated as cells differentiate and commit to particular lineages. Establishment of **induced pluripotent stem cells** (iPSCs) has shown the transcriptional activity of these key pluripotency factors.

Recent evidence highlights the important role of post-translational modifications, including ubiquitination, sumoylation, phosphorylation, methylation and acetylation, in regulating the levels and activity of pluripotency factors to achieve a balance between pluripotency and differentiation(Figure 10-4).

(2) **Lineage-specific transcription factors**   The identification of transcriptional regulatory networks, which control lineage-specific development and function, is of central importance to the understanding of cell biology.

Lineage commitment and differentiation are likely to be coordinated by the combined effects of multiple transcription factors (TFs) acting on numerous different target genes. The examples of lineage-specific TFs include by GATA-1, GATA2, PU1 in erythroid differentiation, HNF-4α in hepatocytic differentiation, Pax6 in neural differentiation in the

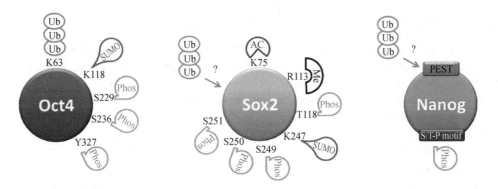

Figure 10-4 Post-translational modifications of pluripotency-associated transcription factors (Ub: ubiquitination; SUMO: sumoylation; Me: methylation; AC: acetylation; Phos: phosphorylation; K118, S229, etc: indicate the amino acid residues modified)

developing cerebral cortex, etc.

The interplay between lineage-specific TFs and chromatin modifying or remodeling complexes allows chromatin modifications at specific lineage gene loci and promotes transcriptionally prone conformations. Molecular and biochemical studies have demonstrated that lineagespecific TFs, even when expressed at basal levels in progeny cells can bind to gene regulatory regions and gene promoters of variety loci. From the very early steps, lineage-specific TFs can recruit chromatin modifying and remodeling complexes to gene regulatory regions, hence favoring the recruitment of general transcription factors (GTFs) and preinitiation complex (PIC) assembly, necessary for the transcription of protein-coding genes in eukaryotes. Therefore, by potentiating lineagespecific gene expression, lineage-specific TFs can influence the identity of progeny cells. Among various differentiations, hematopoiesis represents one of the best-understood models of lineage commitment and cell differentiation (Figure 10-5).

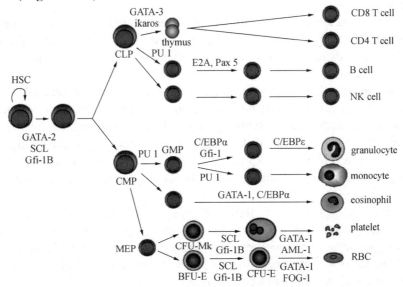

Figure 10-5 Transcription factors regulating hematopoietic differentiation. Transcription factors regulating lineage specification at various differentiation branch points or facilitating differentiation along specific lineage pathways are indicated (From: Keio J Med, 2011, 60: 47-55)

10.4.3 Epigenetic view of cell differentiation

Since a stem cell, progenitor cells, or mature cells, regardless of cell type, possess same genome of that species, determination of cell type must occur at the level of gene expression. As it turns out, epigenetic processes play a crucial role in the control of differentiation. During differentiation, specialized cells take in genetic information from stem cell or progenitor cell, but they exhibit different phenotype (structure and function). Scientists demonstrated that it is the alternation of genes expression pattern, not the change of DNA sequence that results in the different phenotype.

(1) **What is epigenetics** Epigenetics study the heritable changes in gene function that occur without a change in the DNA sequence. Epigenetics describe the alteration of the chromatin architecture to either permit or deny access of transcription factor machinery, and the basis of this control is brought about by **DNA methylation** of cytosine at specific CpG islands or by covalent modification of the N-terminal tails of the histones that form the structure of the nucleosome.

Current researches about epigenetics produce following conclusions: ①The study of stable alterations in gene expression that arises during development and cell proliferation. ②Epigenetic phenomena do NOT change the actual, primary genetic sequence. ③Epigenetic phenomena are important because, together with promotor sequences and transcription factors, they modulate when and at what level genes are expressed. ④The protein context of a cell can be understood as epigenetic phenomenon. ⑤Examples include: DNA methylation, histone acetylation, Chromatin modifications, X-inactivation and genetic imprinting.

(2) **Epigenetic regulation of cell differentiation** Epigenetic mechanisms such as DNA methylation, histone acetylation and methylation, chromatin remodeling, and RNA interference, and their effects in gene activation and inactivation, are increasingly understood to be more than "bit players" in phenotype transmission and differentiation, representing a fundamental regulatory mechanism with consequences for cell differentiation and development.

DNA methylation and histone modifications, are closely related and involved chromatin structure. Successful epigenetic control of gene expression often requires the cooperation and interaction of both mechanisms, and disruption of either of those processes leads to aberrant gene expression.

1) **DNA methylation**: DNA methylation turns off gene expression and DNA methylation profile is reprogrammed during cell differentiation. Establishing a silent chromatin state is achieved through the addition of a methyl-group to (predominately) the cytosine residue of a CpG dinucleotide. The addition of the methyl-group causes a physical change in the state of the chromatin that inhibits the expression of any genes in the methylated region. And, interestingly, this inhibitory chromatin state is passed on to daughter cells during cell division.

During the development of fertilized egg, dynamic changes in both histone modification and DNA methylation can be seen, generating cells with a broad developmental potential. The first true differentiation step in early development is the distinction between the inner cell masses, which go on to form the embryo, and trophectodermal cells, which form extraembryonic tissues.

2) **Histone acetylation**: The nucleus contains the chromosomes of the cell. Each chromosome consists of a single molecule of DNA complexed with an equal mass of proteins. Collectively, the DNA of the nucleus with its associated proteins is called

chromatin.

Each nucleosome consists of eight histone molecules (two each of histones H2A, H2B, H3, and H4) associated with 146 nucleotide pairs of DNA and a stretch of linker DNA about 50 nucleotide pairs in length. The diameter of the nucleosome "bead," or core particle, is about 10 nm. Histone H1 is thought to bind to the linker DNA and facilitate the packing of nucleosomes into 30 nm fibers (Figure 10-6).

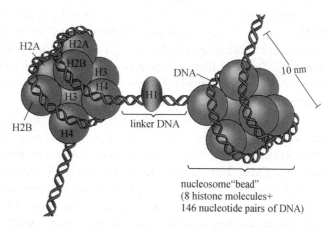

Figure 10-6 Nucleosome. Subunit of chromatin composed of a short length of DNA wrapped around a core of histone proteins (From: Longman AW. 1999)

Histone acetylation and deacetylation are chromatin-modifying processes that have fundamental importance for transcriptional regulation. Transcriptionally active chromatin regions show a high degree of histone acetylation, whereas deacetylation events are generally linked to transcriptional silencing. Acetylation and deacetylation of histone tails are catalyzed by histone acetyltransferases and deacetylases (HDACs), respectively. Histone acetyltransferases have been shown to increase the activity of several transcription factors, including nuclear hormone receptors, by inducing histone acetylation, which facilitates promoter access to the transcriptional machinery. Conversely, HDACs reduce levels of histone acetylation and are associated with transcriptional repression.

3) **Histone methylation**: Post-translational addition of methyl groups to the aminoterminal tails of histone proteins was discovered more than three decades ago. Only now, however, is the biological significance of lysine and arginine methylation of histone tails being elucidated. Recent findings indicate that methylation of certain core histones is catalyzed by a family of conserved proteins known as the histone methyltransferases (HMTs).

The basic amino acid side chains of arginine and lysine residues on histones can be methylated by HMTs. Most modified residues are located in the N-terminal tails of histone H3 and H4 (Figure 10-6).

Arginines can either be mono or dimethylated (symmetric or asymmetric methylation), while lysines can be mono-, di-or trimethylated. Different modification levels of multiple residues give high combinatorial complexity, enabling histone methylation to be involved in different DNA-templated events.

4) **Chromatin remodeling**: Chromatin remodeling refers to numerous *in vitro* ATP-dependent changes in a chromatin substrate, including disruption of histone-DNA contacts within nucleosomes, movement of histone octamers *in cis* and *in trans*, loss of negative supercoils from circular minichromosomes, and increased accessibility of nucleosomal DNA to transcription factors and restriction endonucleases.

Histone modification induces chromatin remodeling by regulating the way in which nucleosomes are packaged into: ① Open chromatin conformations which permit gene expression; ②Inaccessible chromatin conformations in which genes are repressed. Specific modifications define the activity state of a gene.

Two types of chromatin remodeling complexes have been extensively studied, both contribute to many cellular functions, including differentiation. ① ATP-dependent chromatin remodeling complexes; ②Histone-modifying enzymes — which include histone acetyltransferases, histone deacetylases, and **histone kinases**. Transcriptional activators and repressors are responsible for recruitment of one or more of these large, multisubunit chromatin remodeling complexes.

During cell differentiation, DNA methylation is important in the control of gene transcription and chromatin structure. Emerging evidence suggests that chromatin remodeling enzymes and histone methylation are essential for proper DNA methylation patterns. Other histone modifications, such as acetylation and phosphorylation, in turn, affect histone methylation and histone methylation also appears to be highly reliant on chromatin remodeling enzymes. Finally, the genes for survival and genes associated with specialized structure and functions are activated, while other genes are suppressed.

5) **MicroRNA**: MicroRNAs (miRNAs) are small, often phylogenetically conserved, non-protein coding RNAs that mediate posttranscriptional gene repression by inhibiting protein translation or by destabilizing target transcripts. miRNAs provide an additional level of gene regulation by functioning at the posttranscriptional level.

Studies demonstrated the requirement of mature miRNAs for the maintenance of ES cell self-renewal and pluripotency. Additionally, several independent studies have identified distinct miRNAs expressed in ES cells and their differentiated counterparts, reinforcing the role of ES cell-specific miRNAs in regulating ES cell identity(Figure 10-7).

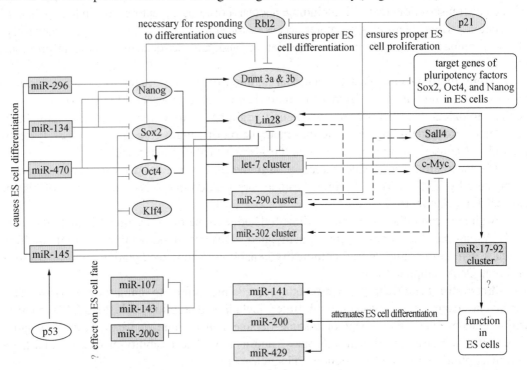

Figure 10-7 Regulatory networks of miRNAs and proteins involved in control of ES cell self-renewal and differentiation (From: Developmental Biology. 2010, 344(1):16-25)

(3) Polycomb proteins function as epigenetic switches **Polycomb group (PcG)** were

originally identified as part of an epigenetic cellular memory system that controls gene silencing via chromatin structure. However, recent reports suggest that they are also involved in controlling dynamics and plasticity of gene regulation, particularly during differentiation, by interacting with other components of the transcriptional apparatus. Global analysis of PcG proteins binding in ES cells identified promoters of a large number of genes associated with differentiation, suggesting that PcG proteins help to maintain the silencing of these genes in undifferentiated ES cells.

1) **PcG proteins repress gene transcription**: PcG proteins are transcriptional master regulators that maintain repressed states of thousands of genes. This function of PcG proteins is crucial during cellular differentiation to establish cell-type specific transcription profiles. PcGs form several complexes that are thought to collaborate to repress gene transcription. To date, at least 15 PcG genes have been identified in Drosophila and even much more PcG genes exist in mammalians.

These complexes dynamically define cellular identity through the regulation of key developmental genes and are responsible for transcriptionally repressing differentiation- and development- promoting genes. Alternately, upon receiving differentiation signals, PcG proteins are recruited to the promoters of transcription factor genes, exerting derepressive roles on the expression of differentiation-and development-promoting genes and repressive roles on those related to maintenance of the ability to self-renew and pluripotency, respectively. Genome- wide studies demonstrate that genes targeted by PcG are predominantly developmental transcription factors.

2) **PcG proteins complex**: A remarkable property of PcG system is that it consists of multimeric complexes and each complex also contains a variety of components. PcG proteins form at least three distinct key complexes, including polycomb repressive complex 1 (PRC1), polycomb repressive complex 2 (PRC2), and pleiohomeotic repressive complex (PhoRC). They reform the chromatin that delineate both repressive and derepressive gene signals, thereby maintaining a particular cell identity or allowing differentiation (Figure 10-8).

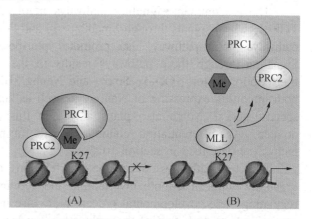

Figure 10-8 **Reversal of Polycomb-group silencing by demethylases.** (A) In the silenced state, the PRC2 complex binds to target genes and adds three methyl groups (Me) to histone H3 at lysine 27 (K27). PRC1 is recruited to histone H3 trimethylated at K27, preventing transcription of the genes associated with this lysine. (B) Such gene silencing might be rapidly reversed by the MLL (mixed lineage leukemia) complexes, which contain either of two histone demethylase enzymes, permitting gene transcription and additional modifications that also promote transcription. (From: Nature. 2007, 450:357)

Recent examples suggested that PRC1 catalyzes the mono-ubiquitinylation of histone H2A at lysine 119 (H2AK119Ub1), blocking RNA polymerase Ⅱ activity and resulting in transcriptional suppression; PRC2 catalyzes the di-and tri-methylation of histone H3 lysine 27 (H3K27me2/me3); while PhoRC selectively interacts with methylated histones in the chromatin flanking PREs to maintain a Polycomb-repressed chromatin state.

3) **PcG proteins interact with epigenetic proteins**: In addition, PcG proteins, together

with a battery of components including sequence-specific DNA binding/accessory factors, chromatin remodeling factors, signaling pathway intermediates, noncoding small RNAs, and RNA interference machinery, generally define a dynamic cellular identity through tight regulation of specific gene expression patterns.

10.4.4 Signal pathways for differentiation

The induction and patterning of tissues and organs in multicellular organisms are regulated by extracellular signals derived from neighboring cells and tissues that control the proliferation, differentiation and death of the cells.

The signals, either soluble or membrane-bound, interact with receptors or receptor complexes that stimulate intracellular signaling pathways, which in most cases lead to the reactions in nucleus, resulting in changes in gene expression. A limited number of different families of extracellular signals appear to be sufficient to induce and pattern most tissues and organs. Each family of signals and receptors consists, however, of multiple members that interact with different specificity.

Most of these signals and pathways were initially identified and analyzed genetically in Drosophila and Caenorhabditis elegans, but have subsequently been shown to perform the same or similar functions in vertebrate organisms. The concerted actions of signal pathways are observed among follows: Bone morphogenetic protein (BMP) pathway, transforming growth factor-β (TGF-β) pathway, fibroblast growth factor (FGF) pathway, epidermal growth factor (EGF) pathway, wingless (Wnt/Wg) pathway. Notch pathway.

The predominant signaling pathways involved in pluripotency and self renewal are TGF-β, which signals through Smad2/3/4, and FGFR, which activates the MAPK and Akt pathways. Wnt pathway also promotes pluripotency through activation of β-catenin. Signaling through these pathways results in the expression and activation of three key transcription factors: Oct-4, Sox2, and Nanog. BMP pathway promotes differentiation by both inhibiting expression of Nanog, as well as activating the expression of differentiation specific genes. Notch also plays a role in this process through the notch intracellular domain. As differentiation continues, cells further differentiate along lineage specific pathways, described in section 10.5.

Inductive pathways control the proliferation, differentiation and death of cells during organ development in both invertebrate and vertebrate organisms. The signaling pathways regulating these developmental processes are also involved in controlling the homeostasis of organs, and perturbation of the pathways often leads to disease.

10.4.5 Nuclear pore complexes and differentiation

Nuclear pore complexes (NPCs) are built from at most 30 different proteins called nucleoporins (Nups). In addition to regulating transport between the nucleus and cytoplasm, they have roles in gene regulation and chromatin organization and exhibit cell-type-specific expression. Current studies confirmed that specific change in NPC composition is required for both myogenic and neuronal differentiation, indicating that NPC might play an important role in differentiation and development by regulating the activity of genes critical for these processes, and that NPC composition is an essential step in cell differentiation.

10.5 The major cell differentiation systems

Adult stem cells, similar to embryonic stem cells, have the ability to differentiate into more than one cell type, but unlike embryonic stem cells they are often restricted to certain lineages. The production of adult stem cells does not require the destruction of an embryo and can be isolated from a tissue sample obtained from an adult. They have mainly been studied in humans and model organisms such as mice and rats.

10.5.1 Haematopoietic stem cell

Haematopoietic stem cells (HSCs) are well characterized adult stem cells, and have been isolated from mice and humans, and have been shown to be responsible for the generation and regeneration of the blood-forming and immune (haematolymphoid) systems. Stem cells from a variety of organs might have the potential to be used for therapy in the future, but HSCs — the vital elements in bone-marrow transplantation — have already been used extensively in therapeutic settings.

(1) **The properties of HSCs** The haematopoietic stem cell is characterized by two cardinal properties. First, they are multipotential; that is, individual stem cells can clonally produce all of the different cell types found in the blood. Second, at least operationally, they are self-renewing; that is, during the proliferation of stem cells commitment/differentiation is balanced by the production of additional stem cells. This self-renewal ability follows logically from the lifelong production of mature cell populations.

(2) **Self-renewal of haematopoietic stem cells** In the haematopoietic system, stem cells are heterogeneous with respect to their ability to self-renew. Multipotent progenitors constitute 0.05% of mouse bone-marrow cells, and can be divided into three different populations: ①longterm self-renewing haematopoietic stem cells (HSCs), short-term self-renewing HSCs, multipotent progenitors without detectable self-renewal potential.

These populations form a lineage in which the long-term HSCs give rise to short-term HSCs, which in turn give rise to multipotent progenitors. As HSCs mature from the long-term self-renewing pool to multipotent progenitors, they progressively lose their potential to self-renew but become more mitotically active. Whereas long-term HSCs give rise to mature haematopoietic cells for the lifetime of the mouse, short-term HSCs and multipotent progenitors reconstitute lethally irradiated mice for less than eight weeks.

(3) **HSCs produces mature blood cells** The vertebrate haematopoietic system produces at least eight distinct lineages of mature blood cells in a continuous manner throughout adult life. These lineages include red blood cells, monocytic, granulocytic and basophilic myeloid cells, the B and T cell compartments of the antigen-specific immune system, as well as the megakaryocytes of the thrombopoietic system. Continuous replenishment of the mature cell populations is necessitated by their finite half-lives. In order to meet daily demands, the rate of mature blood cell production is very high; in the human, many billions of new cells are produced every day.

(4) **CD34 as the universal marker of HSCs** CD34 was discovered originally as the result of a strategy to develop antibodies that recognize small subsets of human marrow cells, but not mature blood and lymphoid cells. Cell surface expression of the CD34 antigen has rapidly become the distinguishing feature used as the basis for the enumeration, isolation, and manipulation of human stem cells, because CD34 is down-regulated as cells differentiate into more abundant mature cells.

In addition to being expressed selectively on stem cells and early progenitors during human and murine haematopoiesis, both mouse and human CD34 are expressed outside the haematopoietic system on vascular endothelial cells and some fibroblasts. This distribution suggests a function outside haematopoiesis. Transplant studies in several species, including baboons and mice, have shown that long-term marrow repopulation can be provided by CD34 + selected cells. However, several recent studies have suggested that there may be human and murine haematopoietic stem cells that do not express CD34.

(5) **The differentiation capacity of HSCs** HSCs have the ability to balance self-renewal against differentiation cell fate decisions; they are multipotent, a single stem cell producing at least eight to ten distinct lineages of mature cells; they have an extensive proliferative capacity that yields a large number of mature progeny; HSCs are rare, with a frequency of 1 in 10,000 to 100,000 total blood cells; they are slowly cycling in a steady-state adult haematopoietic system(Figure 10-7).

1) **HSCs are committed to the haematopoietic lineage**: HSCs can be subdivided into long-term self-renewing HSCs, short-term self-renewing. HSCs and multipotent progenitors can differentiate to the cells in haematopoietic lineage as described in figure 10-7. They differentiate to common lymphoid progenitors (CLPs; the precursors of all lymphoid cells) and common myeloid progenitors (CMPs; the precursors of all myeloid cells). Both CMPs/GMPs (granulocyte macrophage precursors) and CLPs can give rise to all known dendritic cells.

2) **HSCs are capable of differentiation to cells in non-haematopoietic lineage**: The bone marrow of adult rodents contains cells with the capacity to give rise to hepatocytes, muscle tissue, or even neurons. Some haematopoietic stem cells present in adult bone marrow co-purified with stem cells that gave rise to hepatocytes, supporting the new concept that somatic adult stem cells can change cell fate. However, these reports were based on adult rodent stem cell populations, which may differ from human stem cells in their capacity to give rise to multiple tissues.

10.5.2 Neural stem cells

Neural stem cells (NSCs) have the ability to self-renew, and are capable of differentiating into **neurones**, **astrocytes** and **oligodendrocytes**. Such cells have been isolated from the developing brain and more recently from the adult central nervous system.

After newly produced neurons and glial cells reach their final destinations, they differentiate into mature cells. For neurons, this involves outgrowth of dendrites and extension of axonal processes, formation of synapses, and production of neurotransmitter systems, including receptors and selective reuptake sites. Most axons will become insulated by myelin sheaths produced by oligodendroglial cells. Many of these events occur with a peak period from 5 months of gestation onward.

(1) Advances in NSCs research Some important considerations have emerged recently.
- The nervous tissue contains bona fide stem cells that support neuronal cell turnover throughout life.
- Despite their origin from one of the most quiescent tissues in the body, NSCs can undergo effective long-term culturing, proliferation and expansion while retaining stable functional characteristics.
- When properly challenged, the overall developmental potential appears to be broader than that observed under physiological conditions *in vivo*.
- NSCs may differentiate *in vitro*, and their possibility of transdifferentiation are

confirmed.

(2) **Plasticity of NSCs within the CNS**   Although *in vivo* NSCs appear to generate almost exclusively neuronal cells, their actual developmental potential as observed *ex vivo* is much broader than expected. In fact, cultured NSCs do give rise to neurons, astrocytes, and oligodendrocytes and are therefore normally classified as being multipotent in nature.

The concept of NSC plasticity and of their dependence on environmental cues is strengthened by transplantation and manipulation/recruitment studies *in vivo*. For example, intrahippocampus transplantation of hippocampal stem cells results in the generation of the neuronal types normally found within this region. However, when transplanted in adult brain regions in which no neurogenesis takes place, NSCs produce exclusively glial cells.

(3) **Plasticity of NSCs outside the CNS**   NSCs (neuroectodermal in origin) can give rise to cells that normally derive from germ layers other than the neuroectoderm.

One of the most intriguing cases is the virtual pluripotency of bone marrow-derived cells; however, multiple examples of stem cells (SCs) giving rise to cells normally found in other tissues have become available. In some cases, both the original SCs and the cells to which they give rise derive from the same embryonic germ layer (intragerm layer conversion).

Nonetheless, in the presence of sustained damage as caused by lethal irradiation, adult bone marrow cells were found to integrate and differentiate within the brain tissue after systemic injection, that is "blood-into-brain" conversion. In a study, bone marrow cells from wild-type male mice were injected intraperitoneally in PU1-mutant, immunodeficient females. In these animals, 0.2% to 2.3% of the male cells found within the brain parenchyma expressed neuronal markers.

10.5.3 Cancer stem cells

Cancer cell is a cell that divides and reproduces abnormally with uncontrolled growth. This cell can break away and travel to other parts of the body and set up another site, referred to as metastasis. Though a few cancer stem cells and the large number of cancer cells constituting the tumor are morphologically similar, they are functionally heterogeneous.

(1) **Cancer cells have many of the characteristics of normal stem cells**   For decades, cancer researchers have wrestled with two competing visions of tumors: ①All the cells of a tumor are pretty much the same. That is, they have an equal capacity to divide and form new tumors. ②Only a few select cells from a tumor have the capacity to initiate new, full-fledged tumors. These bad seeds are the cancer stem cells.

Cancer cells have many of the characteristics of stem cells. They share the ability for: Self-renewing cell division; resistance to apoptosis (programmed cell death). several of the molecular signaling pathways associated with normal stem cell development, such as Wnt, Shh and Notch pathways, are also active in cancer development. ATP-binding cassette transporters, which remove drugs from the cell. Cancer stem cells maintain this characteristic. Oncogenic pathways, several long-known oncogenic pathways, which are pivotal to maintenance of normal stem cell self-renewal and to better characterization of tissue stem cells.

(2) **The concept of cancer stem cells**   The hypothesis of the cancer stem cells is from the observation that not all the cells within a tumor can maintain tumor growth, and that large numbers of tumor cells are needed to transplant a tumor, even in an autologous context. Moreover, most cancers are not clonal, but consist of heterogeneous cell populations, similar to the hierarchical tree of stem cell lineages. Also, a hypothetical long-lived stem

cell at the base of tumor outgrowth would allow progression towards malignancy through accumulation of mutations and epigenetic changes.

Researchers believe that leukemia is fueled by a small population of blood cells that divide continually, or self-renew, when they harbor certain genetic abnormalities. Continuous self-renewal prevents these cells from developing into the specialized blood cells that the body needs to function normally. According to Gilliland, previous research had established that leukemia and some other cancers were not made up of a homogeneous population of cells.

The concept of cancer stem cells could change notions of how cancers spread and how tumors should be treated. For example, some current cancer drugs may turn out to kill most cells in a tumor but to spare the stem cells, thereby setting the stage for a relapse. If we don't eliminate those cells, then they will just re-form tumors.

(3) Origination of cancer stem cells There is overwhelming evidence that only a subset of cells — referred to as cancer stem cells within a tumour clone, are tumorigenic and possess the metastatic phenotype. The identification of human cancer initiating cells provided a major step forward in this field.

Recently, by applying techniques used in the stem cell field to identify self-renewing populations, has it become possible to prospectively isolate cancer stem cells. The prospective isolation and transplantation of defined cells demonstrate that only a few cells within a tumor can generate heterogeneous tumors, whereas transplantation of the rest of the cells within a tumor did not give rise to tumors upon transplantation.

1) **Cancer stem cell may come from normal stem cells**: Chronic tissue injury can result from cigarette smoke, alcohol, stomach ulcers, acid reflux, inflammatory bowel disease, and other causes. These conditions result in a continuous need to replace damaged tissues. Thus the stem cells are in a continuous state of cell division and signaling pathway activation, which resembles the conditions seen in cancers. The inability to return to a quiescent state may result in the stem cell progressing to a cancer stem cell.

2) **Abnormal activation of genes results in cancer**: Cancer cells have many of the same surface features as normal stem cells. Genes that are abnormally active in several kinds of cancers are the same ones that drive stem cells' self-renewing ability. This suggests that self-renewal, the hallmark of stem cells, may be essential to cancer. Either a normal stem cell goes awry to produce cancer, or an abnormal mature cancer cell may somehow regain stem cells special trait.

On the other hand, previous work had determined that in **chronic myeloid leukemia** (CML), hematopoietic stem cells may not be involved, but rather a progenitor cell that reacquired self-renewal capacities. Although CML is classified as a HSC disorder — the *bcr/abl* fusion gene is expressed in HSCs — direct evidence for HSC malignancy in CML has been lacking.

(4) The discovery of cancer stem cells The existence of cancer stem cells had been hypothesized for many decades, but was not until 1997 that they were isolated from patients with acute myeloid leukemia. Subsequently, cancer stem cells have been isolated from breast and brain cancers, and cancer cell lines. Cancer stem cells may derive from mutations in normal stem cells. Alternatively, differentiated tumor cells may acquire the characteristics of stem cells. Cancer stem cells are present in only very small numbers in tumors, and may not be present in all tumors.

Dysregulation of fundamental processes in developmental biology such as stem cell differentiation and proliferation may be the basis of many neoplastic diseases. Cancer stem cells have recently been identified in a number of malignancies, including leukemia, breast tumors, gliomas and cancers initiated from other tissues.

Researchers are now racing to identify tumor-forming stem cells in skin, lung, pancreatic, ovarian, prostate, and many other cancers. And researchers suspect that the dangerous cells may arise from mutations in the normal stem cells that sustain various tissues.

(5) **Induced differentiation of cancer stem cells** It is recognized that cancers are likely to arise from disruption of differentiation of early progenitors that fail to give birth to cell lineage restricted phenotypes. An understanding of the mechanism that controls growth and differentiation in normal cells would seem to be an essential requirement to elucidate the origin and reversibility of malignancy. But a few reports provide evidences to support the partial differentiation of cancer cell by chemicals, such as all-trans-retinoic acid (t-RA), cyclin-dependent kinase inhibitors, inhibitors of HDACs.

Drugs that affect differentiation, such as retinoic acid, have found therapeutic niches for the treatment of various forms of leukemia. As new understanding of stem cell biology and its role in cancer emerges, the development and application of drugs that target stem cell differentiation and cell fate will provide new therapeutic modalities for the management and treatment of malignant diseases. All-trans retinoic acid administration leads to complete remission in acute promyelocytic leukemia (APL) patients by inducing growth arrest and differentiation of the leukemic clone.

Pluripotential human embryonal carcinoma (EC) cell lines undergo differentiation programs resembling those occurring in embryonal stem cells during development. Expression profiling was performed during the terminal differentiation of the EC cell line also by all-trans-retinoic acid.

10.5.4 Differentiation in plants cells

Dedifferentiation also occurs in plants cells in cell culture can lose properties they originally had, such as protein expression, or change shape. This process is also termed dedifferentiation.

(1) **Plant cell types** Plants have about a dozen basic cell types that are required for everyday functioning and survival. Additional cell types are required for sexual reproduction. While the basic diversity of plant cell types is low compared to animals, these cells are strikingly different.

1) **Epidermal cells**: are the most common cell type in the epidermis. These cells are often called "pavement cells" because they are flat polygonal cells that form a continuous layer, with no spaces between individual cells. Epidermal cells are transparent, because their plastids remain small and undifferentiated; hence light readily penetrates through to the photosynthetic tissues beneath the epidermis. Two more specialized cell types are also found in the epidermis: guard cells and trichomes.

2) **Parenchyma cells**: are relatively unspecialized cells that make up the bulk of the soft internal tissues of leaves, stems, roots, and fruits. Parenchyma cells have thin, flexible cell walls and their cytoplasm typically contains a large, water-filled vacuole that fills 90 percent of the cell's volume. The vacuole may also contain compounds such as sugars.

3) **Collenchyma cells**: are long narrow cells with thick, strong, yet extensible, cell walls. Collenchyma cells are found in strands or sheets beneath the epidermis and function to provide support while a stem or leaf is still expanding.

4) **Sclerenchyma cells**: have extremely hard, thick walls. Two general types of

sclerenchyma cells are recognized: sclereids, which may be more or less isodiametric or may be branched, and fibers, which are greatly elongate cells.

5) **Vascular tissue cells**: are highly specialized cells and aligned end to end within the xylem. The part of the cell wall between adjacent cells is degraded, so that the interior of all the vessel elements in a file becomes continuous, forming a vessel. Vessel elements are dead at maturity, leaving a hollow tube for the flow of water upward from the roots to the shoot system.

(2) Cell differentiation and development   Cell differentiation is only part of the larger picture of plant development. As plant organs develop (the process of organogenesis), the precursors of the tissue systems form in response to positional signals. Then, within each tissue system precursor, cell types must be specified in the proper spatial pattern. For instance, the spacing of trichomes and stomates within the protoderm must be specified before their precursor cells begin differentiation. Exchange of signals among neighboring cells is an important aspect of the processes of spatial patterning and cell differentiation. In addition, long distance signals are required so that the strands of xylem and phloem cells within the leaf vascular bundles connect perfectly with those in the stem.

(3) Hormons control cell differentiation in plants   Many of normal stem segments differentiation are controlled by hormons by regulating the proportion of auxin and cytokinin. The control of cell proliferation and differentiation during development depends, in most cases, on the concerted action of plant hormones. Among them, auxins and cytokinins are the best documented and they can impinge directly on cell cycle regulators. In addition, other hormones, e. g., abscisic acid, ethylene, jasmonic acid and brassinosteroids, whose action is much less well characterized, also have an impact on cell cycle progression and(or) arrest.

1) **Auxin**: plays an important role in the differentiation of vessel elements, both in intact and wounded plants. Auxin produced by the apical meristem and young leaves above the wound induces parenchyma cells to regenerate the damaged vascular tissue. Auxin is involved, since transdifferentiation is blocked when the sources of natural auxin (young leaves and buds) are removed or when auxin transport inhibitors are applied. If natural sources of auxin are removed, and artificial sources added, transdifferentiation of parenchyma cells will occur, regenerating the vascular bundle.

2) **Cytokinins** (CKs) : are a class of plant growth substances and act in concert with auxin. More cytokinin induces growth of shoot buds, while more auxin induces root formation. CKs are implicated in essential plant growth-related processes, e. g. induction of cell growth and differentiation, shoot formation and development, activation of dormant lateral buds, delayed senescence, among others. Evidence has been provided to show that in the root meristem, cytokinin acts in defined developmental domains to control cell differentiation rate, thus controlling root meristem size.

Summary

Cell differentiation is the process by which cells acquire the specialized structures and functions. This process occurs through changes in gene expression, which controls cell differentiation. According to the potency of differentiation, cells could be classified as totipotent cells, pluripotent cells, multipotent cells and terminal differentiated cell. Stem cells have the potency of differentiation, and consist of embryonic stem cells, embryonic germ cells and adult stem cells. The differentiation processes of haematopoietic stem cells, lymphocyts, neural stem cell and liver stem cell are well described. Cancer stem cell as a special sample for cell differentiation is under the investigation.

During differentiation, specialized cells take in genetic information from stem cell or progenitor cell, but they exhibit different phenotype (structure and function). It is the alternation of genes expression pattern, not the change of DNA sequence that results in the different phenotype. Cell differentiation related transcriptional factors determine the fate of cells. PcG proteins are key molecules linked with epigenetic regulation. Epigenetic mechanisms underlying the cell differentiation include DNA methylation, histone acetylation as well as methylation, and chromatin remodeling.

Questions
1. What does the differentiation mean in cell biology?
2. What do you know the epigenetic control in cell differentiation?
3. How is cell differentiation regulated?
4. What are stem cells? What are their common features?
5. Please describe established cell differentiation systems.
6. Can cancer cells differentiate into normal cells? Why?

(吴超群)

Chapter 11

Senescence and apoptosis

Senescence (or aging) and **apoptosis** (or programmed cell death, PCD) are two common cell fates. While apoptosis leads to death and removal of the cells, senescence cells persist and survive for extended periods of time (unable to maintain their normal functions). Both senescence and apoptosis are important not only for development, but also for a cancer treatment perspective.

11.1 Senescence

In biology, senescence is the state or process of aging. Cellular senescence is a phenomenon where isolated cells demonstrate a limited ability to divide in a culture dish (Figure 11-1), while organismal senescence is the aging of organisms.

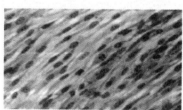
young cells after 20 population doublings(PDs)

older cells entering a senescent phase after 50 population doublings

Figure 11-1 Human fibroblast before and after replicative senescence

Senescence is one of the most intensely studied processes in biology. Currently, senescence is defined as progressive deterioration in physiological functions and metabolic processes of a cell or organism. It is a program activated by normal cells in response to various types of stress.

Cellular senescence is induced by telomere uncapping, DNA damage, **oxidative stress**, **oncogene** activity and others. Senescence can occur following a period of cellular proliferation or in a rapid manner in response to acute stress. Once cells have entered senescence, they cease to divide and undergo a series of dramatic morphologic and metabolic changes. Cellular senescence is thought to play an important role in tumor suppression and to contribute to organismal senescence.

11.1.1 Senescence is a complicate phenomenon

A broad biological approach makes it possible to understand why senescence exists and also why different mammalian species have very different maximum longevities. The adult organism is maintained in a functional state by at least ten major mechanisms, which together constitute a substantial proportion of all biological processes. Although we now have a biological understanding of the senescence process, much future research will be needed to uncover the cellular and molecular changes that give rise to age-associated diseases.

(1) **Some aspects of senescence may be universal** This tentative conclusion has been reached from the discovery that genes which are important for modulating aspects of the rate of senescence in humans play similar roles for such diverse organisms as *Saccharomyces cerevisiae* (baker's yeast), *Drosophila melanogaster* (fruitfly) and *Caenorhabditis elegans* (roundworm).

(2) **Senescence occurs at multiple levels** Cell senescence is thought to contribute to organismic and system senescence. Senescence occurs at multiple levels, such as molecules level, cells level, tissues level and organ systems level. Possible mechanisms for induction of cell senescence involve in telomere shortening, double-strand breaks, oxidative stress, terminal differentiation and DNA methylation.

(3) **Cellular senescence contributes to individual aging** Scientists have identified specific cellular events that enable a cell to make the transition from a state of active growth to an irreversible state of growth arrest, called senescence. Understanding the mysterious process of senescence is of great importance to human biology, because a failure to enter this normal stage in the life cycle of a cell may lead to uncontrolled growth and cellular immortality, the hallmark of cancer.

(4) **Senescent cells undergo three phenotypic changes** Normal cells do not divide indefinitely due to a process termed cellular or replicative senescence. Cell senescence is thought to contribute to organismic senescence. Senescent

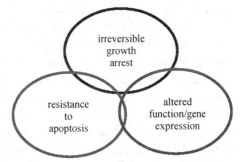

Figure 11-2 Senescent/"aged" cells: many characteristics change

cells undergo three phenotypic changes (Figure 11-2): cell irreversibly arrest growth、cell acquire resistance to apoptotic death、cell acquire altered differentiated functions.

11.1.2 Different theories elaborated about senescence

Several avenues to studying senescence have placed us on the threshold of understanding basic underlying mechanisms. These approaches include the identification of key genes and pathways important in senescence; genetic studies of heritable diseases that cause the appearance premature senescence in affected people; physiological experiments that relate the pace of senescence to caloric intake; and advances in human genetics, as well as cell and molecular biology leading to an understanding of the basis of many diseases of senescence. Strikingly, single gene mutations have been found to significantly extend the lifespan in *C. elegans*, yeast.

Even the exact mechanisms of senescence process are not known, there are at lease five hypotheses that have been proposed to explain senescence: mitochondria ageing、reactive oxygen species (ROS)、replicative ageing and telomere shortening、senescence processes and apoptosis、the evolutionary theories of senescence.

11.1.3 Mitochondria ageing

Senescence is a complex physiological phenomenon and several different theories have been elaborated about its origin. Among such theories, the "mitochondrial theory of senescence", which has gained a large support, indicates the accumulation of somatic mutations of mitochondrial DNA leading to the decline of mitochondrial functionality as one of the driving forces for the process itself.

Within eukaryotic cells mitochondria provide most of the ATP by oxidative phosphorylation. In addition, mitochondria are critical for other aspects of cell function, such as haem and iron sulphur centre biosynthesis and modulating calcium levels. Consequently mitochondrial dysfunction arising from oxidative damage and mutations to mitochondrial and nuclear genes, includes the cumulative degeneration associated with ageing. This mitochondrial dysfunction causes cell damage and death by compromising ATP production, disrupting calcium homeostasis and increasing oxidative stress. Furthermore mitochondrial damage can lead to apoptotic cell death by causing the release of cytochrome C and other pro-apoptotic factors into the cytoplasm.

The mitochondrial theory of ageing states include two aspects: ① Alterations of structure and expression of mitochondrial genome; ②Alterations of structure and expression of cytochrome C oxidase activity.

(1) **Alterations of structure and function of mitochondria** The mitochondrial theory of ageing states that the slow accumulation of impaired mitochondria is the driving force of the ageing process and ageing "phenotype". In healthy individuals, somatic mtDNA mutations accumulate with age, in the form of a wide range of different mtDNA deletions in postmitotic tissues such as skeletal muscle, myocardium, and the brain. Age associated mtDNA mutations occur more frequently and accumulate much faster in the tissues of high energy demand and low mitotic activity. When the proportion of mutant mtDNA exceeds a critical threshold concentration, a defect of mitochondrial oxidative phosphorylation is results.

Oxidative stress is a particularly important factor in mitochondrial dysfunction because the respiratory chain continually leaks the free radical superoxide. Another radical that interacts with mitochondrial metabolism is nitric oxide which can react with superoxide to form the damsenescence by product peroxynitrite. These reactive oxygen species lead to generalized oxidative damage to all mitochondrial components, including an increased mutation rate for mitochondrial DNA relative to nuclear DNA.

(2) **Alterations of structure of cytochrome C oxidase** Cytochrome C oxidase (COX) is the terminal enzyme of the mitochondrial respiratory chain, located in the inner membrane of mitochondria and bacteria, catalyzing the transfer of electrons from reduced cytochrome C to molecular oxygen. The three catalytic subunits, COX I, II and III are coded by the mitochondrial DNA and are synthesized within mitochondria. The mammalian COX IV, Vb, VIa, VIb, VIc, VIIa, VIIb and VIII are encoded by the nuclear genome, synthesized in the cytosol and imported into mitochondria.

Function of COX is often affected by the mutation, resulting in cellular dysfunction. The vast majority of COX disorders are linked to mutations in nuclear-encoded proteins referred to as assembly factors, or assembly proteins. Defects involving genetic mutations alter the COX functionality or structure. Important finding in tissues from elderly individuals is the presence of COX deficiency within individual cells.

11.1.4 Replicative senescence and telomere

Higher organisms contain two types of cells: postmitotic cells, which never divide, and mitotic (or mitotically competent) cells, which can divide. Postmitotic cells include mature nerve, muscle, and fat cells, some of which persist for life. Mitotic cells include epithelial and stromal cells of organs such as the skin. Because postmitotic and mitotic cells differ in their proliferative capacity, they may age by different mechanisms. Normal somatic mitotically competent cells do not divide indefinitely. The process that limits the cell division number is termed cellular or replicative senescence.

Senescent cells differ from their pre-senescent counterparts in three ways: ① They arrest growth and cannot be stimulated to reenter the cell cycle by physiological mitogens; ② They become resistant to apoptotic cell death; ③ They acquire altered differentiated functions.

(1) Telomere shortening *in vitro* The association between cellular senescence and **telomere** shortening in vitro is well established. In the laboratory, telomerase-negative differentiated somatic cells maintain a youthful state, instead of senescence, when transfected with vectors encoding telomerase. Many human cancer cells demonstrate high telomerase activity. Evidence is also accumulating that telomere shortening is associated with cellular senescence *in vivo* (Figure 11-3). What causes changes in expression of telomerase in different cell types and premature senescence syndromes? Does the key to "youthfulness" lie in our ability to control the expression of telomerase? We have reviewed the contemporary literature to find answers to these questions and explore the association between senescence, telomeres, and telomerase.

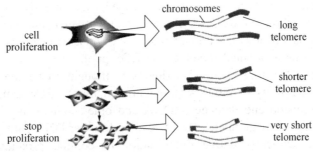

Figure 11-3 Telomeres. The linear chromosomes of all of the eukaryote cells have the telomeres at both ends. As the cells continue to proliferate, the length of the telomeres shortens and ultimately the cells stop dividing (From: PNAS. 2000, 97(23): 12407)

Normal cells do not divide indefinitely due to a process termed cellular or replicative senescence. Several lines of evidence suggest that replicative senescence evolved to protect higher eukaryotes, particularly mammals, from developing cancer.

(2) Replicative senescence is a fail-safe mechanism Replicative senescence occurs because, owing to the biochemistry of DNA replication, cells acquire one or more critically short telomere. Recent findings suggest that certain types of DNA damage and inappropriate mitogenic signals can also cause cells to adopt a senescent phenotype. Thus, cells respond to a number of potentially oncogenic stimuli by adopting a senescent phenotype. These findings suggest that the senescence response is a fail-safe mechanism that protects cells from tumorigenic transformation.

(3) Evidence for the telomere hypothesis of senescence The telomere-based model of

cell senescence has proven to among been among the most enduring hypotheses in cell biology. This model, suggesting that the gradual loss of telomere sequences during the proliferation of cultured human somatic cells imposes a barrier on cellular replicative potential, has been strongly supported by recent genetic and biochemical studies. Evidence in support of the telomere hypothesis of senescence: ① Telomeres are shorter in most somatic tissues from older individuals compared to younger individuals. ② Telomeres are shorter in somatic cells than in germline cells. ③ Children born with progeria (early senescence syndrome) have shortened telomeres compared to age-matched controls. ④ Telomeres in normal cells from young individuals progressively shorten when grown in cell culture. ⑤ Experimental elongation of telomeres extends proliferative capacity of cultured cells.

11.1.5 Senescence by reactive oxygen species

Over recent years evidence has, however, been accumulating in favour of the free radical theory of senescence. This concept is supported by a wealth of recent observations confirming the genomic instability of mitochondria and suggesting that animal and human ageing is accompanied by mtDNA deletions and other types of injury to the mitochondrial genome. Despite this, an understanding of the mechanism by which cells might succumb to the effects of free radicals has proved elusive and the mitochondrial theory of ageing, however attractive in principle, is still supported by very little hard evidence.

(1) **Sources of ROS** *in vivo*   In a cell, approx 90% of cellular reactive oxygen species (ROS) is generated by the mitochondria. Electrons from the react with oxygen in mitochondria generate free radicals.

Superoxide that is not immediately scavenged can directly react with oxidized cytochrome c and cytochrome c oxidase. Despite the highly efficient mechanisms to mop up harmful ROS, there is still a small amount that cannot be accounted for by SOD and other peroxidases-contributing as a small factor to the ROS pool.

Proximity of mtDNA to sources of oxidants is thought to be the reason behind the observed increased sensitivity of mtDNA to oxidative damage. ROS generated in mitos can feed back on the organelle and damage mtDNA and other components in a vicious cycle. Similarly, oxidants can damage nuclear DNA leading to other DNA damage pathways (such as $p53$ activation). Direct oxidative modification of cellular proteins may also be an important element of ageing (this is a good way to relate to chew-wey's topic).

(2) **ROS is increased during ageing**   Some researchers showed that the production of reactive oxygen species (ROS) and free radicals in mitochondria is increased during ageing as a result of the ever-increasing electron leak from the defective respiratory system, and that the ensuing oxidative damage and mtDNA mutations cause further decline in respiratory function.

During ATP production normally about 1% ~ 5% of the consumed oxygen by mitochondria are converted to superoxide anions, hydrogen peroxide, and other ROS. These ROS do have certain important physiological functions, although in excess they can be harmful to cells. Although human cells can counteract the effects of ROS, through the expression of an array of free radical scavenging enzymes, these antioxidant defense systems are by no means perfect especially in the case of excess ROS. Thus there is a known age-dependent increase in the fraction of ROS and free radicals that may escape the cellular defense mechanisms, exerting oxidative damage on cellular constituents, including DNA, RNA, proteins, and membrane lipids.

11.1.6 Cellular senescence involved in genetic errors

Somatic mutation equals usage damages genes which causes faulty information which, in turn, causes senescence. A key prediction of the somatic mutation theory of senescence is that there is an invariant relationship between life span and the number of random mutations. A number of studies at a number of gene loci have shown that somatic mutations of a variety of types accumulate with age. Dietary restriction, which prolongs life span, results in slowed accumulation of HPRT mutants in mice. Conversely, senescence-accelerated mice, which have been bred to have a shortened life span, show accelerated accumulation of somatic mutations. Error catastrophe equals damage in genetic transcription to RNA or translation to protein causes a cascade of errors which, in turn, causes senescence.

(1) The role of genetics in senescence   The role of genetics in determining life-span is complex and paradoxical. Although the heritability of life-span is relatively minor, some genetic variants significantly modify senescence of mammals and invertebrates, with both positive and negative impacts on age-related disorders and life-spans. In certain examples, the gene variants alter metabolic pathways, which could thereby mediate interactions with nutritional and other environmental factors that influence life-span. Given the relatively minor effect and variable penetrance of genetic risk factors that appear to affect survival and health at advanced ages, life-style and other environmental influences may profoundly modify outcomes of senescence.

(2) The rate of error accumulation   According to the model proposed, the rate of error accumulation is intrinsic and species specific. An initially low level of errors could increase exponentially and randomly through the interaction of several self-perpetuating error generating systems, cross-feeding errors into each other, resulting in the accumulation of random somatic mutations and the production of altered proteins with random amino acid substitutions at an increasing frequency with age. Environmental factors can influence the rate of error accumulation by altering DNA, RNA, and proteins.

(3) Initial alteration in DNA sequence   The cellular researches about senescence have indicated the possibility that cellular senescence is conceived as the consequence of an accumulation of genetic errors. Several error-producing mechanisms have been proposed, such as DNA misduplication, gene mistranscription and mistranslation, and DNA misrepair. They have been incorporated into one model, consisting of DNA, RNA, and a few key enzymes, i. e., DNA polymerases, RNA polymerases, and aminoacyl-tRNA synthetases.

The initial alteration in DNA sequence can be introduced in a number of ways:
- Natural errors in replication. These occur at a frequency of about 1 in 1010 base pairs synthesized. For the human genome at 4×10^9 bp, about one cell division in three will contain a replication error.
- UV radiation induces dimers where there are 2 consecutive pyrimidines in the same DNA strand.
- X-radiation. Generates unstable ionic species and free radicals. High proportion of cells are killed, or carry multiple mutations. Not favoured for trying to get single mutations in cells.
- Mutagenic chemicals are both generators of CH^{3+}, potent alkylating agents, and form methyl adducts with bases that alter the H-bonding pattern of bases.
- Deaminating agents nitrous acid HONO, nitrites convert amino groups to OH groups/keto groups.

(4) **Faulty repair of genetic errors**   Efficient repair mechanisms normally eliminate lesions in the DNA. Permanent mutations are the results of errors in repair, and this can be promoted by the use of error prone strains of organisms. Mismatch repair replaces mismatched bases resulting from errors in replication. Cells containing defective components of the mismatch repair process have high spontaneous rates of mutation. Faulty repair equals a decline in DNA repair which causes genetic errors which causes faulty information which, in turn, causes senescence.

11.1.7 Epigenetic regulation of senescence

Epigenetic entails the study of the switching on and off of genes during development, cell proliferation, senescence and also by environmental insults. Epigenetic changes can result in the stable inheritance of a given spectrum of gene activities in specific cells. Genome modifications resulting from epigenetic changes appear to play a critical role in the cellular senescence. Scatter experimental evidence suggests that epigenetic changes could also be critical determinants of cellular senescence and organismal senescence.

(1) **Histone deacetylases participate in senescence**   In mammals, the proteins histone deacetylase SIRT1, a Sir 2 homologue, and the class I ~ II of histone deacetylases (HDACs) control the cellular response to stress. The scientists observed that as cells senesce, there is increased association of HDAC1 with the retinoblastoma protein RB, and a parallel decrease in the levels of the histone acetyltransferases p300 and CBP proteins. Elevated HDAC activity appears causally related to cellular senescence, as overexpression of a p300 mutant protein, or treatment with a specific chemical inhibitor of p300, results in irreversible growth arrest and senescence of normal human cells.

(2) **Euchromatin formation**   Heterochromatin represses gene activity, but many genes in senescent cells actually display higher expression levels, therefore euchromatin, which facilitates gene transcription, must also be involved. Data from a cDNA microarray study provide tentative support for this contention. The study identified transcriptional fingerprints unique to senescence, not those merely correlated with cell-cycle arrest. It also found that upregulated senescence-specific genes were physically clustered, an arrangement consistent with euchromatin formation.

(3) **Chromatin remodeling**   Chromatin remodeling is thought to mediate senescence in human melanocytes. According to one hypothesis, the histone acetyltransferase p300 is removed from promoters of certain cell-cycle regulatory genes, and HDACs are added. The resulting repression of those genes leads to a halt in mitosis.

11.1.8 The evolutionary theories of senescence

Empirical observations have demonstrated that senescence has a substantial genetic component. Most often cited is markedly variable lifespan potentials for different species, even within a single class such as mammals. Therefore many researchers consider senescence to be a byproduct of evolution due to lack of selective pressure for the post-reproductive individual. Simply stated, if evolution does not select for you it will ultimately work against you, perhaps by allowing the accumulation of genes that are harmful to postreproductive individuals. Ironically, some of these genes could be important for reproductive fitness but exert a deleterious effect later in life.

11.1.9 Senescence pathway

p53 and Rb, paradigmatic tumor suppressor proteins, have been shown to play critical roles in the induction of senescence. Both p53 and Rb are activated upon the entry into senescence (Figure 11-4). The p53 protein is stabilized and proceeds to activate its transcriptional targets, such as p21CIP1/WAF1. Rb is found at senescence in its active, hypophosphorylated form, in which it binds to the E2F protein family members to repress their transcriptional targets. These targets constitute the majority of effectors required for cell-cycle progression.

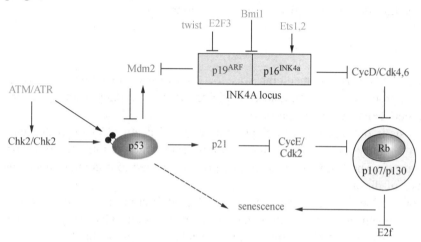

Figure 11-4 The molecular circuitry of senescence. p53 and Rb are the main activators of senescence (From: Int J Biochem & Cell Biol, 2005, 37 : 961-976)

Senescence requires activation of the Rb and/or p53 proteins, and expression of their regulators such as $p16^{INK4a}$, $p21^{CIP}$, and $p14^{ARF}$. P53 can activate senescence by activating Rb through p21 and other unknown proteins, and also, in human cells, can activate senescence independently of Rb. P53 activation is achieved by phosphorylation, performed by the ATM/ATR and Chk1/Chk2 proteins, and by the p19ARF product of the INK4a locus, which sequester Mdm2 in the nucleolus. The transcriptional control of the INK4a products is not fully elucidated, indicated are some of these regulators.

(1) **The $p19^{ARF}/p53/p21^{Cip1}$ pathway** One form of senescence is mediated by p53, which induces a variety of anti-proliferative activities including transcriptional activation of $p21^{CIP}$, a non-INK4 CDKI. The activity of p53 is regulated predominantly at the post-translational level and is induced by inhibiting its MDM2-mediated degradation in response to diverse stimuli including DNA damage signals and oncogene activation. The stabilization of p53 by oncogenes is in part mediated by ARF.

The INK4a/ARF locus encodes two tumor suppressors, $p16^{INK4a}$ and $p19^{ARF}$, that are transcriptionally activated by the accumulation of cell doublings. The induction of cell-cycle arrest by $p19^{ARF}$ requires the functional activity of p53, being $p21^{Cip1}$ one of the most important targets of p53. Two oncogenes are known to interfere with the $p19^{ARF}/p53$ pathway: overexpression of Bmi-1 constitutively represses the $p19^{ARF}$ gene, and overexpression of MDM2 destabilizes p53.

(2) **The $p16^{INK4a}/Rb$ pathway** $p16^{INK4a}$ also increases markedly in senescent cells, and correlates with increasing Rb hypophosphorylation during this process. While the regulation

of p16^{INK4a} in senescence is not as well understood as that of p53, it appears to be induced by several stimuli including MAP kinase signaling, oncogene activation, and growth in culture. Thus, the p16/pRB pathway appears to be particularly important for ensuring that the senescence growth arrest is essentially irreversible and re-fractory to subsequent inactivation of p53, pRB, or both.

The p16^{INK4a} gene seems to be a critical effectors of senescence. The effective establishment of cell-cycle arrest by p16^{INK4a} requires the functional activity of Rb. Several oncogenes may interfere with this pathway, delaying or disabling senescence. Specifically, Bmi-1 overexpression constitutively represses the p16^{INK4a} gene, whereas overexpression of CDK4 or cyclin D1 bypasses p16^{INK4a} by inactivating Rb.

11.2 Apoptosis

2002 year's Nobel Prize in medicine was for discoveries about how genes regulate organ growth and a process of programmed cell suicide. Britons Sydney Brenner, John E. Sulston and H. Robert Horvitz shared the prize.

Apoptosis is often referred to as programmed cell death. Cells can die due to external factors (i.e., injury, infection) or by a natural death which is intrinsic in the programming of the cell itself. The study of the apoptotic process has led to a greater understanding of the pathogenesis of disease.

A coordination and balance between cell proliferation and apoptosis is crucial for normal development and tissue-size homeostasis in the adult. Cancer results when clones of mutated cells survive and proliferate inappropriately, disrupting this balance.

11.2.1 Apoptosis or necrosis

Cells undergo two primary types of death, apoptosis (programmed cell death) and necrosis (accidental cell death). Apoptosis is quite different from **necrosis** and begins when a cell activates it's own destruction by initiating a series of complex cascading events.

Apoptosis is the ordered disassembly of the cell from within. Disassembly creates changes in the phospholipid content of the plasma membrane outer leaflet. **Phosphatidylserine** (PS) is exposed on the outer leaflet and phagocytic cells recognizing this change, may engulf the apoptotic cell or cell-derived, membrane-limited apoptotic bodies. The apoptotic cells are removed by tissue phagocyte by phagcytosis, and no inflammation is initiated.

Necrosis normally results from a severe cellular insult. Internal organelle and plasma membrane integrity are lost, resulting in spilling of cytosol and organella contents into the surrounding environment (Figure 11-5). Immune cells are attracted to the area and begin producing cytokines that generate an inflammatory response.

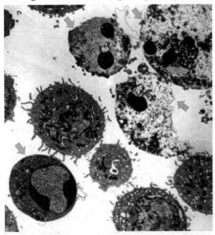

Figure 11-5 **Dexamethasone induced apoptosis of Burkitt cells** (arrows indicate the apoptotic cells)

(1) **Apoptosis** Down regulation of Bcl-2 proteins、 activation of caspases、 cleavage of the DNA repair enzyme PARP、 plasma membrane blebs and the

translocation of phosphatidylserine from the inner cytoplasmic membrane to the cell surface、 endogenous endonuclease release promoting DNA degradation、chromatin condensation forming crescent bodies , decrease in cell volume、increase in cell surface side scatter properties、no inflammation、ATP is needed.

(2) Necrosis By contrast, necrosis is not self initiated and results from an insult or injury to the cell such as cytotoxicity. The characteristics of necrosis are: cell swells、 mitochondria dilate、loss of membrane integrity、significantly less DNA degradation as compared with apoptosis、organelles dissolve and the plasma membrane ruptures、releasing cytoplasmic material、elicits an inflammatory response、no ATP needed.

11.2.2 The extrinsic pathway

In deed, the extrinsic death pathways trigger receptor-mediated apoptosis. Much is known about the biochemical action of many apoptotic players, and key components are being assembled into pathways.

The major components of the extrinsic apoptotic signal transduction pathway are: A proapoptotic ligand; A receptor, such as those membrane bound receptors with homology to the TNF receptor that recognize and bind the apoptotic ligand; The adapter protein tumor necrosis factor receptor (TNFR)1 - associated death domain (TRADD) plays an essential role in recruiting signaling molecules to the TNFRI receptor complex at the cell membrane; Adaptor proteins that bind at the cytoplasmic face of the receptor and recruit additional molecules; Executioners (caspases) to carry out the apoptotic process. The intracellular domain of these receptors contains an 80 amino acid stretch termed the death domain. A death-inducing signaling complex(DISC) formed by the death domain, the adaptor protein Fas-associating protein with death domain (FADD), and procaspase 8 produces active caspase 8. Active caspase 8, in turn, activates downstream or executioner caspases.

The extrinsic death pathway is death receptor-dependent, which activated by the binding of secreted ligands such as Fas ligand (FasL) to "death receptors" of the tumor-necrosis-factor-receptor family, such as Fas. Receptor aggregation recruits other proteins, including the adaptor protein FADD and pro-caspase-8, into the "death-inducing signaling complex" (DISC) (Figure 11-6).

(1) Death receptors Death receptors are cell surface receptors that transmit apoptosis signals initiated by specific ligands. They play an important role in apoptosis and can activate a caspase cascade within seconds of ligand binding. Induction of apoptosis via this mechanism is therefore very rapid. Death receptors belong to the tumour necrosis factor (TNF) gene superfamily and generally can have several functions other than initiating apoptosis. The best characterized of the death receptors are CD95(or Fas), TNFR1 (TNF receptor-1) and the TRAIL (TNF-related apoptosis inducing ligand) receptors DR4 and DR5.

Death receptor members are defined by similar, cysteine-rich extracellular domains. The death receptors contain in addition a homologous cytoplasmic sequence termed the "death domain". Death domains typically enable death receptors to engage the cell's apoptotic machinery, but in some instances they mediate functions that are distinct from or even counteract apoptosis. Some molecules that transmit signals from death receptors contain death domains themselves.

(2) Signaling by CD95 CD95 and CD95 ligand (also termed as Fas and Fas ligand) to "death receptors" of the tumor-necrosis-factor-receptor family, such as Fas. Receptor aggregation recruits other proteins, including the adaptor protein Fas-associated death-

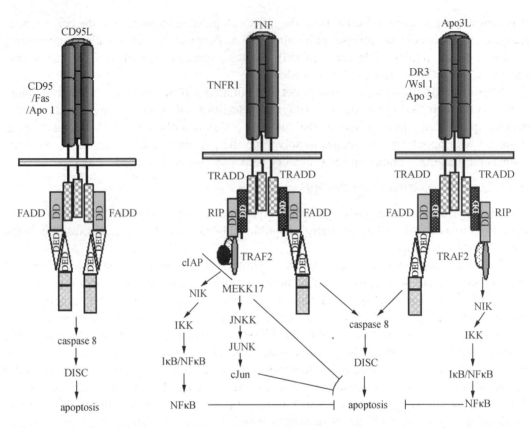

Figure 11-6 Simplified model of extrinsic apoptosis pathways (→indicates activation; ⊣ means inhibition)

domain protein (FADD) and pro-caspase-8, into the death-inducing signaling complex (DISC). Cleavage and activation of caspase-8 in the complex triggers the effecter caspase cascade, which leads to cell death. Caspases are inactivated by inhibitor of apoptosis (IAP) proteins, which can bind to and block the active site on the caspase.

(3) **Signaling by TNFR1** Upon binding, TNF-α trimerizes its ligand TNFR1 and results in the subsequent recruitment of the signal transducing molecules tumor necrosis factor receptor 1-associated death domain protein (TRADD) through conserved protein interaction regions known as "death domains". TRADD recruits RIP and TNF receptor-associated factor (TRAF-2), leading to activation of nuclear factor κB (NF-κB), which suppresses TNF-α-induced apoptosis. The recruitment of FADD by TRADD results in apoptosis through activation of a cell death protease, caspase-8. Activated caspase-8 initiates a protease cascade that cleaves cellular targets and results in apoptotic cell death. Hence, disruption of FADD can prevent activation of caspase-8, thereby producing defects in receptor-mediated cell death.

(4) **Signaling by TRAIL receptors** In a number of ways TRAIL is similar in action to CD95. Binding of TRAIL to its receptors DR4 or DR5 triggers rapid apoptosis in many cells, however unlike CD95, its expression has been shown to be constitutive in many tissues. The DR4 and DR5 receptors contain death domains in their intracellular domain, but as yet no adapter molecule (such as FADD or TRADD) has been identified that associates with the receptor to initiate apoptosis. Work in FADD-deficient mice has indicated that FADD is not essential for triggering apoptosis via these receptors.

(5) **The death-inducing signaling complex (DISC)** The DISC refers to a complex of proteins that is assembled around the cytosolic domains of certain TNF-family death receptors that contain the death-domain structure. Invariant proteins of the DISC include the adaptor protein FADD and caspase-8. Other proteins can be found in some circumstances, depending on the TNF-family receptor and the cell type. Binding of the proenzyme pro-caspase-8 to the DISC results in its activation and the release of active caspase-8 into the cytoplasm where it can cleave a number of death substrates including caspase-3, BID and proteins of the cytoskeleton such as plectin.

Cleavage and activation of caspase-8 in the complex triggers the effector caspase cascade, which leads to cell death. Caspases are inactivated by inhibitor of apoptosis (IAP) proteins, which can bind to and block the active site on the caspase. IAPs are targeted for degradation by the activity of pro-apoptotic proteins like second mitochondria-derived activator of caspase(SMAC)/Diablo and its functional homologues in flies, which include Grim, Reaper and Hid.

11.2.3 The intrinsic pathway

The intrinsic death pathway is the mitochondrial pathway, which is activated by the release of cytochrome C from mitochondria in response to various stresses and developmental-death cues.

The major components of the intrinsic apoptotic signal transduction pathway are(Figure 11-7): ①A extra or/and intracellular proapoptotic signal. ②Cytochrome C release from the mitochondria into the cytosol. ③Cytochrome C binds with apoptotic protease activating factor-1(Apaf-1) proteins. ④A apoptosome where procaspase-9 inactivated, and hydrolysis of adenosine triphosphate/deoxyadenosine triphosphate (ATP/dATP) and binding of cytochrome C to Apaf-1 drives oligomerization of Apaf-1 and therefore complex formation. ⑤Executioners (caspases-3, 6, 7) to carry out the apoptotic process.

This pathway is activated by a variety of extra and intracellular stresses, including oxidative stress and treatment with cytotoxic drugs. The apoptotic signal leads to the release of cytochrome C from the mitochondrial intermembrane space into the cytosol, where it binds to the Apaf-1, a mammalian CED-4 homologue. The resulting complex, termed the apoptosome, recruits and activates pro-caspase-9, which in turn activates downstream caspases. Bcl-2, the mammalian homolog of Ced-9, prevents apoptosis by inhibiting the release of cytochrome C from mitochondria. Inhibitors of apoptosis proteins (IAPs), second mitochondrial activator of caspases (smac), apoptosis-inducing factor(AIF), and Omi/Htr A2 also have important roles in the apoptotic process.

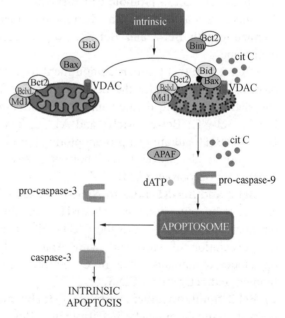

Figure 11-7 Simplified model of intrinsic apoptosis pathways

11.2.4 DNA damage-induced apoptosis

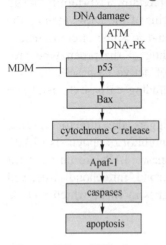

Figure 11-8 DNA damage initiates pathway involved in apoptosis (→indicates activation; ⊣ indicates suppression)

DNA damage-induced apoptosis is common form of apoptosis, which belongs to the intrinsic death pathway. DNA damage can also induce apoptosis through a central player, p53. DNA strand breaks are sensed by kinases such as the DNA dependent protein kinase (DNA-PK) or the ataxia-telangiectasia mutated gene product (ATM), leading to phosphorylation and activation of p53. p53 transmits the apoptotic signal by a complicated mechanism that involves, at least in part, its ability to transactivate target genes such as *bax* and a series of p53-inducible genes. The next step may involve changes in mitochondrial membrane potential and release of cytochrome c. In any case, once in the cytosol, cytochrome c can interact in a apoptosome with Apaf1 and pro-caspase-9, leading to caspase-9 activation and the initiation of a protease cascade similar to that described above (Figure 11-8).

11.2.5 Important role of mitochondria in apoptosis

Mitochondria have the ability to promote apoptosis through release of cytochrome C, which together with Apaf-1 and ATP forms a complex with pro-caspase-9, leading to activation of caspase-9 and the caspase cascade. On the other hand, anti-apoptotic members of the Bcl-2 family of proteins, such as Bcl-2 and Bcl-XL, are located in the outer mitochondrial membrane and act to promote cell survival. Many of the pro-apoptotic members of the Bcl-2 family, such as Bad and Bax also mediate their effects though the mitochondria, either by interacting with Bcl-2 and Bcl-XL, or through direct interactions with the mitochondrial membrane.

(1) **Bcl-2 family** Cellular commitment to apoptosis is partly regulated by the Bcl-2 family proteins, which includes the death antagonists Bcl-2 and Bcl-XL, and death agonists Bax and Bad. They were divided into three subfamilies: ①Bcl-2 subfamily (anti-apoptotic): Bcl-2, Bcl-XL, Bcl-w, Mcl-1 and A1. ②Bax subfamily (pro-apoptotic): Bax, Bak and Bok. ③ BH3 subfamily (pro-apoptotic): Bad, Bid, Bik, Blk, Hrk, BNIP3 and BimL.

Additionally, several Bcl-2 homologs have been identified in viruses, among others the adenovirus oncoprotein E1B-19K.

1) Bcl-2 and Bcl-XL bind to Apaf-1: A central checkpoint of apoptosis is the activation of caspase-9 by mitochondria. The BH4 domain of Bcl-2 and Bcl-XL can bind to the C terminal part of Apaf-1 (to the CED-4 like part and the WD-40 domain), thus inhibiting the association of caspase-9 with Apaf-1. This process seems to be conserved from nematodes to humans since in *C. elegans* CED-9 binds to CED-4, preventing it from binding and activating CED-3.

2) Bcl-2 family members regulate cytochrome C release: The pro-survival proteins also seem to maintain organelle integrity since Bcl-2 directly or indirectly prevents the release of cytochrome C from mitochondria. On the other hand, the pro-apoptotic BH3 subfamily member BID was reported to mediate the release of cytochrome C (without evoking mitochondrial swelling and permeability transition). Interestingly, BID is able to bind to pro-apoptotic members of the Bcl-2 family (e. g. Bax) as well as to pro-survival members Bcl-2 and Bcl-XL.

3) Bax maybe involved in caspase-independent death: The BH1 and BH2 domains of Bcl-2 family members (Bcl-2, Bcl-XL and Bax) resemble membrane insertion domains of bacterial toxins: hypothetically they can form pores in organelles such as mitochondria, what at least was demonstrated in lipid bilayers *in vitro*. In yeast which lack Bcl-2 like proteins, CED-4, caspases -Bax and Bak were shown to induce cell death, while Bcl-2 can protect, apparently by preventing mitochondrial disruption. Bax and Bax-like proteins might mediate caspase-independent death via channel-forming activity, which would promote the mitochondrial permeability transition.

(2) Cytochrome C   Suppression of the anti-apoptotic members or activation of the pro-apoptotic members of the Bcl-2 family leads to altered mitochondrial membrane permeability resulting in release of cytochrome C into the cytosol. In the cytosol, or on the surface of the mitochondria, cytochrome C is bound by the protein Apaf-1 (apoptotic protease activating factor), which also binds caspase-9 and dATP. Binding of cytochrome C triggers activation of caspase-9, which then accelerates apoptosis by activating other caspases. Release of cytochrome C from mitochondria has been established by determining the distribution of cytochrome C in subcellular fractions of cells treated or untreated to induce apoptosis. Cytochrome C was primarily in the mitochondria-containing fractions obtained from healthy, non-apoptotic cells and in the cytosolic non-mitochondria-containing fractions obtained from apoptotic cells. Using mitochondria-enriched fractions from mouse liver, rat liver, or cultured cells it has been shown that release of cytochrome C from mitochondria is greatly accelerated by addition of Bax, fragments of BID, and by cell extracts.

(3) Apoptosome   The **apoptosome** has been considered the cytosolic equivalent of a multi-protein complex, DISC assembled at the plasma membrane. This complex also activates caspases and causes receptor mediated cell death by apoptosis.

The apoptosome is a large, multimeric, caspase-9-activating complex formed after the release of cytochrome C from mitochondria. It contains cytochrome C, Apaf1, and caspase-9. Hydrolysis of adenosine triphosphate/deoxyadenosine triphosphate (ATP/dATP) and binding of cytochrome C to Apaf-1 drives oligomerization of Apaf-1 and therefore complex formation.

The exact mechanism of caspase activation is still uncertain although two possibilities have been proposed. In one case the Apaf-1, cytochrome C and pro-caspase-9 complex can act as a stage to activate cytosolic pro-caspase-9 as it is recruited to the apoptosome. In the other scenario two apoptosome have been proposed to interact with each other and to activate the caspase-9 located on the other apoptosome.

11.2.6 Caspases

Apoptosis relies on a cascade of enzymes. Most of these are caspases (cytosolic aspartate-specific proteases), enzymes that are expressed as inactive precursors (procaspases) and are activated by being cleaved at specific peptide bonds. Likewise, activated caspases in turn activate procaspases at the next point in the enzyme chain by cleaving them at specific bonds.

(1) Caspase family   Caspases are synthesized as pro-caspases with an N-terminal pro-domain. The active caspase is a heterotetramer of two large and two small subunits. Caspases involved in apoptosis. "Initiator" caspases are responsible for kick-starting the process of cell death in response to some external signal, and for activating the "effector"

caspases. Effector caspases actually execute the process — they are responsible for breaking down the cell. It is important to note that caspases may not be the only enzymes involved in cell death, so even if all the caspases are blocked, cell death may still occur.

Caspases have been found in organisms ranging from C elegans to humans. The 14 identified mammalian caspases (named caspase-1 to caspase-14) have distinct roles in apoptosis and inflammation. All caspases are produced in cells as catalytically inactive zymogens and must undergo proteolytic processing and activation during apoptosis.

There are three major groups of caspases: group 1-inflammatory caspases: [caspase 1, 4, 5, 11(murine), 12, 13, 14(mice)], group 2-initiator caspases [caspase 2, 8, 9, 10], group 3-effector caspases [caspase 3, 6, 7].

The initiator caspase activation takes place in large protein complexes bringing together several caspase zymogens. The effector caspases are activated by initiator caspases. Active effector caspases cleave substrates escalating the death signal and executing apoptosis. Caspase-3 is the main effector caspase that cleaves the majority of the cellular substrates in apoptotic cells, while caspase-7 is highly similar to caspase-3. Inflammatory caspases are part of the innate immune system, and have also been suggested to bridge the innate immune responses to the adaptive immune system.

(2) **Activation of caspase** Caspases are responsible for the deliberate disassembly of a cell into apoptotic bodies. *In vivo*, caspases are present as inactive pro-enzymes, most of which are activated by proteolytic cleavage. Caspase-8, caspase-9, and caspase-3 are situated at pivotal junctions in apoptotic pathways. Caspase-8 initiates disassembly in response to extracellular apoptosis-inducing ligands and is activated in a complex associated with the receptors cytoplasmic death domains. Caspase-9 activates disassembly in response to agents or insults that trigger release of cytochrome C from the mitochondria and is activated when complexed with dATP, APAF-1, and extramitochondrial cytochrome C. Caspase-3 appears to amplify caspase-8 and caspase-9 signals into full-fledged commitment to disassembly. Both caspase-8 and caspase-9 can activate caspase-3 by proteolytic cleavage and caspase-3 may then cleave vital cellular proteins or activate additional caspases by proteolytic cleavage. Many other caspases have been described.

Shown is the crystal structure of caspase-3 in complex with a tetrapeptide aldehyde inhibitor. The active enzyme is composed of a large (20 kD) and small (10 kD) subunit, each of which contributes amino acids to the active site. In the two structures that are available, two heterodimers associate to form a tetramer. In common with other proteases, caspases are synthesized as precursors that undergo proteolytic maturation. The NH_2-terminal domain, which is highly variable in length (23 to 216 amino acids) and sequence, is involved in regulation of these enzymes (Figure 11-9).

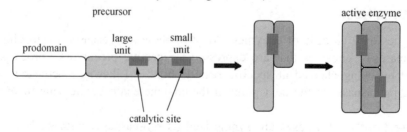

Figure 11-9 **A model for caspase activation**

11.2.7 Suppressors of apoptosis

Apoptosis, or programmed cell death, is an active process of self-destruction, described a long time ago. However, the understanding of the molecular pathways which regulate programmed cell death is more recent. Apoptosis occurs during embryonic and foetal development, and tissue remodeling, and its purpose is to assure homeostasis of cells and tissues. Several genes affecting various steps in programmed cell death must be expressed to trigger apoptosis, but the genes suppressing apoptosis were also identified.

(1) **Inhibitors of apoptosis (IAPs)** The induction of apoptosis or progression through the process of apoptosis is inhibited by a group of proteins called IAPs. These proteins contain a BIR (baculovirus IAP repeat) domain near the amino-terminus. The BIR domain can bind some caspases. Many members of the IAP family of proteins block proteolytic activation of caspase-3 and -7. XIAP, cIAP-1 and cIAP-2 appear to block cytochrome C-induced activation of caspase-9, thereby preventing initiation of the caspase cascade. Since cIAP-1 and cIAP-2 were first identified as components in the cytosolic death domain-induced complex associated with the TNF family of receptors, they may inhibit apoptosis by additional mechanisms.

(2) **Survivin** Survivin (also termed *Birc*5) belongs to the family of genes known as inhibitors of apoptosis, and it has been implicated in both prevention of cell death and control of mitosis.

Survivin contains a baculovirus inhibitor of apoptosis repeat (BIR) protein domain that classifies it as a member of the IAP family. Survivin inhibits apoptosis, via its BIR domain, by either directly or indirectly interfering with the function of caspases. Survivin is also a chromosomal passenger protein that is required for cell division. Survivin is expressed in embryonic tissues as well as in the majority of human cancers.

(3) **Bcl-2 members with anti-apoptotic potentials** The Bcl-2 family of proteins regulates apoptosis, some antagonizing cell death and others facilitating it. Bcl-2 subfamily member Bcl-2, Bcl-XL, Bcl-w, Mcl-1 and A1 are anti-apoptotic proteins.

Human Bcl-2 gene and related Bcl-2 genes were described as anti-apoptotic genes. It is clear that Bcl-2 protects many cell lines from induced apoptosis. Other proteins, like Bcl-XL, A1 or Mcl-1 have the same anti-apoptotic function. It is known that Bcl-2 can interact with other proteins. For example, Bax, which can exist as a homodimer, is also able to form a heterodimer with Bcl-2. A surexpression of Bax in several cell lines allows to counteract the effect of Bcl-2.

R-ras p23 is another example, among others, of a protein interacting with Bcl-2, and this results in an interruption of the apoptotic signal transduction pathway when Bcl-2 is overexpressed. Many interesting results suggest that Bcl-2 is a death repressor molecule functioning in an anti-oxydant pathway, but other data seem to claim the contrary. Recently, the demonstration was made that apoptosis may require the activation of several classes of proteases. It seems now that Bcl-2 has also a function of protease(s) inhibitor.

(4) **Flip/Flame** FLIP/FLAME is highly homologous to caspase-8. It does not, however, contain the active site required for proteolytic activity. FLIP appears to compete with caspase-8 for binding to the cytosolic receptor complex, thereby preventing the activation of the caspase cascade in response to members of the TNF family of ligands. The exact *in vivo* influence of the IAP family of proteins on apoptosis is still under investigation.

11.3 Senescence or apoptosis

As described above, senescence and apoptosis share numerous characteristics at molecular and cellular levels, and one of the characteristics of the senescent cell is apoptotic resistance, indicating complicate linkage between these two phenotypes.

11.3.1 Dysregulation of the apoptosis involves in senescence

Recent evidence shows that dysregulation of the apoptotic process may be involved in some senescence processes. Apoptosis and susceptibility to apoptosis in several types of intact cells are enhanced by senescence. Apoptosis is the process of programmed cell death or cell suicide. It is an active process of cellular self-destruction, a regulated energy-dependent process involving the activation of a cascade of proteases and endonucleases (i. e. caspases).

While senescence is characterized as a relatively undefined process, apoptosis on the other hand is well defined and conserved from *C. elegans* to mammals. Apoptosis was initially described as a series of morphological steps leading to the elimination of the cell without inflammation. During apoptosis, the nuclear DNA is condenced and fragmented, the cell shrinks and is fragmented into several peaces that are phagocytosed *in vivo*.

11.3.2 Senescence and apoptosis result from antagonistic pleiotropy

The cellular and molecular changes that occur with senescence or apoptosis, specifically changes that occur as a result of cellular responses to environmental insults, are evolved. Because the force of natural selection declines with age, it is likely that these processes were never optimized during their evolution to benefit old organisms. That is, some age-related changes may be the result of gene activities that were selected for their beneficial effects in young organisms, but the same gene activities may have unselected, deleterious effects in old organisms, a phenomenon termed antagonistic pleiotropy. These two cellular processes, apoptosis and cellular senescence, may be examples of antagonistic pleiotropy. Both processes are essential for the viability and fitness of young organisms, but may contribute to aging phenotypes, including certain age-related phenotypes.

11.3.3 Mitochondria mediates both senescence and apoptosis

Apoptosis and senescence share common mechanisms in oxidative stress and mitochondrial involvement. The mechanism involved in the aging process is free radical attacks. Mitochondria are the primary source of free radicals due to inevitable leakage of electrons from the electron transport chain. Since free radicals are highly reactive, they may interact with DNA and proteins to alter cellular functions. A very destructive target is the voltage-dependent anion channel (VDAC) which plays the central role in mitochondria-mediated apoptosis.

Several diseases associated with senescence appear to be the direct result of cells containing dysfunctional mitochondria. The typical human cell has several hundred mitochondria. Mitochondria are the sites of three important processes which are involved in the progression of senescence: energy conversion (ATP), production of reactive oxygen species (ROS), initiation of cell death.

Mitochondria function to convert the chemical energy from food into ATP, which cellular enzymes use to do work. As a by product of energy conversion, ROS are produced.

Mitochondria are also involved in sensing when a cell is sick enough to initiate programmed cell death (apoptosis or necrosis).

11.3.4 Oxidative stress causes both senescence and apoptosis

Oxidative stress is a major form of endogenous and exogenous damage to cells. The role of oxidative stress in cell senescence and cell death (apoptosis) is under study. Interestingly, the same genes (e.g., p53 and Rb) are involved in cell senescence and cell death. A better understanding of the molecular mechanism of signal transduction leading to cell senescence or cell death through divergent pathways is required to understand cellular responses to environmental stresses, particularly oxidative damage.

11.3.5 Opting for senescence or apoptosis

Cell could chose senescence or apoptosis under certain stresses. For example, one obvious mechanism that could regulate the choice between senescence and apoptosis is if the apoptotic process was inhibited and senescence occurred instead as a default mechanism. The mitochondria anti-apoptotic protein Bcl-2 has been shown to induce senescence, judged by SA-βGAL staining, yet this may more resemble quiescence as p27 was overexpressed. Bcl-2 can also accelerate RAS-induced senescence to some extent. In support of the hypothesis, Bcl-2 has been described to shift the response from apoptosis to senescence when artificially overexpressed in rat cells.

(1) **p21 overexpression refers to senescence** When cell shifted from apoptosis to senescence upon Bcl-2 overexpression, p21 was also found to be upregulated. This should be the reason for the shift from apoptosis to senescence as p21 expression after DNA damage leads to senescence while absence of p21 induction after DNA damage leads to apoptosis. Similarly, apoptosis was associated with low p21 levels whereas senescence was associated with high p21 levels in a cancer cell treated with interferon-γ.

p21 is a molecule downstream of p53 induction, then an important question is why p53 sometimes induces p21 expression and sometimes not. Some of the regulation could be a result of the convergence of several pathways that directly regulate p21. Such as cisplatin, a chemotherapy that is given as treatment for certain types of cancer, can induce growth arrest at low concentration and this growth arrest is associated with translational activation of p21 by an unknown mechanism.

(2) **Higher levels of p53 facilitate apoptosis** It is also possible that the decision, of whether or not to induce p21, occurs at the activation level of p53. Interestingly the phosphorylation pattern of p53 is different during induction of senescence and DNA damage-induced apoptosis. It appears that a large DNA damage response leads to apoptosis while a low but persistent activation of p53 induces senescence. For example, upon hydrogen peroxide treatment, both senescence and apoptosis are possible outcomes; apoptosis was associated with higher levels of p53 and lower levels of p21, while senescence was associated with lower levels of p53 and higher levels of p21. Similarly, a TRF2 (telomeric-repeat binding factor 2) mutant that causes telomere dysfunction induced apoptosis or senescence, depending upon the expression level and thereby the extent of telomere damage; substantial overexpression of p53 induced apoptosis while lower overexpression induced senescence. Therefore, it seems like that the nature of p53 response is the major determinant whether p53 response will induce apoptosis or senescence.

(3) **Bmi1 links senescence and apoptosis** Polycomb group (PcG) genes are involved in the maintenance of cellular memory through epigenetic chromatin modifications. The

polycomb group protein BMI1 (B lymphoma Mo-MLV insertion region 1 homolog) has been shown to inhibit the INK4A-ARF tumor suppressor locus that encodes two tumor suppressors, $p16^{INK4a}$ and $p19^{ARF}$, which inhibit cell proliferation by activating Rb and p53, respectively. Therefore BMI1 has been linked to cell proliferation, senescence and apoptosis.

Summary

Senescence and apoptosis are two common cell fates. While apoptosis leads to death and removal of the cells, senescence cells persist and survive for extended periods of time. Both senescence and apoptosis are important not only for development, also for a cancer treatment perspective.

Senescence is defined as progressive deterioration in physiological functions and metabolic processes of a cell or organism. It is a program activated by normal cells in response to various types of stress. Several hypotheses explain senescence, such as mitochondria ageing, reactive oxygen species stress, replicative ageing and telomere shortening. Apoptosis is often referred to as programmed cell death, which mainly results from the extrinsic apoptotic signal transduction, the intrinsic death signal transduction and DNA damage. The mitochondria plays important role in cellular apoptosis. Cell senescence and apoptosis share numerous characteristics at molecular and cellular levels, and there is complicate linkage between these two phenotypes.

Questions

1. What does the senescence mean in cell biology?
2. What do you know the mechanisms in cell apoptosis?
3. What are the roles of mitochondria in senescence and apoptosis?
4. How does cell chose senescence or apoptosis?
5. How can cancer cells undergo senescence or apoptosis?

(吴超群)

Chapter 12

Cells in immune response

Loving organisms provide ideal habitats in which other organisms can grow. It is not surprising that animals are subject to infection by viruses, bacteria, protists, fungi, and animal parasites. Vertebrates have evolved several mechanisms that able to recognize and destroy these infectious agents. As a result, vertebrates are able to develop immunity against invading pathogens. Immunity results from the combined activities of immune response with many different cells, lymphoid organs. It is often divided into the two sections of innate and adaptive immunity, the former encompassing unchanging mechanisms that are continuously in force to ward off noxious influences, and the latter responding to new influences by mounting an immune response. In this chapter, we will focus on the basic events in the body's response to the presence of an intruding microbe.

12.1 The immune system

The immune system is the organ system that protects an organism from outside biological influences. It is composed of many interdependent cell types that collectively protect the body from bacterial, parasitic, fungal, viral infections and from the growth of tumor cells. Many of these cell types including the **B cells**, **T cells**, **macrophages**, **nature killer cells** (**NK cells**), **neutrophils**, **basophils**, **eosinophils**, **endothelial cells**, or **mast cells**, have distinct roles in the immune system, and communicate with other immune cells by cytokines, which control proliferation, differentiation and function of cells in the immune response. They are also involved in processes of inflammation and in the neuronal, haematopoietic and embryonal development of an organism.

12.2 The organs of the immune system

The immune system is composed of related organs which harbor immune cells, and where the cells undergo development, differentiation and maturation. Immune cells arise from the same type of precursor cell, but they differentiate along different pathways in different lymphoid organs (Table 12-1).

Table 12-1 Central and peripheral lymphoid organs

Central lymphoid organs	Peripheral lymphoid organs
Bone marrow	Spleen
Thymus	Lymph nodes
	Mucosal associated lymphoid tissues

12.2.1 Central lymphoid organs

(1) **Bone marrow**   Bone marrow is the tissue comprising the center of large bones. It is the place where new blood cells are produced. Bone marrow contains two types of stem cells: hemopoietic (which can produce blood cells) and stromal (which can produce fat, cartilage and bone). **Stromal stem cells** have the capability to differentiate into many kinds of tissues, such as nervous tissue. **Hematopoietic stem cells** give rise to the three classes of blood cell that are found in the circulation: leukocytes, red blood cells (erythrocytes), and platelets (thrombocytes). All the immune cells are initially derived from the bone marrow by hematopoiesis. During hematopoiesis, the bone marrow produces B cells, natural killer cells, granulocytes and immature thymocytes, in addition to red blood cells and platelets (Figure 12-1).

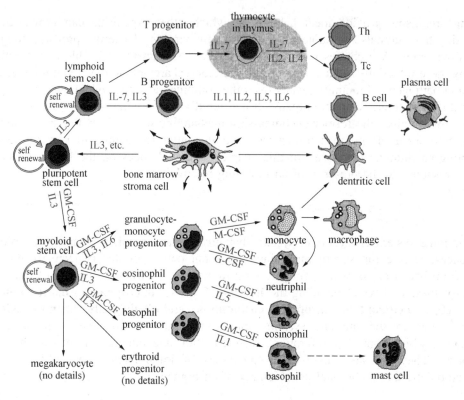

Figure 12-1   **The immune cells are differentiated from plurpotent stem cell.** The stem cell can give rise to two different progenitor cells: a myeloid progenitor cell that can differentiate into most of the various blood cells (e. g., erythrocytes, basophils and neutrophils), macrophages, or dendritic cells; a lymphoid progenitor cell that can differentiate into any of the various types of lymphocytes (NK cells, T cells, or B cells). T-precursor cells migrate to the thymus where they undergo differentiation into T cells. In contrast, B cells undergo differentiation in the bone marrow.

　　In addition to self-renewing stem cells and their differentiating progeny, the marrow contains numerous antibody-secreting plasma cells. These plasma cells are generated in peripheral lymphoid tissues as a consequence of antigenic stimulation of B cells and then migrate to the marrow, where they may live and continue to produce antibodies for many years. Some long-lived memory T lymphocytes also migrate to and may reside in the bone marrow.

(2) **Thymus**   In human anatomy, the **thymus** is a ductless gland located in the upper anterior portion of the chest cavity. It is most active during puberty, after which it shrinks

in size and activity in most individuals and is replaced with fat. The thymus plays an important role in the development of the immune system in early life, and its cells form a part of the body's normal immune system.

The thymus is the site for T cell maturation. Its function is to produce mature T cells. In general, the most immature — T cell leaves the bone marrow and enter the thymic cortex through the blood vessels. Through a remarkable maturation process sometimes referred to as thymic education, T cells that are beneficial to the immune system are spared, while those T cells that might evoke a detrimental autoimmune response are eliminated. In the thymus lobules lymphocytes mature into T cells (where T stands for "thymus") that behave in different ways according to their type. Maturation begins in the cortex, and as thymocytes mature, they migrate toward the medulla, so that the medulla contains mostly mature T cells. Only mature T cells exit the thymus and enter the blood and peripheral lymphoid tissues.

12.2.2 Peripheral lymphoid organs

(1) **Spleen** Spleen is an organ derived from mesenchyme and lying in the mesentery. It is the major site of immune responses to blood-borne antigens. The organ consists of masses of lymphoid tissue of granular appearance located around fine terminal branches of veins and arteries. The spleen is an immunologic filter of the blood. In addition to capturing foreign materials (**antigens**) from the blood that passes through the spleen, migratory macrophages and dendritic cells bring antigens to the spleen via the bloodstream. An immune response is initiated when the macrophage or **dendritic cells** present the antigen to the appropriate B or T cells. This organ can be thought of as an immunological conference center. In the spleen, B cells become activated and produce large amounts of antibody. Also, old red blood cells are destroyed in the spleen.

(2) **Lymph nodes and the lymphatic system** Lymph nodes are small nodular organs and components of the lymphatic system. Clusters of lymph nodes are found in the underarms, groin, neck, chest, and abdomen. Lymph nodes act as filters for antigens, with an internal honeycomb of connective tissue filled with T cells, B cells, dendritic cells and macrophages, that collect and destroy bacteria and viruses. When the body is fighting an infection, these lymphocytes multiply rapidly and produce a characteristic swelling of the lymph nodes.

The lymphatic vessels (or lymphatics) are a network of thin tubes that branch, like blood vessels, into tissues throughout the body (Figure 12-2). Lymphatic vessels carry lymph, a colorless, watery fluid originating from inter-

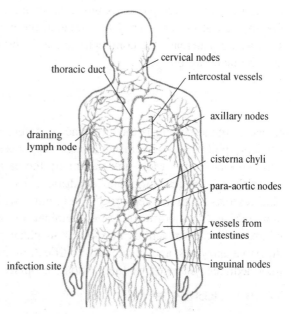

Figure 12-2 The lymphatic system. The major lymphatic vessels and collections of lymph nodes are illustrated. Antigens are captured from a site of infection, and the draining lymph node to which these antigens are transported and where the immune response is initiated.

stitial fluid (fluid in the tissues). The lymphatic system transports infection-fighting cells-lymphocytes, is involved in the removal of foreign matter and cell debris by phagocytes and is part of the body's immune system. It also transports fats from the small intestine to the blood.

Unlike the circulatory system, the lymphatic circulatory system is not closed and has no central pump; the lymph moves slowly and under low pressure. Like veins, lymph vessels have one-way valves and depend mainly on the movement of skeletal muscles to squeeze fluid through them. Rhythmic contraction of the vessel walls may also help draw fluid into the lymphatic capillaries.

(3) **Other important lymphoid tissues** The skin is the largest organ in the body and is an important physical barrier between an organism and its external environment. The principal cell populations within the epidermis are keratinocytes, melanocytes, epidermal Langerhans cells, and intraepithelial T cells. Many foreign antigens gain an entry into the body through the skin, **Langerhans cells** is an important immature dendritic cell for cutaneous immune system.

The bulk of the body's lymphoid tissue (>50%) is found associated with the mucosal system. Mucosal epithelia are barriers between the internal and external environments and are an important site of entry of microbes. The mucosal immunity is essential for life and requires continual and effective protection against invasion. Mucosal associated lymphoid tissue (MALT) play crucial role in the immune system. The organized lymphoid tissues in the gut are called the gut-associated lymphoid tissues (GALT), which including the Peyer's patches and solitary lymphoid follicles of the intestine, the appendix, the tonsils and adenoids in throat, and the mesenteric lymph nodes. The organized lymphoid tissues in the wall of the upper respiratory tract are called the bronchus-associated lymphoid tissues (BALT), and in the lining of the nose are called the nasal-associated lymphoid tissues (NALT). These lymphoid tissues contain large numbers of effector lymphocytes even in the absence of disease. They also include the exocrine glands associated with these organs, such as the pancreas, the conjunctivae and lachrymal glands of the eye, the salivary glands, and the lactating breast.

12.3 Cells in the innate immune system

Innate immune system is the first line of defense against infections. It comprises epithelial barriers, phagocytes and inflammatory responses, natural killer cells, circulating proteins and cytokines, etc. Most notable among the early defensive cells are phagocytes — mainly macrophages, neutrophils, and dendritic cells — which destroy an invading organism by phagocytosis, a multi-step process beginning with the complete engulfment of the organism and ending with modification and chemical breakdown of its structural components. A host of toxic chemical reactions may occur in phagocytes including biomolecular breakdown by digestive enzymes and chemical modification by highly reactive oxygen (and nitrogen) intermediates generated by the "oxidative burst" reaction.

12.3.1 Epithelial barriers

The first defense includes barriers to infection such as skin and mucus coating of the gut and airways, physically preventing the interaction between the host and pathogen. The skin and the extensive epithelial lining of the mucosal tissues act as nonspecific barriers impeding, if not preventing, infectious agents from entering the body. Pathogens which penetrate these barriers encounter constitutively expressed antimicrobial molecules such as lysozyme that restrict the infection. However, infectious agents usually encounter the activities of "defensive" cells, either resident to the site or recruited there from the blood by chemotactic signals which promote diapedesis, or extravasation.

12.3.2 Phagocytes

Phagocytes are bone marrow-derived cells. A range of phagocytic cells have evolved in humans, each with specific functions. Neutrophils are the most numerous white cells in blood. They migrate rapidly into the sites of infection, where they kill pathogens. Monocytes are very different from neutrophils. They also migrate from blood into tissues where they mature into macrophages and take on a number of specialized forms (see later). Phagocytes can recognize danger signals by using receptors on their cell membranes during their journey through the tissues and their encounters with pathogens or damaged cells. Once they are stimulated through their receptors, phagocytes kill and clear pathogens through phagocytosis, respiratory burst, and the release of proteolytic enzymes and various cytokines. More details will be discussed later.

12.3.3 Macrophages

Macrophages (termed monocytes when in the blood stream) are cells found in tissues that are responsible for phagocytosis of pathogens, dead cells and cellular debris. As part of the innate immune system, macrophages are derived from monocytes. Once through the endothelium, the monocytes are distributed to different parts of tissue to form monocyte phagocytic system, then tissue bound and, as a result, are further sub-divided depending on their location (Table 12-2).

Table 12-2 Cells of the mononuclear phagocytic system and their respective locations

Cells	Locations
Monocytes	Blood stream
Alveolar macrophages	Lungs
Sinus macrophages	Lymph nodes and spleen
Kupffer cells	Liver
Osteoclast	Bone
Microglia	Nerve system

Macrophages have a horseshoe-shaped nucleus and are large cells. Its manly functions are phagocytosis and antigen presentation to T cells (see later). Macrophages typically respond to microbes nearly as rapidly as neutrophils do, but macrophages survive much longer at sites of inflammation. Unlike neutrophils, macrophages are not terminally differentiated and can undergo cell division at an inflammatory site. Therefore, macrophages are the dominant effector cells of the later stages of the innate immune response, 1 or 2 days after infection.

Activated macrophages serve many functions in defense against infections. Membrane receptors on the surface of macrophage are important for identification of "danger signals". Mannose receptor (MS), scavenger receptor (SR) and Toll like receptor (TLR) belong to Patter-recognition receptors (PRR), which are crucial receptors for macrophage recognize pathogen associated molecular patterns (PAMP). Mammalian TLRs are involved in responses to widely divergent types of molecules that are commonly expressed by microbial but not mammalian cells, e. g., lipopolysaccharide (LPS). TLRs expressed on innate immune cells are critical in the activation of adaptive immunity.

12.3.4 Dendritic cells

Dendritic cells (DC) are immune cells and form part of the mammal immune system. They are present at a low frequency in those tissues which are in contact with the environment, such as in the skin (called Langerhans cells) and the lining of nose, lungs, stomach and intestines. Especially in immature state, they can also be found in blood. They have long spiky arms, so called dendritic cells. Dendritic cells start out as immature dendritic cells which are characterized by high endocytic activity and low T cell activation potential. Dendritic cells constantly sample the surroundings for viruses and bacteria. Once they have come into contact with a pathogen, they become activated into mature dendritic cells. Immature dendritic cells phagocytose pathogens and degrade its proteins into small pieces and upon maturation present those fragments at their cell surface using MHC molecules. Simultaneously, they upregulate cell-surface receptors that act as co-receptors in T cell activation, and greatly enhance their ability to activate T cells (see later).

Dendritic cells act as antigen-presenting cells. They can travel through the blood stream to the spleen or through the lymphatic system to a lymph node, and activate $CD4^+$ T cell and $CD8^+$ T cells by presenting antigen-peptide derived from the pathogen. Every T cell only recognizes one particular antigen-peptide. Only dendritic cells are able to activate a naïve T cell which has never encountered its antigen before. Dendritic cells form an important bridge between innate and adaptive immunity as the cells present the antigenic peptide to the T helper cell (adaptive immunity). They are the most potent of all the antigen-presenting cells and served a critical function in adaptive immune responses by capturing and displaying microbial antigens to T lymphocytes.

12.3.5 Natural killer cells

Nature killer (NK) cells are a lineage of cells related to lymphocytes that recognize infected and/or stressed cells and respond by directly killing these cells and by secreting inflammatory cytokines. NK cells are derived from bone marrow precursors and appear as large lymphocytes with numerous cytoplasmic granules, so sometimes called large granular lymphocytes. By surface phenotype and lineage, NK cells are neither T nor B lymphocytes, and they do not express somatically rearranged, clonally distributed antigen receptors like immunoglobulin or T cell receptors. Many receptors expressed on the surface of NK cell are responsible for NK cell activation, which is regulated by a balance between signals that are generated from activating receptors and inhibitory receptors.

Nature killer cells play several important roles in defense against intracellular microbes. They kill virally infected cells during the first few days after viral infection. When NK cells are activated, granule exocytosis releases these proteins adjacent to the target cells. One NK cell granule protein, called perforin, facilitates the entry of other granule proteins, called granzymes, into the cytoplasm of target cells. The granzymes are enzymes that initiate apoptosis of the target cells. By killing cells infected by viruses and intracellular bacteria, NK cells eliminate reservoirs of infection. The IFN-γ secreted by NK cells activates macrophages to destroy phagocytosed microbes. This IFN-γ-dependent NK cell-macrophage reaction can control an infection with intracellular bacteria such as *Listeria monocytogenes* for several days or weeks.

12.3.6 Granulocytes

(1) **Neutrophils**　　Neutrophils are a class of white blood cells. They are much more numerous than the longer-lived monocyte/macrophages. Microscopically, these cells possess a characteristic, salient feature-a multilobular nucleus, which has been referred to as polymorphonuclear leukocytes (PMNs). They have a pivotal role to play in the development of acute inflammation as a part of the immune response. Neutrophils activate as phagocytes and capable of only one phagocytic event, expending all of their glucose reserves in an extremely vigorous respiratory burst. The respiratory burst involves the activation of an NADPH oxidase enzyme, which produces large quantities of superoxide, a reactive oxygen species. Being highly motile, neutrophils quickly congregate at a focus of infection, attracted by cytokines expressed by activated endothelium, mast cells and macrophages. In addition, neutrophils release their granules including acidic and alkaline phosphatases, defensins and peroxidase-all of which represent the requisite molecules required for successful elimination of the unwanted microbe (s). However, neutrophil antimicrobial products can also damage host tissues at site of inflammation.

(2) **Eosinophils**　　Eosinophils are granulocytes named because of their intense staining with "eosin". Under the microscope, eosinophils typically have a bilobed nucleus and contain many basic crystal granules in their cytoplasm. The granules are eosinophil mediators that are toxic to many organisms and also to tissues as in Asthma. Eosinophils are motile and phagocytic and are particularly active in parasitic infection, such as schistosome infection, hay fever etc.

(3) **Basophils and Mast cells**　　Basophils are found in low numbers in the blood. Their functions are not well understood but they are known to be involved in Type I hypersensitivity responses. These cells have high affinity Fc receptors for IgE on their surface. Crosslinking of the IgE causes the basophils to release pharmacologically active mediators such as heparin and histamine. Basophils, therefore act very much like mast cells except that they are in the blood instead of the tissues.

Mast cells are derived from blood-borne precursors and differentiated in the local tissues. They often reside near surfaces exposed to pathogens and allergens, including the submucosal tissues of the gastrointestinal and respiratory tracts and the dermis that lies just below the surface of the skin. Mast cells contain large distinctive cytoplasmic granules in their cytoplasm that contain a mixture of chemical mediators, including histamine, chymase, tryptase, and serine esterases, etc. High affinity Fc receptors specific for IgE and IgG are expressed on the surface of mast cells and can be activated to release their granules that act rapidly to make local blood vessels permeable, swelling and mucus secretion. On activation, mast cells also synthesize and release cytokines, chemokines, lipid mediators such as prostaglandins, leukotrienes, etc. These mediators contribute to both the acute and the chronic inflammatory response.

12.3.7 Other important molecules involved in innate immunity

There are many molecules, which work in concert with the cells of the innate immune system and which also foster close functional links with their adaptive counterpart. The three major molecules are complement, acute phase proteins (APP), and interferons (IFNs).

(1) **Complement**　　Complement comprises over 20 different serum proteins that are produced by a variety of cells including, hepatocytes, macrophages and gut epithelial cells. Some complement proteins bind to immunoglobulins or to membrane components of cells.

Others are proenzymes that, when activated, cleave one or more other complement proteins. Upon cleavage some of the complement proteins yield fragments that activate cells, increase vascular permeability or opsonize bacteria.

The complement system is a very complex group of serum proteins which is activated in a cascade fashion. Three different pathways are involved in complement activation: ①the classical pathway recognizes antigen-antibody complexes; ② The alternative pathway spontaneously activates on contact with pathogenic cell surfaces; ③The mannose-binding lectin pathway recognizes mannose sugars, which tend to appear only on pathogenic cell surfaces. The cascade of complement system can result in a variety of effects including opsonization of the pathogen, destruction of the pathogen by formation and activation of the membrane attack complex and inflammation. The functions of the complement system are including lysis bacteria(destruction of cells through damage/rupture of plasma membrane), chemotaxis cells (directed migration of immune cells), and initiation of active inflammation via direct activation of mast cells.

The complement system can be regulated by a series of regulatory proteins to protect host cells from damage and/or their total destruction. These regulatory proteins are expressed on the host cells themselves. The complement system takes part in both specific and non-specific resistance and generates a number of products of biological and pathophysiological significance.

(2) Acute phase proteins Acute phase proteins are a class of serum proteins that are synthetized in the liver by hepatocytes and are produced in high numbers in response to cytokines released from macrophages. For example, in response to injury, local inflammatory cells (neutrophil granulocytes and macrophages) secrete a number of cytokines into the bloodstream, in which the most notable are the interleukins IL-1, IL-6 and IL-11 and TNF-alpha. The liver responds by producing a large number of acute phase reactants, and the most notable are C-reactive protein, Alpha 1-antitrypsin, Alpha 1-antichymotrypsin, Alpha 2-macroglobulin, some coagulation factors (Fibrinogen, prothrombin, factor VIII, von Willebrand factor, plasminogen), complement factors, and serum amyloid protein. This response is called the acute phase reaction.

(3) Interferons (IFNs) Interferons (IFNs) are natural proteins produced by the cells of the immune systems of most animals in response to challenges by foreign agents such as viruses, bacteria, parasites and tumour cells. Interferons belong to the large class of glycoproteins known as cytokines. IFNs are a group of molecules, which limit the spread of viral infections. There are two categories of IFNs, namely type I and type II.

Type I IFNs contains IFN-α, IFN-β, IFN-ε, IFN-ω and IFN-κ isoforms. They are produced by leukocytes in response to viruses and regulate innate immune response. Type II IFN only consists of one isoform, IFN-γ. It is produced by activated helper T cells (CD4$^+$), cytotoxic T cells (CD8$^+$), and NK cells. The main functions of IFN-γ are: ①Enhances the microbicidal function of macrophages through formation of nitric oxide and reactive oxygen intermediates (ROI); ②Stimulates the expression of class I and class II MHC molecules and co-stimulatory molecules on antigen presenting cells; ③Promotes the differentiation of naive helper T cells into Th1 cells; ④ Activates polymorphonuclear (PMN) leukocytes and cytotoxic T cells and increases the cytotoxicity of NK cells. IFN-γ is an important cytokine for both innate and adaptive immune responses.

12.4 Cells in the adaptive immune system

The principal cells of the adaptive immune system are lymphocytes, antigen-presenting cells, and effector cells. Lymphocytes are the cells that specifically recognize and respond to foreign antigens and are therefore the mediators of humoral and cellular immunity. There are distinct subpopulations of lymphocytes that differ in how they recognize antigens and in their functions.

12.4.1 B and T cells development

As discussed earlier in this chapter, the cells of the lymphoid lineage — B cells, T cells, and NK cells — are derived from common lymphoid progenitors (CLP) (Figure 12-1). CLPs give rise mainly to B and T cells but may also contribute to NK cells and some dendritic cells. Pro-B cells can eventually differentiate into follicular(FO)B cells, marginal zone (MZ) B cells, and B-1 B cells. Pro-T cells may commit to either the $\alpha\beta$ or $\gamma\delta$ T cell lineages. The $\alpha\beta$ T cell lineages further developed into $CD4^+$ T cells and $CD8^+$ T cells. The $\gamma\delta$ T cell acts as a part of the first line of defense and involved in innate immune response.

The development of a B lineage cell proceeds through several stages marked by the rearrangement and expression of the immunoglobulin genes, and B cells receptors. Before reaching maturity, the developing B cells undergo a selection process (referred to as negative selection) in an attempt to ensure that their antigen receptor does not display self-reactivity. Those B cells surviving negative selection disperse to the peripheral lymphoid organs. Once encountering the foreign antigen, naïve B cells become activated, and ultimately differentiate into antibody-producing cells. The stages for B cell development include stem cells, pro-B cells, pre-B cells, immature B cells, mature B cells, plasma cells, and memory B cells.

T lymphocytes are derived from bone marrow and generated in the thymus. T cell development in the thymus is a multistep process with several built-in check points to ensure that the appropriate differentiation has taken place. As a result of selection (including positive and negative selection), only a small proportion of progenitor T cells, actually exit the thymus to the periphery as different subsets of mature T cells, such as $CD8^+$ T cell, $CD4^+$ T cell. The stages for T cell development include stem cells, thymocytes, mature T cells, effector T cells, and memory T cells.

12.4.2 B cells

B cells are produced in the bone marrow and migrate to secondary lymphoid organs for their functions. The molecules on the surface of B cells are responsible for B cell activation, including BCR, $CD79\alpha/\beta$, and other co-stimulators. The B cell receptors (BCRs) are unique in the early cellular events that are induced by antigen initiate B cell proliferation and differentiation. B cells are activated by antigen and, usually with T cell help, proliferate and mature into plasma or memory cells. Each B cell clone secretes antibody with single antigen specificity; diversity in the B cell population is generated by immunoglobulin gene re-arrangements that occur at the Pro-B cell stage. The affinity of the antibody for the cognate antigen can be modified by further gene re-arrangements that take place during the maturation of the humoral response. The isotype (IgM, IgG, IgE, IgA or IgD) is influenced by direct and indirect (i.e., mediated via cytokines) interactions with Th1 and

Th2 cells.

Not all proliferating B cells develop into plasma cells. Indeed, a significant proportion remains as memory B cells through a process known as clonal selection. This process is vital in eliminating the antigen and the body should become re-exposed to the antigen in the future. T cells are also clonally selected and this confers to the production of memory T cells.

12.4.3 T cells

The mature T cells that leave the thymus have not yet encountered the antigen that they have specificity for. These cells are called naïve T cells and mainly circulate from the bloodstream to the peripheral lymphoid organ for years before they die or encounter antigen. Important molecules on the surface of T cells are TCR, CD4/CD8, and other co-stimulators. The T cell receptors (TCRs) are unique in that it is only able to identify antigen when it is associated with a MHC molecule on the surface of the cell (see later). T cells described by cluster determination (CD) can be broadly divided into T helper (Th, or $CD4^+$ T cells) and cytotoxic T cells (Tc, or $CD8^+$ T cells).

$CD4^+$ T cells recognize antigen displayed on MHC class II molecules. $CD4^+$ T cells may differentiate into one of four subsets of T cells (Th1, Th2, Th17 or Treg), depending on the type of immune response taking place and in particular the cytokines present in the environment of the T cells. Th1 cells are pro-inflammatory T cells and stimulate macrophages activation whilst Th2 cells orchestrate B cell differentiation and maturation and hence are involved in the production of humoral immunity (antibody mediated). Th17 cells help to recruit neutrophils to sites of infection early in adaptive immune response, which is also a response aimed mainly at extracellular pathogens. Regulatory T cells (called Treg) have inhibitory activity that limits the extent of immune activation and prevent autoimmunity.

$CD8^+$ T cells recognize antigen displayed on MHC class I molecules. Because MHC class I molecules are found on essentially all uncleated cells of the body, $CD8^+$ T cells can monitor all cells for signs of infection. Hence, activated $CD8^+$ T cells (also called CTL), unlike Th cells, are primarily involved in the destruction of infected cells, notably viruses.

12.4.4 Other important molecules involved in adaptive immunity

(1) Co-stimulatory cascade T-cells are stimulated via the TCR by peptide antigen bound to MHC molecules. Additional signals are generated via engagement of co-stimulatory molecules, including the interaction of B7 molecules on antigen presenting cells with CD28 on T cells, CD40 on B cells with CD40 ligand (CD40L) on T cells. These can establish a cycle of reciprocal activation of T and B cells, resulting in the expansion of antigen-specific plasma cells that secrete a multi-clonal antibody response, resulting in the ability of immune system to distinguish between self and non-self.

(2) Cytokines Cytokines (also termed interleukins, IL; meaning "between white blood cells") are small molecules that act as a signal between cells and have a variety of roles including chemotaxis, cellular growth and cytotoxicity. Owing to their ability to control immune activity, they have been described as the "hormones" of the immune. Unlike hormones, cytokines are not stored in glands as preformed molecules, but are rapidly synthesized and secreted by different cells mostly after stimulation. Cytokines are pleiotropic in their biological activities and play pivotal roles in a variety of responses, including the immune response, hematopoiesis, neurogenesis, embryogenesis, and

oncogenesis. They frequently affect the action of other cytokines in an additive, synergistic or antagonistic manner.

Cytokines have been classified on the basis of their biological responses into: chemokines subfamily, interleukins subfamily, interferon subfamily, granulocyte-colony stimulating factors, tumor necrosis factor subfamily, etc. They play an important role in the communication between cells of multicellular organisms. As intercellular mediators, they regulate survival, growth, differentiation and effector functions of cells. Besides their pleiotropic effects, cytokine actions are often redundant and they exert their actions, which can be autocrine, paracrine or endocrine, via specific cell-surface receptors on their target cells. They are key players in the regulation of the immune response, particularly during infections, inflammatory joint, kidney, vessel and bowel diseases, or neurological and endocrinological autoimmune diseases.

(3) **Specific antibodies** The specific antibodies are circulating proteins that are produced by activated B cells in response to exposure to foreign antigens. More details will be described later.

12.5 Innate and adaptive immune responses

12.5.1 Innate immune response

The "innate" or "nonspecific" immunity refers to various physical and cellular attributes that collectively represent the first lines of defense against an intruding pathogen and infectious disease. The response evolved is therefore rapid, and is unable to "memorize". The body can be exposed to the same pathogen again with out any memory in the future.

The study of the innate immune response has recently flourished. Earlier studies of innate immunity utilized model organisms that lack adaptive immunity such as the plant *Arabidopsis thaliana*, the fly *Drosophila melanogaster*, and the worm *Caenorhabditis elegans*. Recent advances have been made in the field of innate immunology with the discovery of the Toll-like receptors, which are the receptors in mammal cells that are responsible for a large proportion of the innate immune recognition of pathogens. There is strong evidence that these Toll-like receptors are responsible for sensing the "pathogen-associated molecular patterns" and/or providing the "danger signal" to immune system.

Innate immunity serves two important functions: ①The initial response to microbes that prevents, controls, or eliminates infection of the host; ②Microbes stimulates adaptive immune responses and can influence the nature of the adaptive responses to make them optimally effective against different types of microbes. Thus, innate immunity not only serves defensive functions early after infection. but also provides the "warning" that an infection is present against which a subsequent adaptive immune response has to be mounted. Moreover, different components of the innate immune response often react in distinct ways to different microbes (e. g., bacteria versus viruses) and thereby influence the type of adaptive immune response that develops.

12.5.2 Role of innate immunity in stimulating adaptive immunity

Splitting the innate and adaptive immunity has served to simplify discussions of immunology. However, the systems are quite intertwined in a number of important respects. There are two aspects that stress the intersections between these two immune systems. Antigen capture and antigen presentation are important steps for linking them together.

(1) Antigens and their recognition receptors

1) **Antigen**: Antigens are macromolecules that elicit an immune response in the body. Antigens can be proteins, polysaccharides, conjugates of lipids with proteins (lipoproteins) and polysaccharides (glycolipids), but also can be small molecules (haptens) coupled to a protein (carrier). Antigen can be classified by different manner.

Ⅰ. According to their functional effect, there are three types of antigens: ①Immunogen: any substance that provokes the immune response when introduced into the body. An immunogen is always a macromolecule (protein, polysaccharide). Its ability to provoke the immune response depends on its foreignness to the host, molecular size, chemical composition and heterogeneity (e. g., different amino acids in a protein). ②Tolerogen: an antigen that invokes a specific immune non-responsiveness due to its molecular form. If its molecular form is changed, a tolerogen can become an immunogen. ③Allergen : an allergen is a substance that causes the allergic reaction. It can be ingested, inhaled, injected or comes into contact with skin.

Ⅱ. According to their origin, antigens can be classified as two groups: ①Exogenous antigens: antigens that enter the body from the environment; these would include inhaled macromolecules (e. g., proteins on cat hairs that can trigger an attack of asthma in susceptible people) ; ingested macromolecules (e. g., shellfish proteins that trigger an allergic response in susceptible people); molecules that are introduced beneath the skin (e. g., on a splinter or in an injected vaccine). The antigen-peptides are presented on the cell surface in the complex with MHC class Ⅱ molecules. ②Endogenous antigens:antigens that have been generated within the cell, as a result of normal cell metabolism, proteins encoded by the genes of viruses that have infected a cell, aberrant proteins that are encoded by mutant genes; such as mutated genes in cancer cells. The antigen-peptides are then presented on the cell surface in the complex with MHC class Ⅰ molecules.

2) **Cellular pattern recognition receptors for antigens**: As discussed earlier, macrophage and dendritic cell can express many receptors for identification of "danger signals", such as Mannose receptor (MS), scavenger receptors (SR), and Toll-like receptor (TLR) to directly recognize antigen from pathogen. B cells and T cells also have surface receptors for antigens. Each cell has thousands of receptors of a single specificity; that is, with a binding site for a particular epitope. T cell receptors (TCRs) enable the cell to bind to and, if additional signals are present, to be activated by and respond to an epitope presented by another cell called the antigen-presenting cell (APC). B cell receptors (BCRs) enable the cell to bind to and, if additional signals are present, to be activated by and respond to, an epitope on molecules of a soluble antigen. The response ends with descendants of the B cell secreting vast numbers of a soluble form of its receptors, called "specific antibodies".

A gene that is assembled from separate gene segments during the differentiation of the cell encodes each chain of BCR or TCR. The resulting gene is transcribed into a mRNA to be translated into one chain of the receptor. The number of gene segments from which the variable regions are constructed is sufficiently large that both B cells and T cells can generate as many as 10^{10} different antigen-binding sites. There is probably no epitope that could exist for which BCRs and TCRs able to bind it are not built.

(2) Major histocompatibility complex (MHC)   Major histocompatibility complex (MHC) is cell's surface proteins which expressing by MHC genes. MHC gene expresses a very high degree of polymorphism over 150 different HLA alleles (haplotypes). Polymorphism is most prominent in the regions that code for the peptide binding domains of HLA. The function of MHC polymorphism is to generate diversity at the population level.

The primary immunological function of MHC molecules is to bind and "present" antigenic peptides on the surfaces of cells for recognition (binding) by the antigen-specific T cell receptors (TCRs).

There are two main types of MHC molecules, called MHC class I and MHC class II. They have slightly different structures but both have an elongated cleft in the extracellular surface of the molecule, in which a single peptide is trapped during the synthesis and assembly of the MHC molecule inside the cell. The MHC molecule bearing its cargo of peptide is transported to the cell surface, where it displays the peptide to T cells. The most important differences between the two classes of MHC are summarized in the table 12-3.

Table 12-3 Differences between the MHC class I and II molecules

MHC class I	MHC class II
Compose with polymorphic α chains, β_2-microglobulin	Compose with polymorphic α and β chains
MHC class I α subunits are encoded by polymorphic MHC genes while the common β subunit is encoded by the conserved, non-MHC gene for β_2-microglobulin (β_2m)	Both MHC class II α and β subunits are encoded by polymorphic MHC genes. Class II α/β dimers also tend to associate as $(\alpha\beta)_2$ "dimers of dimers"
Expressed on all uncleated cells	Expressed on dendritic cells, mononuclear phagocytes, B lymphocytes, endothelial cells, thymic epithelium
Bind smaller antigenic peptides, about 8~12 amino acid residues in length	Bind larger antigenic peptides, about 9~25 amino acid residues in length
Source of protein from cytosolic proteins (mostly synthesized in the cell; may enter cytosol from phagosomes)	Source of protein from endosomal/lysosomal proteins (mostly internalized from extracellular environment)
Loading of peptide at endoplasmic reticulum	Loading of peptide at vesicular compartment
Presented to $CD8^+$ T cells	Presented to $CD4^+$ T cells

(3) **Antigen-presenting cells (APC)**   The initiation and development of adaptive immune responses require that antigens would be captured and displayed to specific lymphocytes. The cells that serve this role are called antigen-presenting cells (APCs). Antigen-presenting cells are cells that can "present" antigens in a form that T cells can recognize it. They are heterogeneous population of leukocytes with an enormous capacity for immuno-stimulation, including all nucleated cells that express MHC class I and MHC class II. The most specialized APCs are called "professional APCs", including dendritic cells, macrophages and B cells, express MHC class II antigens on their surface. These are important for communication with T cells and subsequent T cell activation. Although many cells do not normally express MHC class II antigens, certain cytokines can induce their expression and thereby potentiate antigen presentation. Such induction of inappropriate MHC class II antigen expression may contribute to autoimmune disease.

The most specialized APCs are dendritic cells. They are more efficient APCs than macrophages. The primary function of dendritic cells is to capture and present protein antigens to naïve T cells (those cells have not yet encountered an antigen). The MHC class II molecules with bound peptide on the dendritic cell are recognized by complementary shaped TCR and CD4 molecules on the naïve $CD4^+$ T cells. The MHC class I molecules with bound peptide on the dendritic cell are recognized by complementary shaped TCR and

CD8 molecules on naïve CD8⁺ T cells. These interactions enable the CD4⁺ T cell or CD8⁺ T cell to become activated, proliferate, and differentiate into effector T cells.

Macrophages are important in the regulation of immune responses. The primary function of macrophages is to capture and present protein antigens to effector T cells (these cells have encountered an antigen, proliferated, and matured into a form capable of actively carrying out immune defenses). The MHC class Ⅱ molecules with bound peptide on the macrophage are recognized by complementary shaped TCR and CD4 molecules on an effector CD4⁺ T cell. This interaction results in the activation of that macrophage. Like dendritic cells, macrophages are also capable of capturing and presenting protein antigens to naïve T cells although they are not so important as in this function. Stimulated macrophages exhibit increased levels of phagocytosis and secretion.

B cells capture and present protein antigens to effector CD4⁺ T cells as well. The MHC class Ⅱ molecules with bound peptide on the B cells are recognized by complementary shaped TCR and CD4 molecules on the CD4⁺ T cell. This interaction eventually triggers the CD4⁺ T cell to produce and secrete various cytokines that enable that B cell to proliferate and differentiate into antibody-secreting plasma cells.

(4) **Antigen processing and presentation** Antigen presentation is defined as a process whereby a cell expresses antigen on its surface in a form capable of being recognized by a T cell. The protein antigens typically undergoes some form of processing in which it is degraded into small peptides that are capable of associating with MHC class Ⅰ antigen to become a target for CD8⁺ T cells, or in association with MHC class Ⅱ antigen for presentation to CD4⁺ T cells (Figure 12-3).

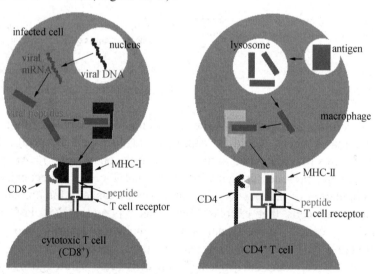

Figure 12-3 **The antigen presentation processes.** Endogenous antigen presentation pathway with MHC class Ⅰ molecule (left) and exogenous antigen presentation pathway with MHC class Ⅱ molecule (right).

1) The MHC class Ⅰ pathway of antigen presentation: MHC class Ⅰ molecules are transmembrane proteins expressed at the cell surface. Like all transmembrane proteins, they are synthesized by ribosomes on the rough endoplasmic reticulum (RER) and assembled within its lumen. Binding a peptide is an important step in the assembly of a stable MHC class Ⅰ molecule in the endopasmic reticulum. These peptides are transported from cytosol through the transporter associated with antigen processing (TAP). With out peptide, the MHC class I α subunit with β-2 microglobulin will not associate and transported to the cell surface. There are three subunits in each MHC class I complex including the transmembrane polypeptide (α subunit), the antigenic peptide and β-2 microglobulin. All of these must be present within the lumen of the endoplasmic reticulum if they are to assemble correctly and move through the Golgi apparatus to the cell surface.

2) **The MHC class II pathway of antigen presentation**: The MHC class II molecule's path toward the cell surface is somewhat different from MHC class I. In the rough endoplasmic reticulum (RER), the MHC class II α and β subunits associate with each other and a third protein, called the "invariant (Ii) chain", stabilizes the complex. Without the invariant chain, the α and β subunits will not associate together. So that, the MHC class II consist of two transmembrane polypeptides and a third molecule nestled in the groove they form. All three components of this complex must be present in the endoplasmic reticulum for proper assembly. But antigenic peptides are not transported to the endoplasmic reticulum, so an Ii protein temporarily occupies the groove.

The assembled MHC class II molecules containing α and β subunits and invariant chain leave rough endoplasmic reticulum forming vesicles, then transported through the Golgi apparatus and fuse with vesicles called lysosomes. Engulfed foreign antigen by endocytosis forming endosomes with proteases to degrade antigen into peptide fragments also fuse with the vesicles of lysosomes. At this point, the complex of MHC class II α and β subunits with Ii pass through such acidified vesicles, Ii is cleaved by acid protease. The groove is free for binding peptides derived from self protein or foreign antigen, transporting the peptides to the cells surface.

(5) **Antigen presentation provides necessary signal for adaptive immune responses**

The innate immune response provides signals that function in concert with antigen to stimulate the proliferation and differentiation of antigen-specific T and B lymphocytes. As the innate immune response is providing the initial defense, it also sets in motion the adaptive immune response. The activation of lymphocytes requires two distinct signals, antigen recognition by lymphocytes provides 1st signal for the activation of the lymphocytes, and molecules induced on host cells during innate immune responses to microbes provide 2nd signal. This idea is called the two-signal hypothesis for lymphocyte activation. The first requirement for antigen (so-called signal 1) ensures that the ensuing immune response is specific. The second requirement for additional stimuli triggered by microbes or innate immune reactions to microbes (signal 2) ensures that adaptive immune responses are induced when there is a dangerous infection and not when lymphocytes recognize harmless antigens, including self antigens. The molecules produced during innate immune reactions that function as second signals for lymphocyte activation include co-stimulators (for T cells), cytokines (for both T and B cells), and complement breakdown products (for B cells).

12.5.3 Adaptive immune response

Defense against microbes is mediated by the early reactions of innate immunity and the later responses of adaptive immunity. Innate immune response provides the early line of defense against microbes. It consists of cellular and biochemical defense mechanisms that are in place even before infection and are poised to respond rapidly to infections. Adaptive immune response is exquisite specificity for distinct molecules and an ability to "remember" and respond more vigorously to repeated exposures to the same microbe. It is able to recognize and react to a large number of microbial and nonmicrobial substances.

Innate and adaptive immune responses are components of an integrated system of host defense in which numerous cells and molecules function cooperatively. The mechanisms of innate immunity provide an effective initial defense against infections. However, many pathogenic microbes have evolved to resist innate immunity, and their elimination requires the more powerful mechanisms of adaptive immunity. The innate immune response to microbes also stimulates adaptive immune responses and influences the nature of the adaptive responses. The differences between the innate and the adaptive immune systems are summarized in the table 12-4.

Table 12-4 Differences between the innate and the adaptive immune systems

Innate Immunity	Adaptive Immunity
Antigen independent	Antigen dependent
No time lag	A lag period
Not antigen specific	Antigen specific
No Immunologic memory	Development of memory
Cells involved: natural killer (NK) cells, mast cells, phagocytes, dendritic cells	Cells involved: T and B cells
Molecules involved: cytokines, acute phase proteins, complement	Molecules involved: cytokines, specific antibodies

As mentioned previously, there is a great deal of synergy between the adaptive immune system and its innate counterpart. The adaptive immune system frequently incorporates cells and molecules of the innate system in its fight against harmful pathogens. For example, complement (molecules of the innate system) may be activated by antibodies (molecules of the adaptive system) thus providing a useful addition to the adaptive system's armamentaria. Although the cells and molecules of the adaptive system possess slower temporal dynamics, they possess a high degree of specificity and evoke a more potent response on secondary exposure to the pathogen.

The adaptive immune system may take days or weeks after an initial infection to have an effect. However, most organisms are under constant assault from pathogens, which must be kept in check by the faster-acting innate immune system. There are two types of adaptive immune responses, called humoral immunity and cell-mediated immunity, that are mediated by different components of the immune system and function to eliminate different types of microbes.

(1) The humoral immune response   Humoral immunity is mediated by secreted antibodies, produced in the cells of the B lymphocyte lineage (B cell), thus it is also named as antibody-mediated immunity. The humoral immune response combats microbes in many ways. Many polysaccharide and lipid antigens have multiple identical antigenic determinants that are able to engage many antigen receptor molecules on each B cell and initiate the process of B cell activation. The response of B cells to protein antigens requires activating signals ("help") from $CD4^+$ T cells (which is the historical reason for calling these T cells as "helper" cells). B cells ingest protein antigens, degrade them, and display peptides bound to MHC class II molecules for recognition by helper T cells, which then activate the B cells. The activated B cells secrete different classes of antibodies with distinct functions. Some of the progeny of the expanded B cell clones differentiate into antibody-

secreting plasma cells. Each plasma cell secretes antibodies that have the same antigen binding site as the cell surface antibodies (B cell receptors) that first recognized the antigen. Antibodies bind to the antigens of extracellular microbes and function to neutralize and eliminate these microbes. The elimination of different types of microbes requires several effector mechanisms, which are mediated by distinct classes, or isotypes, of antibodies.

1) **The types of antibodies (immunoglobulins, Ig)**: Antibodies have two roles to play, the first is to bind antigen and the second is to interact with host tissues and effector systems in order to ensure removal of the antigen. The basic antibody unit consists of a glycosylated protein consisting of two heavy and two light chains with Y-shaped structure (Figure 12-4). The region which binds to the antigen is known as the variable region, while the constant region, not only determines the isotype but is the region responsible for evoking effector systems, such as binding with Fc receptors present on the surface of phagocytic cells to facilitate the phagocytosis. Antibodies can protect the body in a variety of ways. For instance, by binding tightly to a toxin or virus, an antibody can prevent it from recognizing its receptor on a host cell. Antibody also can activate complements and induce cell-mediated cytotoxicity (ADCC).

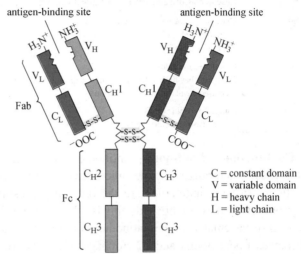

Figure 12-4 Schematic structure of an antibody molecule.
An antibody is composed of two identical heavy chains (blue) and two identical light chains (red). Each chain has a constant part and a variable part. The variable region is created by a light chain and a heavy chain of variable parts, which contains the antigen-binding site. The stem is formed constant region from the constant parts of the two heavy chains, which is involved in the elimination of the bound antigen.

There are five different types (known as isotypes) of antibodies in the human immune system — namely IgM, IgG, IgA, IgE and IgD. Each antibody can occur as transmembrane antigen receptors or secreted antibodies. IgG is the most important immunoglobulin and can be further subdivided into four subclasses (IgG 1~4). IgA antibodies are found as two subclasses (IgA 1~2). IgM forms pentamers in serum, which accounts for its high molecular weight. IgE is associated with immediate-type hypersensitivity when fixed to tissue mast cells. More characteristics of these antibodies are summarized in the table 12-5.

Table 12-5 Antibody isotypes and corresponding functions

Antibody	Characteristics
IgG	Predominant antibody in blood and tissue fluid With a high affinity Activate complement Cross placenta thus providing newborn with useful humoral immunity

Antibody	Characteristics
IgM	Large pentameric structure in circulation
	Present in monometric form on B cell surface
	Secreted form is predominant antibody in early immune response against antigen
	Reach 75% of adult levels at 12 months of age
IgA	Exist in both a monometric and dimeric form
	Secretory IgA (dimeric form) represents 1st line of defence against microbes invading the mucosal surface (e. g., tears)
IgE	Low level in circulation
	Fc region has high affinity for mast cell and basophil thus involved in allergy
	Increased levels in worm infections
IgD	Antigen receptor on B cells
	Absent from memory cells

2) **The functions of antibody**: Antibodies bind to microbes and prevent them from infecting cells, thus "neutralizing" the microbes. In this way, antibodies are able to prevent infections. In fact, antibodies are the only mechanisms of adaptive immunity that block an infection before it is established; this is why eliciting the production of potent antibodies is a key goal of vaccination. IgG antibodies coat microbes and target them for phagocytosis, since phagocytes (neutrophils and macrophages) express receptors for the tails of IgG. IgG and IgM activate the complement system, by the classical pathway, and complement products promote phagocytosis and destruction of microbes. Some antibodies serve special roles at particular anatomic sites. IgA is secreted from mucosal epithelia and neutralizes microbes in the lumens of the respiratory and gastrointestinal tracts (and other mucosal tissues). IgG is actively transported across the placenta, and protects the newborn until the immune system becomes mature. Most antibodies have half-lives of about 3 weeks. However, some antibody-secreting plasma cells migrate to the bone marrow and live for years, continuing to produce low levels of antibodies. The antibodies that are secreted by these long-lived plasma cells provide immediate protection if the microbe returns to infect the individual. More effective protection is provided by memory cells that are activated by the microbe.

3) **Diversity of antibodies**: B cells can express any class of Ig as a membrane-bound antigen receptor or as secreted antibody. This is achieved by differential splicing of mRNA to include exons that encode either a hydrophobic membrane anchor or a secretable tailpiece. The primary repertoire of Ig takes place without interaction of B cells with antigen. Although this primary repertoire is large, further diversification can occur that enhances both ability of Ig to recognize and bind to foreign antigen and the effector capacities of the expressed antibodies. This secondary phase of diversification occurs in activated B cells and is largely driven by antigen. The diversification is achieved through three mechanisms: somatic hypermutation, gene conversion, and class switching. After secondary diversification, estimates of the total number of distinct antibody variable regions or antigen combining sites that an individual's B cells can theoretically produce ranged from 10 million upward. It is not surprising then that antibody responses to even simple antigens are generally very heterogeneous in terms of the numbers of different antibodies made, a bonus for the individual as far as immune protection goes but a hindrance to the immunologist trying to sort out and manipulate this intricate process. However, each different antibody is produced only by a single clone of B cells, which is the "biological key" for creating hybridomas producing monoclonal antibodies. The fine-specificity of antigen recognition by monoclonal antibodies coupled with the relative ease of producing them has

resulted in widespread use of monoclonal antibodies in both research and medicine.

4) **General features of humoral immune responses**: In a primary immune response, naïve B cells are stimulated by antigen, become activated, and differentiate into antibody-secreting cells that produce antibodies specific for the eliciting antigen. Some of the antibody-secreting plasma cells survive in the bone marrow and continue to produce antibodies for long periods. Long-lived memory B cells are also generated during the primary response. A secondary immune response is elicited when the same antigen stimulates these memory B cells, leading to more rapid proliferation and differentiation and production of greater quantities of specific antibody than they are produced in the primary response. The principal characteristics of primary and secondary antibody responses are summarized in the table 12-6. These features are typical of T cell-dependent antibody responses to protein antigens.

Table 12-6 The comparison of primary and secondary humoral immune responses

feature	primary response	secondary response
time lag after immunization	usually 5 ~ 10 days	usually 1 ~ 3 days
peak response	smaller	larger
antibody isotype	usually IgM > IgG	IgG > IgM
antibody affinity	lower average affinity more variable	higer average affinity
induced by	all immunogens	only protein antigens
required immunization	high doses of antigen optimally with adjuvants	low doses of antigen Adjuvants not necessary

(2) **The cell-mediated immune response** T cells mainly involved in cell mediated immunity. These immune responses are directed principally at intracellular pathogens. Unlike B cells, T cells involve the destruction of infected cells (also called target cells) or intracellular pathogens in macrophages.

1) **T cells activation and clonal expansion**: Naïve T cells leave thymus and home to secondary lymphoid organs, where they may encounter antigens presented by mature dendritic cells with MHC class I or II molecules. TCR on the surface of T cells is an important membrane receptor for binding MHC on the surface of mature dendritic cells. $CD4^+$ T cells only bind with MHC II-antigen peptide while $CD8^+$ T cells bind with MHC I-antigen peptide. Antigen-stimulated T cells that have received both "signal one" through the antigen receptor and "second signals" via costimulatory receptors may be induced to secrete cytokines and to express cytokine receptors. The cytokine interleukin-2 (IL-2) provides autocrine signals to activated T cells, leading to expansion of antigen-specific clones. Only the T-cell clones that activated by the specific antigen are divided and proliferation. This results in the expansion of the antigen-specific lymphocyte pool and further differentiation into effector and memory T cells.

2) **T cells proliferation and differentiation**: Antigen recognition by T cells induces cytokine (e.g., IL-2) secretion, clonal expansion as a result of IL-2-induced autocrine cell proliferation, and differentiation of the T cells into different types of effector cells or memory cells. In the effector phase of the response, the activated $CD4^+$ T cells differentiate into Th1, Th2 and Th17 subsets of T cells. Some of $CD4^+$ T cells tend to function as down-regulating immune response. These cells are called regulatory T cells (Treg). The activated $CD8^+$ T cells differentiate into $CD8^+$ cytotoxic T lymphocytes (CTLs). These subsets of T cells all involved in cell-mediated immunity (Figure 12-5).

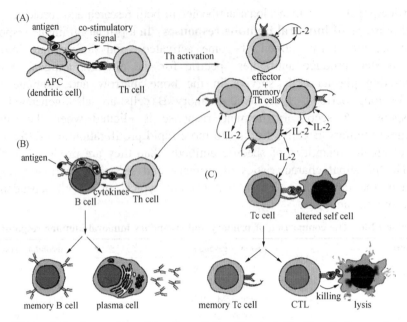

Figure 12-5 Activation and proliferation of T cells. (A) T cells activation require antigen-presenting cell (APCs) provide MHC-antigen peptide; (B) Effector $CD4^+$ T cells help B cell generate specific antibodies for humoral response; (C) Effector $CD8^+$ T cells induce cell-mediated response to altered self-cells.

3) **The functions of subsets T cells:** Unlike B cells, T cells involve the destruction of infected cells and tumor cells that display MHC class I -associated antigens by CTLs, or the destruction of intracellular pathogens in macrophages activated by Th1 cells. Th2 cells contribute to humoral immunity by stimulating the production of antibodies by B cells. TH17 cells help to recruit neutrophils to sites of infection early in adaptive immune response, which is also a response aimed mainly at extracellular pathogens. Regulatory T cells tend to suppress the adaptive immune response and are important in preventing immune response from becoming uncontrolled and in preventing autoimmunity. More functions of effector T cells are summarized in the table 12-7.

Table 12-7 The role of effector T cells in adaptive immunity

effector T cells	CTL	Th1	Th2	Th17	Treg
markers on the surface	CD8	CD4	CD4	CD4	CD4
recognized MHC	class I	class II	class II	class II	class II
effector molecules	perrorin, granzymes granulysin	IFN-γ, NF-α, CD40 ligand	IL-4, IL-5, CD40 ligand	IL-17A, IL-17F, IL-6	IL-10, TGF-β
main functions	kill virus-infected cells, tumor cells	activate infected macrophages, help B cells produce antibody	help B cells produce antibody	enhance neutrophil response	suppress T cell responses
pathogens targeted	viruses; some intracellular bacteria	microbes that persist in macrophage vesicles, extracellular bacteria	helminth parasites	extracellular bacteria	

12.6 Immunological memory

An effective immune response eliminates the microbes that initiated the response. This is followed by a contraction phase, in which the expanded lymphocyte clones die and homeostasis is restored. Memory cells are more effective in combating microbes than are naive lymphocytes because as mentioned earlier, memory cells represent an expanded pool of antigen-specific lymphocytes (more numerous than naive cells specific for the antigen), and memory cells respond faster and more effectively against the antigen than do naive cells. This is why generating memory response is the second important goal of vaccination.

Summary

The immune system is the organ system, which is composed of many interdependent cell types that collectively protect the body from invaders and the growth of tumor cells. Many of these cell types have specialized functions. It is often divided into two sections of innate and adaptive immunity, the former, representing a non-specific (no memory) response to antigen, and the latter, responding to new influences by mounting an immune response, which displays a high degree of memory and specificity.

Among the innate defensive cells the most notable are macrophages, nature killer cells, and dendritic cells. The adaptive immune defense consists of the humoral and cell-mediated responses. The cellular effectors comprise two main types of lymphocytes known as B and T cells. There is a great deal of synergy between the innate and adaptive immunity. Antigen presentation is required to initiate both forms of adaptive immune response. Humoral immunity is mediated by secreted antibodies, produced in the B cells, thus it is also named as antibody-mediated immunity, whilst the cell-mediated immunity protects against foreign organisms, and also protects the body from itself by controlling cancerous cells through different subsets of T cells. Memory lymphocytes also are important cells for protecting body.

Questions

1. What is immune system? What are the components of immune system?
2. What are the different lines of defense against infection?
3. What is innate immune response? Which types of cells are involved in?
4. What is called antigen? How does antigen have been presented?
5. What allows T and B cells to recognize antigens?
6. What is humoral immune response? What type of cells produce antibody?
7. Which lymphocytes are involved in the cell mediated immunity?
8. How does adaptive immunity differ from innate immunity?
9. What is immunological memory?

(戴亚蕾)

Chapter 13

Cancer cells

Cancer is a very common disease which is one of the leading causes of death in the world. Between 100 and 350 of each 100,000 people die of cancers in the world each year.

Since being one of the highest occurrence frequency and mortality rate among all the disease, cancer has been the focus of massive research effort for decades. Though these studies have led to a remarkable breakthrough in our understanding of the cellular and molecular basis of cancer, they have had very little impact on either preventing the occurrence or increasing the chances of surviving most cancers.

13.1 Basic knowledge about cancer cell

All cancer cells share one important characteristic: they are abnormal cells in which the processes regulating normal cell division are disrupted.

How a specific cancer cell behaves depends on which processes are not functioning properly. Some tumor cells simply divide and produce more tumor cells, the tumor mass remains localized, and the disease can usually be cured by surgical removal of the tumor. Other malignant tumors tend to metastasize, that is, to break away from the parent mass, enter the lymphatic or vascular systems, and metastasize to a remote site in the body where they establish lethal secondary tumors that are on longer amenable to surgical removal.

13.1.1 Cancer types

Cancer comprises more than 100 different diseases. Cancer cells within a tumor are the descendants of a single cell, even after it has metastasized. Hence, a cancer can be categorized based on the type of cell in which it originates and by the location of the cell. Generally, there are classified into four major types of tumors.

(1) Carcinoma **Carcinomas** are cancers derived from epithelial (lining) cells, which cover the surface of our skin and internal organs. This is the most common cancer type and represents about 80% ~ 90% of all cancer cases reported.

(2) Adenocarcinoma **Adenocarcinoma** refers to carcinoma derived from cells of glandular origin. One can, for instance, have an adenocarcinoma of the pancreas, or an adenocarcinoma of the lung. These are very different cancers.

(3) Sarcoma **Sarcomas** are cancers of the connective tissue, cartilage, bone, fat, muscle, and so on.

(4) Leukemia **Leukemias** are cancers derived from bone marrow and affects the lymphatic system. In leukaemia, lymphocytes cells do not mature properly and become too

numerous in the blood and bone marrow. Leukaemias may be acute or chronic. The most common type is acute lymphoblastic leukaemia (ALL). There are a number of other less common acute types which may be grouped together as acute non-lymphoblastic leukaemia (ANLL), this group includes acute myeloid leukaemia (AML).

(5) **Myelomas**   Myelomas are cancers of specialized white blood cells, involving the white blood cells responsible for the production of antibodies (B-cells).

13.1.2   Basic properties of cancer cells

Tumor cells display a characteristic set of features that distinguish them from normal cells. These traits allow the individual cells to form a tumor mass and eventually to metastasize to other parts of the body.

The behavior of cancer cells is more easily studied when the cells are growing in culture. Cancer cells can be obtained by removing a malignant tumor, dissociating the tissue into the separate cells, and culturing the cells *in vitro*. Alternatively, normal cells can be transformed *in vitro* to cancer cells by treatment with carcinogens which will be discussed in the next section. These transformed cells can generally cause tumors when they are introduced into host animal. A wide range of changes occur during the **transformation** of a normal cell to a cancerous cell.

There are many differences in properties from one type of cancer cell to another. Meanwhile, there are a number of common basic properties that are shared by cancer cells, regardless of their tissue of origin.

(1) **Loss of growth control**   The most important characteristic of a cancer cell is its unlimited number of cell divisions and loss of growth control. All cancer cells acquire the ability to grow and divide in the absence of stimulatory growth signals and/or in the presence of inhibitory signals.

1) Normal culture cells: Normal culture cells depend on growth factors, such as epidermal growth factor and insulin, that are present in serum added in the medium. Cancer cells can proliferate even in the absence of these external factors and are therefore immune to many of the normal regulations on cell division. The cancer cells no longer function as a part of a larger organism, but behave more like independent entities living without regard for the organism as a whole.

2) Cancer cells failure to respond to "stop" signals: The division of normal cells is restricted. Cells will stop dividing when they are in contact with neighboring cells. When the normal cells grown in tissue culture under suitable conditions proliferate to the point where they convert the bottom of the culture dish, their growth rate decreases dramatically, and they tend to remain as a single layer (monolayer) of the cells. This is because the normal cells respond to inhibitory influences from their environment. The cell-cell contacts send signals into the dividing cells that cause them to stop dividing. In contrast, tumor cells show density and anchorage independent growth. When tumor cells are cultured under the same conditions, they continue to grow, piling on top of one another to form clumps.

(2) **Dedifferentiation of the cancer cells**   An alternative mechanism to stop cell division is differentiation. This is the process by which a precursor cell acquires its final functional capabilities. The maturation process involves the coordinated regulation of gene expression that results in differential morphological and biological properties for the cells. Often, the differentiation process results in cells that have very limited potential for cell division. The cancer cell acquests dedifferentiation in some extension and keep dividing.

(3) **Avoidance of cell death**   In normal tissue, there is a balance between the generation

of new cells via cell division and the loss of cells through cell death. Old cells become damaged over time and are eliminated by a process termed programmed cell death or apoptosis. This process is a normal and necessary means of maintaining ourselves. There are safeguards built into the cell cycle that allow for the identification and elimination of cells that are dividing in an aberrant manner. These safeguards are responsible for preventing the development of cancer. Cancer cells that get through this stage have acquired the ability to avoid the cell death signals triggered by their abnormal behavior. Avoidance of cell death, coupled with continued cell division leads to the growth of the tumor.

(4) **The morphological and chemical changes of cancer cells** Cancer cells can often be distinguished from normal cells by microscopic examination. They are usually less well differentiated than normal cells. There are other detectable changes in the physical and chemical properties of the cells. These main changes include the following.

1) **Cytoskeletal changes:** The distribution and activity of the microfilaments and microtubules may change. These alterations change the ways in which the cell interacts with neighboring cells and alter the appearance of the cells. Changes in the cytoskeleton also affect cell adhesion and movement (motility).

2) **Nuclear changes:** The shape and organization of the nuclei of cancer cells may be markedly different from that of the nuclei of normal cells of the same origin, such as a high nucleus-to-cytoplasm ratio, prominent nucleoli, and relatively little specialized structure. This change in appearance may be useful in the diagnosis and staging of tumors (Figure 13-1).

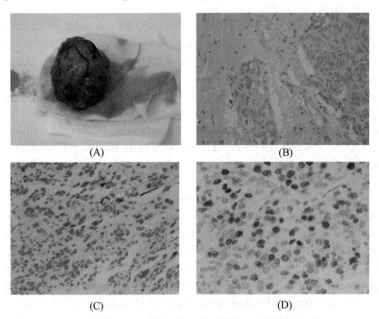

Figure 13-1 Gross and microscopic views of a tumor. (Meningioma).
(A) The gross morphology of meningioma. The dark red protrusion is the tumor (B. C. D). A light micrograph of a section of the tumor. (B) Showing tumor cells invading the surround normal tissue, tumor cells have darker and bigger nucleus. (C) Part of HE (hematoxylin and eosin) stained tumor cells. (D) Immatohistochemical tumor tissue stained with ki-67 antibody. It shows ki-67 existing in the nucleus (Courtesy of Dr. Yingqun Tao in the Department of Neurosurgery, the General Hospital of Shenyang Millitary Region).

3) **Enzyme production:** Cancer cells often secrete enzymes that enable them to invade neighboring tissues. These enzymes digest away the barriers to migration and spread of the

tumor cells.

For normal human cells, only Germ-line cells and rapid dividing somatic cells produce telomerase, and the telomerase will gradually lose its activity during differentiation of somatic cells. Tumor cells keep strong telomerase activity, and telomerase expression is essential for a tumor cell to immortal.

4) Decrease of proteins in cell surface and Loss of gap junction: Decrease of proteins in cell surface and Loss of gap junction allow the reduction of cell-cell and cell-extracellular matrix adhesion, and let cancer cells do not exhibit contact inhibition and are able to continue to grow even when surrounded by other cells. The alterations in cell adhesion also impact on the ability of the cells to move. Cancer cells must be able to move and migrate in order to spread, and cell adhesion plays a major role in regulating cell movement.

13.1.3 Angiogenesis and metastasis

All of the above properties of cancer cells can be studied *in vitro*. They are also practical for *in vivo* cancer cells. Besides these, there are two important properties of cancer *in vivo*: **angiogenesis** and **metastasis**.

(1) Angiogenesis   All living tissues are amply supplied oxygen and nutrients such as glucose (sugar). Nutrients and oxygen are pumped through the body via the circulatory blood system. Our tissues are full of many small blood vessels (capillaries) that can deliver blood within a very short distance to any cell.

1) Tumor cells also need requiring nutrients and oxygen in order to grow to a large mass: As tumors enlarge, the cells in the center no longer receive nutrients from the normal blood vessels. To provide a blood supply for all the cells in the tumor, it must form new blood vessels to supply the cells in the center with nutrients and oxygen. In a process called angiogenesis, tumor cells make growth factors (or cause nearby cells to produce), which activate the blood vessel cells to divide and stimulate formation of new capillary blood vessels. Angiogenesis is a prerequisite for the growth and metastasis of solid tumors.

2) Angiogenesis is an essential step in the growth of a tumor: Dormant tumors are those that do not have blood vessels, they can only grow into a mass of about 10^6 cells, roughly a sphere1 ~ 2 mm in diameter. Several autopsy studies in which trauma victims were examined for such very small tumors revealed that thirty-nine percent of women aged forty to fifty have very small breast tumors, while forty-six percent of men aged sixty to seventy have very small prostate tumors. Amazingly, ninety-eight percent of people aged fifty to seventy have very small thyroid tumors. However, for those age groups in the general population, the incidence of these particular cancers is only one-tenth of a percent (thyroid) or one percent (breast or prostate cancer). The conclusion is that the incidence of dormant tumors is very high compared to the incidence of cancer. Therefore, angiogenesis is critical for the progression of dormant tumors into cancer.

(2) Tumor metastasis   In most cases tumors have little risk to their host because they are localized and small size. Tumors that grow only in their original location are said to be benign. Benign tumors can cause health problems, and they will become serious medical problem only when they become malignant tumors which metastasize to other locations outside its site of origin in the body.

1) Metastasis of the cancer cells: Metastasis is the process by which cancer cells migrate from the initial or primary tumor to a distant location through the blood or lymphatic system or via direct contact to another location, which divide and form a secondary tumor at the new site. The distant growths are termed metastases. Metastatic tumors often interfere with the functions of the organs involved and lead to the morbidity and mortality seen in cancer. The majority of mortality associated with cancer is due to the metastasis of primary tumor.

The tumor cells retain the characteristics they had when they were in their original location. As an example, breast cancer that metastasizes to the liver is still breast cancer, NOT liver cancer.

2) **Bloodstream metastasis**: In order for cells to move through the body, they must first climb over/around neighboring cells. They do this by rearranging their cytoskeleton and attaching to the other cells or extracellular matrix via proteins on the outside of their plasma membrane. By extending part of the cell forward and letting to go at the back end, the cells can migrate forward. The cells can crawl until they hit a blockage which cannot be bypassed. Often this block is a thick layer of proteins and glycoproteins surrounding the tissues, called the basal lamina or basement membrane. In order to cross this layer, cancer cells secrete a mixture of enzymes that degrade the proteins in the basal lamina and allow them to crawl through.

3) **Matrix metalloproteases (MMP)**: The enzymes secreted by cancer cells contain a group of enzymes called MMP. These enzymes digest proteins that inhibit the movement of the migrating cancer cells. Once the cells have traversed the basal lamina, they can spread through the body in several ways. They can enter the bloodstream by squeezing between the cells that make up the blood vessels.

Once in the blood stream, the cells float through the circulatory system until they find a suitable location to settle and re-enter the tissues. The cells can then begin to grow in this new location, forming a new tumor.

4) **Lymphatic metastasis**: An alternative but similar route would be entry of the cancer cells in to the vessels of the lymphatic system. The fluid in this extensive network flows throughout the body, much like the blood supply. It is the movement of cancer cells into the lymphatic system, specifically the lymph nodes, that is used in the detection of metastatic disease. Some cancers may spread through direct contact with other organs, as in the gut cavity.

(3) Angiogenesis and metastasis have close relationship   Angiogenesis contribute largely to metastases. Without a blood supply, tumor cells cannot spread to new tissues. Tumor cells can cross through the walls of the capillary blood vessel at a rate of about one million cells per day. However, not all cells in a tumor are angiogenic. Both angiogenic and non-angiogenic cells in a tumor cross into blood vessels and spread; however, non-angiogenic cells give rise to dormant tumors when they grow in other locations. In contrast, the angiogenic cells quickly establish themselves in new locations by growing and producing new blood vessels, resulting in rapid growth of the tumor.

13.2 Carcinogenesis

Cancer cells behave as independent cells, growing without control to form tumors. The development of a cancer is remarkably rare event. Tumors are invariably found to have arisen from a single cell. They grow in a series of steps. The process of forming cancer cells from normal cells to carcinomas is called **oncogenesis** (onco = cancer), or **tumorigenesis**, or **carcinogenesis**.

13.2.1 Stages of tumor progression

The cells, as the basic components of the tissues and organs, experience several stages from the normal to the cancerous. This process described below is applicable mainly for a solid tumor such as a carcinoma or a sarcoma (Figure 13-2). Blood cell tumors go through a similar process but since the cells are free-floating, they are not limited to one location in the body.

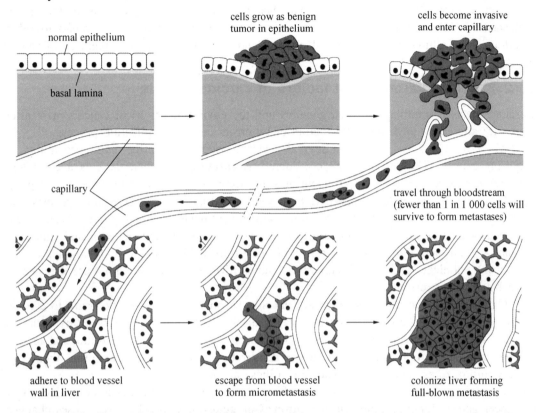

Figure 13-2 Stages of tumor progression (From: Alberts B, et al. Molecular Biology of the Cell. 2008)

(1) **Hyperplasia** This is the first step. These cells appear normal, but genetic changes have occurred that result in some loss of control of growth. leading to an excess of cells in that region of the tissue.

(2) **Dysplasia** In this stage, the additional genetic changes in the hyperplastic cells lead to the even more abnormal growth. The cells and the tissue no longer look normal and become disorganized.

(3) **Carcinoma *in situ*** (benign tumor) Further genetic changes are required in the stage of benign tumor, which result in cells that are even more abnormal and can now spread over a larger area of tissue. These cells often "regress" or become more primitive in their capabilities. They begin to lose their original function, for example a liver cell that no longer makes liver-specific proteins. Such cells are called de-differentiated or anaplastic. At this stage, because the benign tumor is still sited within its original location (called *in situ*), and the tumor cells have not yet crossed the basal lamina to invade other tissues. These growths are often considered to have the potential to become invasive and are treated as malignant growths.

Benign tumors are not considered to be cancerous because they do not progress to the point where they invade distant tissues. Though they are less often lethal than malignant tumors, they can still cause serious health problems in two ways. First, large benign tumors can put pressure on organs and cause other problems. In the case of brain tumors, the limited space within the skull means that a large growth in the brain cavity can be fatal. Secondly, they secrete excess amounts of biologically active substances like hormone which will bring serious healthy problem.

(4) Cancer (malignant tumors)　　The last step occurs when the cells in the tumor metastasize, which means that they can invade surrounding tissue, and spread to other locations. These tumors are the most dangerous and account for a large percentage of cancer deaths.

13.2.2　The environmental factors of carcinogenesis

Carcinogenesis is interplay between genetics and the environment. Most cancers arise after genes are altered by environmental factors or by errors in DNA replication process.

(1) The concept of carcinogens　　Several environmental factors affect one's probability of acquiring cancer. These factors are considered as **carcinogens** when there is a consistent correlation between exposure to an agent and the occurrence of a specific type of cancer.

Carcinogens can induce inherited mutations and finally result in carcinogenesis, so carcinogen is also called mutagen. A known human carcinogen means that there is sufficient evidence of a cause and effect relationship between exposure to the material and cancer in humans. Such determination requires evidence from epidemiologic (demographic and statistical), clinical, and(or) tissue(cell) studies involving humans who were exposed to the substance in question.

Carcinogen-induced cancer generally develops in many years after exposure to a toxic agent. A latency period of as much as thirty years has been observed between exposure to asbestos and incidence of lung cancer. Some cancers associated with environmental factors are preventable. Simply understanding the danger of carcinogens and avoiding them can usually minimize an individual's exposure to these agents.

(2) Carcinogens affect the genes　　The effect of environmental factors is not independent of cancer genes. Sunlight alters **tumor suppressor genes** in skin cells; and causes changes in lung cells, making them more sensitive to carcinogenic compounds in smoke. The environmental factors probably act directly or indirectly on the genes that are already known to be involved in cancer. Individual genetic differences also affect the susceptibility of an individual to the carcinogenic affects of environmental agents. About ten percent of the population has an alteration in a gene, causing them to produce excessive amounts of an enzyme that breaks down hydrocarbons presented in smoke and various air pollutants. The excess enzyme reacts with these chemicals, turning them into carcinogens. These individuals are about twenty-five times more likely to develop cancer from hydrocarbons in the air than others are.

(3) The classification of carcinogens　　Carcinogens are classified into three main types: Radiation carcinogens, chemical carcinogens and virus. Besides carcinogens from environment, physical condition is also contributed to carcinogenesis, such as chronic inflammation and oxygen radicals.

1) **Radiation carcinogens**: Radiation is one of the first known mutagens, it's a potent inducer of mutations. Different types of radiation cause different types of genetic changes. Ultra-violet (UV) radiation causes point mutations. X-rays can cause breaks in the DNA double-helix and lead to translocations, inversions and other types of chromosome damage. Exposure to the UV rays in sunlight has been linked to skin cancer.

2) **Chemical carcinogens**: There are hundreds of distinct types of chemical carcinogens

capable of inducing cancer after prolonged or excessive exposure. Many chemical carcinogens usually exert their effect by binding to DNA or the building blocks of DNA and interfering with the replication or transcription processes. Many industrial chemicals are carcinogenic. The fumes of the metals cadmium, nickel, and chromium are known to cause lung cancer. Vinyl chloride causes liver sarcomas. Exposure to arsenic increases the risk of skin and lung cancer. Aflatoxin, a mutagen most often found on improperly stored agricultural products, induces hepatocellular carcinoma.

3) **Viruses**: In addition to chemicals and radiation, the third source of mutation is viruses. Many viruses infect humans but only a few viruses play a significant role in the development of particular cancers in many different animals, including humans. Viruses associated with cancer include human papillomavirus (genital carcinomas), hepatitis B (liver carcinoma), Epstein-Barr virus (Burkitt's lymphoma and nasopharyngeal carcinoma), human T-cell leukemia virus (T-cell lymphoma); and, probably, a herpes virus HHV-8 (Kaposi's sarcoma and some B cell lymphomas), SV40 (mesothelioma). Viruses can disrupt cell behavior in several different ways. These include both DNA viruses and retroviruses.

Oncogenic DNA viruses contribute to cancer by inserting their genomes into the DNA of the host cell. Insertion of the virus DNA directly into a **proto-oncogene** may mutate the gene into an **oncogene**, resulting in a tumor cell. Insertion of the virus DNA near a gene in the chromosome that regulates cell growth and division can increase transcription of that gene, also resulting in a tumor cell.

The viruses may contain their own oncogenes (v-onc comparing to c-onc) that disrupt the regulation of the infected cell. This process may be beneficial to the virus if it allows for rapid production of progeny, but can be seriously detrimental to the host. For example, human papillomavirus makes proteins that bind to two tumor suppressors, p53 protein and RB protein, transforming these cells into tumor cells. Remember that these viruses contribute to cancer, but they do not by themselves cause it.

The ability of retroviruses to promote cancer is associated with the presence of oncogenes in these viruses. Retroviruses have acquired the proto-oncogene from infected animal cells, and altered the versions of genes by some ways. These altered genes no longer function properly, and when they are inserted into a new host cell, they cause disregulation and can lead to cancerous growth. An example of this is the normal cellular c-SIS proto-oncogene, which makes a cell growth factor. The viral form of this gene is an oncogene called v-SIS. Cells infected with the virus that has v-SIS overproduce the growth factor, leading to high levels of cell growth and possibly to tumor cells.

4) **Chronic inflammation**: Chronic inflammation can lead to DNA damage due to the production of mutagenic chemicals by the cells of the immune system. For example, the long-term inflammation caused by infection with the hepatitis virus will cause hepatocellular carcinoma.

5) **Oxygen Radicals**: During the capture of energy from food, which occurs in our mitochondria, oxygen radicals may be generated, which are very reactive and are capable of damaging cell membranes and DNA itself. These **reactive oxygen intermediates** (ROI) may also be generated by exposure of cells to radiation.

The mutagenic activity of ROI is associated with the development of cancer as well as the activities of several anticancer treatments, including radiation and chemotherapy.

13.3 The genes involved in cancer

There are a lot of genes participating in the initiation, development and metastasis of the cancers. These malfunctioning genes which play key roles in cancer induction can be broadly categorized into two broad classes, depending on their normal functions in the cell.

The first group, called proto-oncogenes in normal cells, produces protein products that normally enhance cell division or inhibit normal cell death. The mutated forms of these genes are called oncogenes. The second group, called tumor suppressors, makes proteins that normally prevent cell division or lead to cell death. Uncontrolled cell growth occurs either when active oncogenes are expressed (dominant) or when tumor suppressor genes (recessive) are lost. In fact, for a cell to become malignant, numerous mutations are necessary. In some cases, both types of mutations may occur.

13.3.1 Proto-oncogenes and oncogenes

An oncogene is any gene that encodes a protein being able to transform cells in culture or to induce cancer in animals. Almost all the oncogenes are derived from normal cellular genes which is called proto-oncogenes.

The proto-oncogenes that have been identified so far have many different functions in the cell, they usually code for the proteins that participate in signal transduction or pathway and promote cell proliferation.

(1) **The proto-oncogenes includes (Figure 13-3)** the extracellular signaling molecules、the membrane receptors for the signal molecules (e.g., HER-2/neu (erbB-2)、the growth factor receptor、The signal-transduction proteins (ras and src)、the transcription factors in the nucleus (myc)、The enzymes that function in DNA replication (hTERT)、the genes that resist apoptosis and promote survival (Bcl-2), a membrane associated protein that functions to prevent apoptosis.

Despite the differences in their normal roles, most proto-oncogenes are responsible for providing the positive signals that lead to cell division. Some proto-oncogenes work to regulate cell death. The defective versions of these genes, known as oncogenes, can cause a cell to divide in an unregulated manner. This growth can occur in the absence of normal pro-growth signals such as those provided by growth factors.

(2) **Activation of a proto-oncogene into an oncogene** The oncogene derived from proto-oncogene is the gene that encodes the protein being able to transform cells in vivo or vitro to induce cancer in animals. Conversion or activation of a proto-oncogene into an oncogene generally involves a *gain-of-function* mutation, because the cells with the mutant form of the proteins have gained a new function which does not present in cells with the normal gene. Most oncogenes are dominant mutations; and a single altered copy is sufficient to cause alterations in cell growth. This is in contrast with tumor suppressor genes which must be defective to lead to abnormal cell division.

The myc protein acts as a transcription factor and it controls the expression of several genes. Mutations in the myc gene have been found in many different cancers, including Burkitt's lymphoma, B cell leukemia, and lung cancer. RAS is another oncogene that normally functions as an "on-off" switch in the signal cascade. Mutations in RAS cause the signaling pathway to remain "on", leading to uncontrolled cell growth. Mutant RAS has been identified in cancers of many different origins, including: pancreas (90%), colon (50%), lung (30%), thyroid (50%), bladder (6%), ovarian (15%), breast, skin,

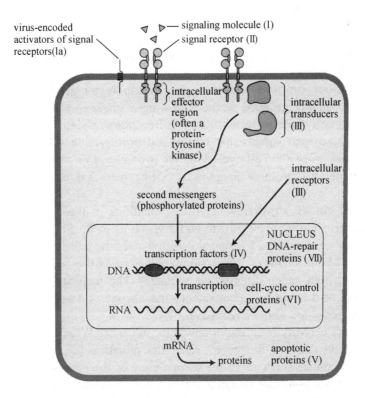

Figure 13-3 Seven types of proteins that participate in carcinogenesis. Mutations of I-IV part of V are in oncogene category. VI, VII and part of V are belong to tumor-suppressor genes (From: Lodish H, et al. Molecular Cell Biology. 2004)

liver, kidney, and some leukemias.

(3) Some oncogenes induce cells to produce angiogenic factors Oncogenes in some tumor cells allow the cells to produce angiogenic factors. An oncogene called BCL-2 has been shown to greatly increase the production of a potent stimulator of angiogenesis. It appears, then, that oncogenes in tumor cells may cause an increased expression of genes that make angiogenic factors. There are at least fifteen angiogenic factors and their production is greatly increased by a variety of oncogenes.

(4) Telomerase The gene that codes for the active component of the **telomerase** enzyme, hTERT, is considered a proto-oncogene because abnormal expression contributes to unregulated cell growth. Cancer cells have the ability to replicate without reaching a state of senescence. In many cancers, the ability to divide without limitation is achieved by the production of telomerase. Telomerase maintains the ends of chromosomes so that they do not shorten. Telomerase is a normal protein that is present in cells during fetal development. In most cells of an adult human, telomerase does not present as the gene for the enzyme is not being expressed (transcribed and translated). However, in some cancer cells the necessary task of achieving unlimited replication is made possible by the reactivation of the gene those code for telomerase.

(5) Oncogenes in the germ line cells The presence of an oncogene in a **germ line cell** (egg or sperm) results in an inherited predisposition for tumors in the offspring. However, a single oncogene is not usually sufficient to cause cancer, so inheritance of an oncogene does not necessarily result in cancer.

13.3.2 Tumor suppressor genes

Tumor suppressor genes generally encode proteins that inhibit cell proliferation and tumor formation in certain ways.

(1) **Groups of tumor suppressor genes** Tumor suppressor genes function in many key cellular processes, mainly grouped in the following 5 classes: ①Intracellular proteins that regulate or restrain cell cycle (p16 and Rb); ②Receptors or signal transducers for secreted hormones or developmental signals that inhibit cell proliferation (TGFβ); ③Checkpoint-control proteins that arrest the cell cycle if DNA is damaged or chromosomes are abnormal (P53); ④Proteins that promote apoptosis; ⑤Enzymes that participate in DNA repair (BRCA).

(2) **Mutations of tumor suppressor genes** Mutations in these genes result in loss-of-function and lead to abnormal cell growth and division as well. Since generally one copy of a tumor-suppressor gene suffices to control cell division, the cell loses the ability to prevent division only when both alleles of tumor-suppressor gene in the cell are lost.

How is it that both genes can become mutated? The majority of cancers are sporadic with no indication of a hereditary component, but oncogenic loss-of-function mutations in tumor-suppressor genes are genetic recessive.

In some cases, heredity is an important factor in cancer When a parent provides a germ line mutation in one copy of the gene, since the mutation is recessive, the trait is not expressed. Later a mutation occurs in the second copy of the gene in a somatic cell. In that cell both copies of the gene are mutated and the cell develops uncontrolled growth. This may lead to a higher frequency of loss of both genes in the individual who inherits the mutated copy than in the general population. However, mutations in both copies of a tumor suppressor gene can occur in a somatic cell, so these cancers are not always hereditary.

(3) **pRB** A very good example is hereditary retinoblastoma, a serious cancer of the retina that occurs in early childhood. When one parent carries a mutation in one copy of the RB tumor suppressor gene, it is transmitted to offspring with a fifty percent probability. Because retinoblasts are rapidly dividing cells and there are thousands of them, about ninety percent of the offsprings who receive the one mutated RB gene from a parent, and who would also develop a mutation in the second allele, producing no functional RB protein. These individuals then develop retinoblastoma in both eyes usually very early in life. Thus the tendency to develop retinoblastoma is inherited as a dominant trait. Not all cases of retinoblastoma are hereditary: sporadic retinoblastoma can also occur by mutation of both copies of RB in the somatic cell of the individual. Because losing two copies of RB is far less likely than losing one, sporadic retinoblastoma is rarely, developed late in life, and usually affects only one eye.

(4) *p53* The *p*53 (or T*p*53) gene has emerged as one of the most important tumor suppressor genes to date. The gene produces a protein product that functions as a transcription factor. The genes controlled by *p*53 are involved in cell division and viability. The P53 protein functions to prevent unregulated cell growth.

The P53 protein interacts directly with DNA. It also interacts with other proteins that direct cellular action. The P53 protein is at the center of a large network of proteins that "sense" the health of a cell and cellular DNA. The P53 protein is the conductor of a well orchestrated system of cellular damage detection and control. When DNA damage or other cellular insults are detected, the activity of the P53 protein aids in the decision between repair and the induction of apoptosis. The *p*53 gene is found to be defective in about half of all tumors, regardless of their type or origin.

13.4 The genetic and epigenetic changes of cancer

Cancer is a highly variable disease with multiple heterogeneous genetic and epigenetic changes. For almost all types of cancer studied up to date, it seems as if the transition from a normal, healthy cell to a cancer cell is step-wise progression that requires an accumulation of genetic changes in several different oncogenes and tumor suppressors. The cells become progressively more abnormal as more genes become damaged.

13.4.1 The genetic changes in cancers

It has been concluded that several independent genetic changes are required to create cells that become cancerous.

To complicate matters, it is clear that a cancer cell can be accomplished in many different ways. The genes involved may differ from each other, and the order in which the genes become abnormal may also vary. As an example, colon cancer tumors from two different individuals may involve very different sets of tumor suppressors and oncogenes, even though the outcome is the same (Figure 13-4).

The genetic changes that lead to unregulated cell growth may be acquired in two different ways. The mutation in somatic cells can occur gradually over a number of years, leading to the development of a "sporadic" case of cancer. Alternatively, inherit dysfunctional genes lead to the development of a familial form of a particular cancer type. Hereditary cancers constitute about 10 percent of human cancers, including breast cancer, colon cancer and retinoblastoma.

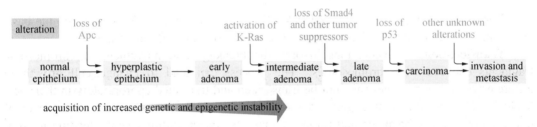

Figure 13-4 **The development and metastasis of human colorectal cancer and its genetic alterations** (From: Alberts B, et al. Molecular Biology of the Cell. 2008)

(1) The ways to cause genetic alterations   There are several ways to cause genetic alterations. It can be placed into two large categories. The first category is comprised of changes that alter only one or a few nucleotides along a DNA strand. A second category involves alterations of larger amounts of DNA, often at the chromosome level, such as translocation, amplification and deletion.

1) **DNA point mutations**: Changes that alter only one or a few nucleotides along a DNA strand are termed point mutations. The altered gene may lead to the production of a protein that no longer functions properly, or alter gene expression levels. This includes both *gain-of-function* mutation of proto-oncogene (Figure 13-5) and *loss-of-function* mutation of tumor-suppressor gene.

The mechanisms by which the changes are induced is varied. It includes spontaneous mutations and induced mutations.

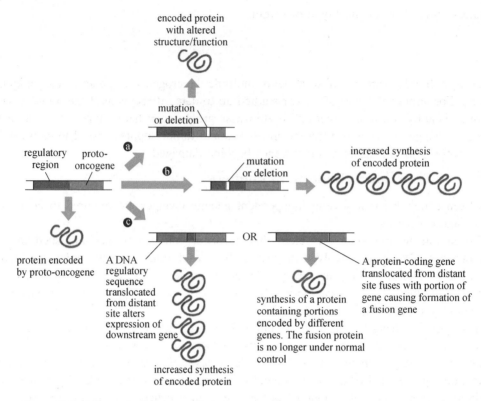

Figure 13-5 The ways for *gain-of-function* mutation of proto-oncogene (From: Kart G, et al. Cell and Molecular Biology. 2005)

2) **Alterations of larger amounts of DNA**: The larger amounts of DNA occur on the basis of chromatin, or fragments of chromatin. It contains several different ways.

Translocations: Translocations involve the breakage and movement of chromosome fragments between two chromosomes. The process can lead to change cell growth in different ways: ①The genes may not be transcribed and translated appropriately in their new location. The movement of a gene can lead to an increase or a decrease in its level of transcription. ②The breakage and rejoining may also occur within a gene, leading to form a new protein with changed structure and activation.

Translocations are common in leukemias and lymphomas and have been less commonly identified in cancers of solid tissues. For example, an exchange between chromosomes 9 and 22 was observed in over 90% of patients with chronic myelogenous leukemia (CML). The exchange leads to the formation of a shortened form of chromosome 22 called the Philadelphia chromosome (after the location of its discovery). This translocation leads to the formation of an oncogene from the *abl* proto-oncogene.

Inversions: In these alterations, DNA fragment is released from a chromosome and then re-inserted in the opposite orientation. As in the previous examples, these rearrangements can lead to abnormal gene expression, either by activating an oncogene or de-activating a tumor suppressor gene.

Duplications/deletions: Through replication errors, a gene or group of genes may be copied more than one time within a chromosome. This is different from gene amplification in that the genes are not replicated outside the chromosome and they are only copied one extra time, not hundreds or thousands of times. Genes may also be lost due to failure of the replication process or other genetic damage.

Aneuploidy: Cancer cells often have highly aberrant chromosome complements. Aneuploidy is genetic changes that involve the loss or gain of entire or part of chromosomes. Aneuploidy is a common feature of cancer cells. Cancer cells often have sometimes more than 100 chromosomes. The presence of the extra chromosomes makes the cells unstable and severely disrupts the controls on cell division (Figure 13-6). There is controversy as to whether aneuploidy occurs at an early stage in tumor formation, and is a cause of genetic instability that characterizes cancer cells, or is a late event and is simply a consequence of abnormal cancer growth.

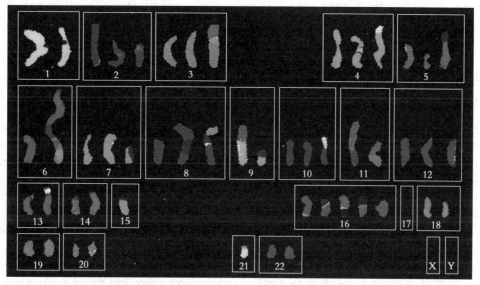

Figure 13-6 Karyotype of a cell from a breast cancer line showing a highly abnormal chromosome complement (From: Kart G, et al. Cell and Molecular Biology. 2005)

(2) Gene amplification When the normal DNA replication process is seriously flawed, it leads to DNA amplification from a single copy of a region of a chromosome to many copies. it occurs quite often in cancer cells.

The genes on each of the copies can be transcribed and translated, leading to an overproduction of the mRNA and encoded protein. If an oncogene is included in the amplified region, then the resulting *gain-of-function* mutation of this proto-oncogene. Examples of this include the amplification of the *myc* oncogene in a wide range of tumors and the amplification of the *ErbB-2* or *HER-2/neu* oncogene in breast and ovarian cancers.

Gene amplification also contributes to one of the biggest problems in chemotherapy. A gene commonly involved in multiple drug resistance (MDR). The protein product of this gene acts as a pump located in the membrane of cells. It is capable of selectively ejecting molecules from the cell, including chemotherapy drugs. In the presence of chemotherapy drugs, and such multiple resistances do occur. The amplification of MDR let the cancer cells make large amounts of this protein, allowing them to keep multiple chemotherapy drugs with similar structure outside the cell.

(3) Loss of heterozygosity of tumor-suppressor genes In hereditary cancers, the progeny receive one mutated copies of tumor-suppressor genes from his parents and it means that he is heterozygous for the mutation. Subsequent loss or inactivation of the normal copy

in somatic cells, referred to as **loss of heterozygosity** (LOH), is a prerequisite for cancer development. There are three ways for LOH, mis-segregation during mitosis; mitotic recombination between two homologous chromatin and subsequent segregation, mutation or deletion of normal copy.

(4) Genetic changes contribute to cancer   Most cancers are thought to arise from a single mutant precursor cell. As that cell divides, the resulting "daughter" cells may acquire different mutations and different behaviors over a period of time. Those cells that gain an advantage in division or resistance to cell death will tend to take over the population. In this way, the tumor cells are able to gain a wide range of capabilities that are not normally seen in the healthy version of the cell type represented. This is also the reason why cancer is much more prevalent in older individuals. The cells in a 70 year old body have had more time to accumulate the changes needed to form cancer cells but those in a child are much less likely to have acquired the requisite genetic changes. Of course, some children do get cancer but it is much more common in older individuals.

13.4.2   The epigenetic changes in cancer

In addition to actual alterations in DNA sequence, gene expression can also be altered by modification of the DNA and chromatin that do not change the sequence. Since these changes do not alter the sequence of the DNA in the genes, they are termed **epigenetic changes** or **epigenetic DNA modification**. The epigenetic mechanisms are described in details in Chapter 10, cell differentiation. Two types are epigenetic changes are described below.

(1) DNA methylation   **DNA methylation** is an important epigenetic alteration in the control of gene transcription, In this alteration, some nucleotides in the DNA are modified by the addition of a methyl ($-CH_3$) group to the base, usually to C in CpG dinucleotides. Almost half of the known genes have CpG islands in promoters and the first exon regions. 80% of CpG dinucleotides are heavily methylated, but some areas remain unmethylated in GC-rich CpG islands that are 0.5~2 kb in length. Methylation of DNA is associated with the inactivation of that particular region of DNA. In cancer cells, DNA methylation abnormality is frequently observed in normally unmethylated CpG islands and silences the function of normally expressed genes. methylation is more common than genetic alterations, the epigenetic-mediated silencing is seen in many tumor-suppressor genes and it is important for cell cycle function and DNA repairing in many cancer types.

While DNA methylation clearly enhances the ability of cells to regulate and package the genetic information, it also adds an additional level of complexity. Genomic methylation patterns are frequently altered in tumor cells, with global hypomethylation accompanying region-specific hypermethylation events. When hypermethylation events occur within the promoter of a tumor suppressor gene, this can silence expression of the associated gene and provide the cell with a growth advantage in a manner somewhat akin to deletions or mutations. Epigenetic controls become misdirected in cancer cells.

(2) Histon acetylation   **Histone acetylation** is another important epigenetic change which cause changes of gene transcription, and has recently been shown to be important in cancer development. In this epigenetic change, the histone proteins around which the DNA is wound become modified by the addition of acetyl ($-CH_3CHO$) groups. This alteration leads to a loosening of the DNA-histone interaction and is associated with increased gene expression.

Histones can be reversibly modified in several ways, and the best character of these modifications is histone acetylation. This is a reversible modification, which is carried out by two families of enzymes, the histone acetyltransferases (HATs), and the histone deacetylases(HDACs). These enzymes have important activities in many cellular processes including transcription, DNA replication and cell cycle progression. Because of their important roles in the regulation of such events, enzymes that affect histone acetylation status are increasingly being associated with tumors.

13.4.3 Aberrant gene expression pattern of cancer cells

Genetic and epigenetic changes in cancer cell result in changes in **expression pattern** of many genes which finally cause aberrant cell behavior.

Because the heterogeneity of cancer, expression profiles are diverse between normal cells and corresponding cancer cells and among different types of tumors, even in different individual of same kind of cancer. Some of these differences can be correlated with biological differences between the tumors or diverse of tumorigenesis. Even though most of the differences can not be explained, they can provide cancer researchers with a list of genes look at more closely as potential targets for therapeutic drugs and **diagnosis biomarkers**.

13.5 Treatment of cancer

Cancer is such a heterogeneous disease, even those of the same kind of cancer, means that diagnosis and treatment are complicated. Current advances in the molecular classification of tumors should allow the rational design of treatment protocols based on the actual genes involved in any given case. New diagnostic tests may involve the screening of hundreds or thousands of genes to create a personalized profile of the tumor in an individual. This technical revolution is accelerating the process of identifying the genes of diagnostic, prognostic or therapeutic significance in cancer.

13.5.1 Traditional treatments

Because cancer comprises many diseases, doctors use many different treatments, mainly including **surgery**, **chemotherapy**, and **radiation.** The course of treatment depends on the type of cancer, its location, and its state of development.

Surgery, often the first treatment, is used to remove solid tumors. It may be the only treatment necessary for early stage cancers and benign tumors. Radiation-based cancer treatments intended to kill all the rapid dividing cancer cells based on the DNA damaging properties of radiation, therefore radiation treatment can causes some of the side effects by non-specifically killing the normal cells. Surgery and radiation treatment are often used together.

Chemotherapy drugs are toxic compounds that target rapidly growing cells by interfering their cell cycle or by promoting apoptosis. The most common class of drugs are designed to interfere S phase with the synthesis of precursor molecules needed for DNA replication, or interfere with the ability of the cell to complete the S phase of the cell cycle, or cause extensive DNA damage, which stops replication. The second class of drugs called spindle inhibitors stops cell replication early in mitosis. During mitosis, chromosome separation requires spindle fibers made of microtubules; spindle inhibitors stop the synthesis of microtubules. Because most adult cells don't divide often, they are less sensitive to these drugs than are cancer cells. The third kind of drugs works by forcing the cancer cells to undergo apoptosis. They kill certain adult cells that divide more rapidly, such as those that line the gastrointestinal tract, bone marrow cells, and hair follicles. This also causes some of the side effects of chemotherapy, including gastrointestinal distress, white blood cell count down, and hair loss.

13.5.2 New strategies for cancer treatment

Different kind of new anticancer strategies and new anticancer drugs appear along with the new discoveries of cancer and development of new technologies, most of them are still in the clinic trials. The main new strategies are separated into 4 groups: ①**Immunotherapy**; ②Gene therapy; ③Inhibiting the activity of oncogenes; ④Preventing angiogenesis.

(1) Immunotherapy   We have all heard or read cancer patients who was expected to survive several months by the doctors remain alive and cancer-free years later, yet they defy the prognosis. This is because of the capability to destroy a tumor with his own immune system. But in most cases, one contributing factor in cancer is the failure of the immune system to destroy cancer cells. Immunotherapy depends on antibodies or immune cells to attack tumor cells, it encompasses several techniques that use the immune system to attack cancer cells or treat the side effects of some types of cancer treatment. In recent years, two broad treatment approaches involving the immune system have been pursued: passive immunotherapy and active immunotherapy.

Passive immunotherapy is an approach that treats cancer patients with the therapeutic antibodies agents which recognize and bind to specific proteins that play role in the activities of the target tumor.

Active immunotherapy is an approach that gets a person's own immune system more involved in the fight against cancer cells. The least specific of these are the **immunostimulants**, such as interleukin-2 and alpha interferon, which enhance the normal immune response.

Although many tumor cells do contain proteins that are not normally expressed by their normal counterparts, or mutated proteins that are different from those presented in normal cells. They are still basically host proteins presented in host cells. As a result, the immune system typically fails to recognize these proteins as "inappropriate". Even if a person does possess T cells that recognize the tumor-associated antigens, tumors evolve mechanisms that allow them to escape immune destruction.

Different strategies have been formulated to stimulate the immune system to achieve a more vigorous response against tumor cells. For example, T cells are isolated from the patient, stimulated in certain way in vitro, allowed to proliferate in culture, and then reintroduced into the patient. The ultimate goal is to produce immunopreventive treatments in which people would be vaccinated with antigens that would prevent them from ever developing cancers.

A technique called **chemo-immunotherapy** attaches chemotherapy drugs to antibodies that are specific for cancer cells. The antibody then delivers the drug directly to cancer cells without harming normal cells, reducing the toxic side effects of chemotherapy. These molecules contain two parts: the cancer-cell-specific antibody and a drug that is toxic once it is taken into the cancer cell. A similar strategy, **radio-immunotherapy**, couples specific antibodies to radioactive atoms, thereby targeting the deadly radiation specifically to cancer cells.

(2) Gene therapy   Gene therapy was first conducted in inherited disease with single gene mutation. In recent years, the focus of gene therapy development has shifted toward the cancer treatment. Since *loss-of-function* change of tumor-suppressor genes in cancer cells,

using gene therapy, a wild-type tumor suppressor gene can be introduced to tumor cells through infection with engineered vector such as virus, so that it either kills tumor cells or converts them to nonmalignant state.

(3) **Inhibiting the activity of oncogenes**   Cancer cells contain *gain-of-function* mutation of oncogenes, if these proteins can be selectively blocked, it should be possible to stop the uncontrolled growth and invasive properties of malignant cells. Oncogenes are usually the valid targets for drug screen, small-molecular-weight compounds that specifically bind the cancer-promoting proteins and inhibit its activity are screened for cancer drug.

Hormone therapy is another strategy for cancer treatment. Many of the factors that affect normal cell growth are hormones. Although cancer cells have lost some of the normal responses to growth factors, some cancer cells still require hormones for growth. Hormone therapy for cancer attempts to starve the cancer cells of these hormones. This is usually done with drugs that block the activity of the hormone, although some drugs can block synthesis of the hormone. For example, some breast cancer cells require estrogen for growth. Drugs that block the binding site for estrogen can slow the growth of these cancers. These drugs are called selective estrogen receptor modulators (SERMs) or anti-estrogens. Tamoxifen and Raloxifene are examples of this type of drug.

(4) **Preventing angiogenesis**   Another promising target for cancer therapy is angiogenesis. As previous introduction, angiogenesis is required to deliver nutrients and oxygen to the rapidly growing tumor cells, or to provide the conduits for cancer cells to spread to other sites in the body.

It was postulated as long ago as 1971 by Dr. Judah Folkman that prevention of angiogenesis could inhibit tumor growth by starving them of vital nutrients. The existence of natural inhibitors of angiogenesis was hinted by an intriguing observation made by surgeons.

Cancer cells promote angiogenesis by secreting growth factors, such as VEGF. There are also inhibitors. The first naturally discovered inhibitor was thrombospondin, identified in 1989 by Dr. Noel Bouck. More agiogenic inhibitors have been developed using different approaches, including antibodies and synthetic compounds directed against the targets.

The anti-angiogenesis treatments share two important advantages: ①Because they are natural products of the body, they should be much less toxic than conventional chemotherapy drugs. ② Because they act on normal (blood vessel) cells instead of attacking the tumors directly, there is less chance than the cancer cells will develop resistance to the drug.

All treatments are far from representing a ultimate cure. It's hoped, over the next few years, the advances in genomics and proteomics will lead to the development of new strategies for cancer diagnosis and therapy.

Summary

In this chapter, we introduce the basic knowledge about cancer cells. All cancer cells share one important characteristic: they are abnormal cells in which the processes regulating normal cell division are disrupted. That is, cancer develops from violations of the basic rules of social cell behavior. Unlike normal cells, cancer cells ignore signals to stop proliferation, to differentiation, or to apoptosis and be shed. Some tumor cells simply divide and produce more tumor cells, the tumor mass remains localized, and the disease can usually be cured by surgical removal of the tumor. Other malignant tumors tend to metastasize.

Cancer is a highly variable disease with multiple heterogeneous genetic and epigenetic changes. Carcinogenesis is interplay between genetics and the environment. Most cancers arise after genes are altered by environmental factors or by errors in DNA replication process. For almost all types of cancer studied up to date, it seems as if the transition from a normal, healthy cell to a cancer cell is step-wise progression that requires an accumulation of genetic changes in several different oncogenes and tumor suppressors. There are several ways to cause genetic alterations, which is placed into two large categories, one is comprised of changes that alter only one or a few nucleotides along a DNA strand; another involves alterations of larger amounts of DNA, often at the chromosome level, such as translocation, amplification and deletion.

Cancer is such a heterogeneous disease, even those of the same kind of cancer, means that diagnosis and treatment are complicated. Current advances in the molecular classification of tumors should allow the rational design of treatment protocols based on the actual genes involved in any given case. Different kind of new anticancer strategies and new anticancer drugs appear along with the new discoveries of cancer and development of new technologies, most of them are still in the clinic trials.

Questions
1. Give short definitions of the following terms: cancer, tumor, transformation, angiogenesis, metastasis, carcinogen, proto-oncogene, oncogene, tumor suppressor gene, gain-of-function mutation, Loss of function mutation.
2. What characteristics distinguish benign from malignant tumor?
3. What's angiogenesis? Explain the mechanism of angiogenesis, and its effect on carcinogenesis.
4. Comparing proto-oncogenes and tumor suppressor genes.
5. *Loss-of-function* occurs in the majority of human tumors. Name two ways to cause loss of $p53$ function.
6. Is cancer hereditary?
7. Example two new strategies to anti-cancer treatments.

(李 瑶)

References

[1] Alberts B, Bray D. and Hopkin K, *et al*. Essential Cell Biology. Second edition. New York and London: Garland Science, 2004

[2] Alberts B, *et al*. Molecular Biology of the Cell. 5th ed. 2008

[3] Alberts B, Lewis J, Johnson A, *et al*. Molecular Biology of the Cell. 5th ed. New York: Gardand Science, 2008: 366-399

[4] Alberts B, Lewis J, Johnson A, *et al*. Molecular Biology of the Cell. 5th ed. New York: Garland Science, 2008

[5] Alizadeh AA, Eisen MB, Davis RE, *et al*. Distinct types of diffuse large B-cell lymphoma identified by gene expression profiling. Nature, 2000, 403(6769): 503-511

[6] Pombo A. Cellular genomics: which genes are transcribed, when and where? Trends Biochem Sci, 2003, 28(1):6-9

[7] Steimer A, Schöb H, Grossniklaus M. Epigenetic control of plant development: new layers of complexity. Curr Opin Plant Biol, 2004, 7:11-19

[8] Antoine HFMP, Dirk S. Methylation of histones: playing memory with DNA. Curr Opin Cell Biol, 2005, 17: 230-238

[9] Asim KB, Ronald MA. Polar Microbiology: The Ecology, Diversity and Bioremediation Potential of Microorganisms in Extremely Cold Environments. Sacramento CA: CRC Press, 2009

[10] Avi As, Vishva MD, Death R. Signaling and modulation. Science, 1998, 281(5381):1305

[11] B Alberts B, Lewis J, Johnson A, *et al*. Molecular Biology of the Cell. Fourth Edition. New York: Garland Science, 2002

[12] Beachy P, et al. Tissue repair and stem cell renewal in carcinogenesis. Nature, 2004, 18(432): 324

[13] Bonfanti ALV. Neural stem cells. Circ Res. 2003, 92: 598

[14] Brown TA. Genomes. 2nd ed. Oxford UK: BIOS Scientific Publishers Ltd, 2002

[15] Brush up on your 'omics. Chemical & Engineering News, 2003, 81(49): 20. http://pubs.acs.org/cen/coverstory/8149/8149genomics1.html

[16] Buchanan B, Gruissem W, Jones R. Biochemistry & Molecular Biology of Plants. Rockville: American Society of Plant Physiologists, 2000

[17] Burt R, Neklason DW. Genetic testing for inherited colon cancer. Gastroenterology, 2005, 128(6):1696-1716

[18] Classon M, Harlow E. The retinoblastoma tumor suppressor in development and

cancer. Nature Rev Cancer, 2002, 2:910-917

[19] Cooper GM. The Cell – A Molecular Approach. 2nd ed. Sunderland (MA): Sinauer Associates Inc, 2000

[20] Cozzolino R, Palladino P, Rossi F, et al. Antineoplastic cyclic astin analogues kill tumour cells via caspase-mediated induction of apoptosis. Carcinogenesis, 2005, 26 (4): 733

[21] Craig LP. Chromatin remodeling enzymes: taming the machines. EMBO Reports, 2002, 3 (4):319-322

[22] David RH, Stephen MC. Connecting proliferation and apoptosis in development and disease. Nat Rev Mol Cell Biol, 2005, 5 : 805

[23] David S. Cancer stem cell refined. Nat Immunol, 2004, 5(7):701

[24] Dedrick J, Emmanuel ML. Signaling networks: the origins review of cellular multitasking. Cell, 2000, 103: 193

[25] Enrique López-Juez. Plastid biogenesis, between light and shadows. Exp Botany, 2007, 58: 11-26

[26] Fulda S, Debatin KM. Sensitization for tumor necrosis factor-related apoptosis-inducing ligand-induced apoptosis by the chemopreventive agent resveratrol. Cancer Res, 2004, 64(1): 337

[27] Gautier J, et al. Cyclin is a component of maturation-promoting factor from *Xenopus*. Cell, 1990, 60:487-494

[28] Gordon K. Embryonic stem cell differentiation: emergence of a new era in biology and medicine. Gen Dev, 2005, 19:1129

[29] Griffiths A, Gelbart W, Miller J, et al. Modern Genetic Analysis. New York: WH Freeman & Co,1999

[30] Hezel AF, Bardeesy N, Maser RS. Telomere induced senescence: end game signaling. Curr Mol Med, 2005, 5(2):145

[31] Hiroo F. Gignals that control plant vascular cell differentiation. Nat Rev Mol Cell Biol, 2004, 5: 379

[32] Isayeva T, Kumar S, Ponnazhagan S. Anti-angiogenic gene therapy for cancer. Int J Oncol, 2004, 25(2):335-343

[33] Jacks T. The expanding role of cell cycle regulators. Science, 1998, 280:1035-1036

[34] Darnell J, Lodish H, Berk A, et al. Molecular cell biology. Fifth Edition. 2004

[35] Johnson F B, David AS, Leonard G. Molecular biology of senescence. Cell, 1999, 96:291

[36] Johnston WK, et al. RNA-catalyzed RNA polymerization: accurate and general RNA-template primer extension. Science, 2001, 292 (5520): 1319-1325

[37] Levenson JM, Sweatt JD. Epigenetic mechanisms in memory formation. Nature, 2005, 6:108-118

[38] Katharine LA, Amanda GF. Epigenetic aspects of differentiation. Cell Sci, 2004, 117, 4355

[39] Katja N. Lineage-specific transcription factors and the evolution of gene regulatory networks. Brief Funct Genomic, 2010, 9 (1): 65

[40] Klug W, Cummings M. Essentials of Genetics. 4th ed. Beijing: Higher Education Press, 2002

[41] Korostelev AA. Structural aspects of translation termination on the ribosome. RNA, 2011, 17 (8): 1409-1421

[42] Lincoln TA, Joyce GF. Self-sustained replication of an RNA enzyme. Science,

2009, 323 (5918): 1229-1232

[43] Lodish H, Berk A, Zipursky SL, et al. Molecular Cell Biology. 4th ed. New York: WH Freeman & Co, 2000

[44] Lodish H, Berk A, Zipursky SL, et al. Molecular Cell Biology. Fourth Edition. New York: W H Freeman Company, 2001

[45] Malumbres M, Barbacid M. To cycle or not to cycle: a critical decision in cancer. Nature Rev Cancer, 2001, 1: 222-231

[46] D' Angelo MA, Gomez-Cavazos JS, Mei A, et al. A change in nuclear pore complex composition regulates cell differentiation developmental. Cell, 2012, 22(2): 446

[47] Ming Guo, Bruce AH. Cell proliferation and apoptosis. Cur Opin Cell Biol, 1999, 11:745

[48] Ning Cai, Mo Li, Jing Qu, et al. Post-translational modulation of pluripotency. J Mol Cell Biol, published online, doi: 10.1093/jmcb/mjs031, 2012

[49] Nurse P. Universal control mechanism regulation onset of M-phase. Nature, 1990, 344:503-507

[50] Paolo SC. Unique chromatin remodeling and transcriptional regulation in spermatogenesis. Science, 2002. 296(5576): 2176

[51] Patrick F C, David CS, Joanna E, et al. Accumulation of mitochondrial DNA mutations in ageing, cancer, and mitochondrial disease: is there a common mechanism? Lancet, 2002, 360(9342):1323

[52] Phillip N, Yau-Huei Wei. Ageing and mammalian mitochondrial genetics. Trends Genetics, 1998, 14(12):513

[53] Purves WK, Sadava D and Orians GH, et al. Life: The Science of Biology. Seventh Edition. New York: Sinauer Associates, Inc & WH Freeman Company, 2003

[54] Reik W. Stability and flexibility of epigenetic gene regulation in mammalian development. Nature, 2007, 447, 425

[55] Robert SB, Shino N, Toren F. Mitochondria, oxidants, and senescence. Cell, 2005, 120(4):483

[56] Beck S, Olek A and Walter J. From genomics to epigenomics. Nature Biotechnol, 1999, 17 (12):1144

[57] Sandeep K, Donna LF, George CT. T cell rewiring in differentiation and disease. Immunol, 2003, 171: 3325

[58] Scott WL, Athena WL. Apoptosis in cancer. Carcinogenesis, 2000, 21(3) :485

[59] Solange D. Mitochondria as the central control point of apoptosis. Trends Cell Biol, 2000, 10:369

[60] Sunil KM, Angie R. Emerging roles of microRNAs in the control of embryonic stem cells and the generation of induced pluripotent stem cells. Dev Biol, 2010, 344(1): 16

[61] Takahashi K, Yamanaka S. Induction of pluripotent stem cells from mouse embryonic and adult fibroblast cultures by defined factors. Cell, 2006, 126: 663

[62] Tannishtha R, Sean JM, Michael FC, et al. Stem cells, cancer, and cancer stem cells. Nature, 2001, 414: 105

[63] Audesirk T, Audesirk G, EB B. Biology: Life on Earth. 5th Edition. 2002

[64] Thompson CB. Apoptosis in the pathogenesis and treatment of disease. Science, 1995, 267: 1456

[65] Tom KK. Polycomb group complexes-many combinations, many functions. Trends

Cell Biol, 2009, 19(12): 692
[66] Wakasugi T, Tsudzuki T and Sugiura M. The genomics of land plant chloroplasts: gene content and alteration of genomic information by RNA editing. Photosynth Res, 2001, 70: 107-118
[67] Wei Yau-Huei, Lu Ching-You, Lee Hsin-Chen, *et al*. Oxidative damage and mutation to mitochondrial DNA and age-dependent decline of mitochondrial respiratory function. Ann NY Acad Sci, 1998, 854: 155
[68] Yusupova G, *et al*. Structural basis for messenger RNA movement on the ribosome. Nature, 2006, 444 (7117): 391-394
[69] Zhai Zhonghe, Wang Xizhong and Ding Mingxiao. Cell Biology. Beijing: Higher Education Press, 2000

Glossary

A

A band 暗带,A 带
横纹肌肌节,中粗肌球蛋白丝所在的区段,具有重折光性。

accessory cell 辅佐细胞
在免疫应答过程中,协助淋巴细胞介导特异性抗原识别的一类细胞的总称。主要包括抗原呈递细胞,如巨噬细胞、中性粒细胞、肥大细胞和自然杀伤细胞等。

acinar cell 腺泡细胞
在腺体(如胰腺)的腔周围形成球状结构的细胞体。

actin 肌动蛋白
肌肉中的一种主要蛋白质;横纹肌的细丝和细胞微丝的主要蛋白质,与细胞运动有关。

actin-binding protein 肌动蛋白结合蛋白
在细胞中与肌动蛋白单体或肌动蛋白纤维结合的,能改变特性的蛋白质。

actin-depolymerizing protein (depactin) 肌动蛋白解聚蛋白
一种可使肌动蛋白解聚的蛋白质。主要存在于肌动蛋白纤维骨架快速变化的部位,与肌动蛋白纤维结合并引起肌动蛋白纤维的快速解聚,形成球状肌动蛋白单体。

actin filament 肌动蛋白丝
由球状肌动蛋白分子构成 7 nm 直径的蛋白质丝,是真核细胞中的主要成分,在肌肉细胞中特别丰富。

actinin 辅肌动蛋白
与肌动蛋白结合的一种小蛋白质,在横纹肌的 Z 线和应力纤维中比较集中,在将肌动蛋白丝横连在一起和使应力纤维与质膜结合中发挥作用。

active transport 主动运输
将分子向着高浓度方向穿膜的耗能运输。

actomyosin 肌动球蛋白
肌球蛋白和肌动蛋白的复合物,是肌肉收缩单元的蛋白质。

adaptive immunity 适应性免疫
又称获得性免疫(acquired immunity)。机体通过与抗原物质接触而由淋巴细胞所产生的免疫力。具有特异性和记忆性。

adenovirus 腺病毒
一种具双链 DNA 的动物病毒。基因组大小约为 36 kb。常用于研究 DNA 复制、转录和作为基因工程载体。

adenylate cyclase (adenylyl cyclase, cAMPase) 腺苷酸环化酶
催化腺苷三磷酸(ATP)形成环腺苷酸(cAMP)的膜结合酶,是细胞内某些信号传递途径中的重要成分。

adhering junction (adherens junction, zonula adherens) 黏着连接
又称锚定连接(anchoring junction)。是上皮细胞间或上皮细胞与细胞外基质间的连接结构。根据连接中所涉及的细胞外基质和细胞骨架的关系分为桥粒、半桥粒、黏着带和黏着斑 4 种类型。

adhesion belt 黏着带
位于上皮细胞紧密连接的下方,靠钙黏蛋白与肌动蛋白相互作用,将两个细胞连接在一起,其质膜内侧与肌动蛋白丝连接。

adhesion protein 黏着蛋白
存在于细胞外基质中的与细胞黏附于基质有关的一类蛋白质,包括纤连蛋白、层粘连蛋白和血纤维蛋白原等。在细胞的黏附、迁移、增殖、分化等活动中起作用。

adult stem cell 成体干细胞
存在于一种组织或器官中的未分化细胞,具有自我更新的能力,并能分化为来源组织的主要类型特化细胞。

amoeba (plural amoebae) 变形虫
广泛用于细胞研究的一种原生动物。

amoeboid movement (amoeboid locomotion) 变形运动
细胞在前端伸出伪足爬行的运动方式。伪足的伸出与细胞趋化性和细胞质的溶胶-凝胶变化有关。

animal pole 动物极
卵母细胞中细胞核偏向的一端。

ankyrin 锚蛋白
红细胞上连接血影蛋白与整合膜蛋白的球

蛋白。

antenna complex　天线复合物
为叶绿体光系统的主要组成部分之一，可将吸收的光能传递给光反应中心。

anther culture　花药培养
将成熟或未成熟的花药从母体植株上取下，放在无菌的条件下，使其进一步生长、发育成单倍体细胞或植株的技术。

antibody（Ab）　免疫球蛋白，抗体
具有抗原结合部位，能与抗原分子上相应表位发生特异性结合的具有免疫功能的球蛋白。

antibody-dependent cell-mediated cytotoxicity（ADCC）　抗体依赖性细胞介导的细胞毒作用
是一种细胞介导免疫反应机制，即免疫系统的免疫效应细胞把结合了特异性抗体的靶细胞裂解的过程。

antigen（Ag）　抗原
又称免疫原（immunogen）。能诱导免疫应答并能与相应抗体或T细胞受体发生特异反应的物质。

antiport　反向运输
膜载体将两种不同的小分子同时或依次向相反的方向穿膜运输。

antisense RNA　反义RNA
与基因的专一RNA转录本互补的RNA，可同此专一的RNA杂交。

apical cell　顶端细胞
植物胚胎发生过程中，合子不等分裂后产生的两个细胞中近合点端的较大的细胞。

apoptosis　〔细胞〕凋亡
即程序性细胞死亡（programmed cell death），为非坏死性细胞死亡，细胞分解成凋亡小体，被邻近的细胞吞噬。

apoptosome　凋亡体
与细胞凋亡有关的多蛋白质复合物。由细胞凋亡蛋白酶激活因子1、胱天蛋白酶9及细胞色素C组成。激活胱天蛋白酶起始物和效应物反应机构，启动细胞凋亡级联反应下游的过程变化。

apoptotic body　凋亡小体
细胞凋亡过程中，细胞萎缩、碎裂形成的有膜包围的含有核和细胞质碎片的小体，可被吞噬细胞所吞噬。

aquaporin（AQP）　水孔蛋白
又称"水通道蛋白"。一个高度保守的膜运输蛋白家族，蛋白质形成同四聚体，每一亚基多次穿膜构成穿膜通道，允许水和亲水性小分子（如甘油）穿过生物膜。

archaea　古核生物
曾称古细菌（archaebacteria）。现今最古老的生物群，为地球原始大气缺氧时代生存下来的活化石。为单细胞生物，无真正的核，染色体含有组蛋白，RNA聚合酶组成比细菌的复杂，翻译时以甲硫氨酸为蛋白质合成的起始氨基酸，细胞壁中无肽聚糖，不同于真细菌，核糖体蛋白与真核细胞的类似。许多种类生活在极端严酷的环境中。与真核生物、原核生物并列构成现今生物三大进化谱系。

archesporium（sporogonium）　孢原细胞
能通过有丝分裂产生孢子母细胞的二倍体细胞。

arm ratio　臂比
染色体短臂与长臂长度的比值，是染色体组型的数据。

asexual reproduction　无性生殖
不经过生殖细胞的结合由亲体直接产生子代的生殖方式。

A site　A部位（受体部位）
核糖体接受氨酰基tRNA的部位。

astrocyte　星形〔胶质〕细胞
脊椎动物中枢神经系统中呈星形的神经胶质细胞，可调节细胞外离子和化学环境，支持脑血屏障，为神经组织提供营养物质，并在大脑瘢痕修复中起重要作用。

asymmetrical division　不对称分裂
细胞分裂可产生两个大小不等的子细胞，所含的细胞组分也有相应差别的一种分裂方式。这种分裂往往与子细胞向不同方向分化有关。

autophagolysosome　自噬溶酶体
又称自〔体吞〕噬泡（autophagic vacuole）。为融入细胞自身多余或衰老细胞器（如线粒体和内质网等）的一类溶酶体。在细胞内起清道夫作用。

autophagosome　自噬体
双层膜结构将衰老或病理的细胞器如线粒体圈住所形成的结构，然后与溶酶体融合形成自噬溶酶体。

autophagy　自〔体吞〕噬
细胞自身一部分内容物被次级溶酶消化的过程。其消化产物可作为营养再利用，或用于细胞分化中的细胞结构重建。

auxin　植物生长素
一类植物激素。常指吲哚乙酸（IAA），有促进细胞壁伸长的作用。

axoneme　轴丝
纤毛和鞭毛中由微管构成的中轴部分。

axoneme dynein　轴丝动力蛋白
又称纤毛动力蛋白（ciliary dynein）。为纤毛或鞭毛中组成外周二联丝侧臂的蛋白质，是外周二联丝相互滑动的动力来源。

B

bacterial artificial chromosome BAC　细菌人工染色体
一类以F因子为基础构建的克隆载体。用来在大肠杆菌中克隆分子量较大的外源DMA片段。

bacteriophage（phage）　噬菌体
以细菌为寄主进行复制的病毒。

basophil　嗜碱性粒细胞
人体血液中含量最少的一种颗粒白细胞。颗粒中含有组胺，激活时分泌组胺、肝素和软骨素，以及蛋白水解酶，且质膜上带有IgE的Fc受体，在速发型超敏反应中可释放过敏毒素反应嗜酸性粒细胞趋化因子。

B cell hybridoma　B 细胞杂交瘤
B 细胞与骨髓瘤细胞融合产生的杂交瘤细胞克隆。

B cell receptor　B 细胞受体
又称膜表面免疫球蛋白（surface membrane immunoglobulin, smIg）。为 B 细胞特异性识别抗原的受体。

biochip　生物芯片
广义的生物芯片是指一切采用生物技术制备或应用于生物技术的微处理器，包括用于研制生物计算机的生物芯片，将健康细胞与电子集成电路结合起来的仿生芯片，缩微化实验即芯片实验室，以及利用生物分子相互间的特异识别作用进行生物信号处理的基因芯片、蛋白质芯片、细胞芯片和组织芯片等。狭义的生物芯片就是微阵列，包括基因芯片、蛋白质芯片、细胞芯片和组织芯片等。

biomembrane　生物膜
围绕细胞或细胞器的脂质双层膜。由磷脂结合蛋白质和胆固醇、糖脂构成，起渗透屏障、物质转运和信号转导的作用。为细胞内的膜系统与质膜的统称。

biotin　生物素
可共价与蛋白质或核酸结合，并可与抗生物素蛋白紧密结合，从而用于蛋白质或核酸检测的一种辅酶。

blastula　囊胚
动物在早期胚胎发育过程中形成的中空球形胚体，是胚胎发育的一个阶段。

B lymphocyte　B 淋巴细胞
在抗原刺激下具有分泌球蛋白能力的淋巴细胞。

C

cadherin　钙黏蛋白，钙黏素
细胞表面黏着因子之一，是一种钙依赖性的细胞黏着膜蛋白。

calcium ATPase　钙 ATP 酶
又称钙泵（calcium pump, Ca^{2+}-pump）。为肌细胞肌质网膜上的整合蛋白。起钙泵 ATP 酶的作用，即将钙离子泵入肌质网腔，降低肌质中的钙离子浓度，终止肌肉收缩。

calcium-binding protein　钙结合蛋白质
可与钙结合的蛋白质。包括两类：一类为 EF 手形蛋白，如钙调蛋白；另一类可与钙和磷脂结合，归类于膜联蛋白，常作为信号传递的中介体，激活下游信号组分，有的为钙离子运输体。

calcium channel　钙通道
专门针对钙离子的膜通道，如肌质网膜上的电压门控离子通道。

calcium store (calcium pool)　钙库
细胞内一些具有钙离子贮存能力的细胞器（如内质网、肌质网及液泡），其钙离子含量很高。

calcium signal　钙信号
当细胞受到各种刺激时，导致细胞外钙离子进入细胞或细胞内钙库钙离子释放，可提高细胞溶质内的游离钙离子浓度，成为引起细胞反应的信号。

calcyclin　钙周期蛋白
与催乳素受体相关的钙结合蛋白（分子量约 1,000）。为含有 EF 手形结构域的小的钙结合蛋白家族成员之一，使细胞周期受到调节。

cancer cell　癌细胞
失去控制而恶性增殖的细胞。

capsid　衣壳
由衣壳粒亚单位构成的病毒外壳。

carcinogen　致癌因素[物]
能诱发细胞癌变的物理和化学因素。

carotenoid　类胡萝卜素
一种脂溶性的光合作用色素，呈黄红色。

carrier protein　载体蛋白
为膜内的运输蛋白，可同穿膜运输的分子进行暂时性结合。

cascade　级联反应
细胞内信号传递途径关联蛋白质的系列反应，即通过多次的逐级放大，使较弱的输入信号转变为极强的输出信号，导致各种生理响应的过程。一般包括磷酸化和去磷酸化反应。

cell aging (cell senescence)　细胞衰老
随着时间的推移，细胞增殖能力和生理功能逐渐下降的变化过程。细胞在形态上发生明显变化，如细胞皱缩、质膜透性和脆性提高、线粒体数量减少、染色质固缩和断裂等。

cell communication　细胞通信
在多细胞生物的细胞社会中，细胞间或细胞内高度精确和高效地发送与接收信息的通信机制，并通过放大机制引起快速的细胞生理反应。

cell culture　细胞培养
细胞的体外人工培养。

cell cycle　细胞周期
分裂细胞由有丝分裂末尾到第二次有丝分裂末尾的顺序变化。在两次相邻的有丝分裂间期又分为 G_1、S、G_2 3 期。

cell differentiation　细胞分化
细胞的后代在结构和功能上发生差异，形成不同细胞的过程。分化细胞获得并保持特征，合成专一性的蛋白质。

cell division　细胞分裂
一个细胞分成两个含一个核的子细胞的变化。

cell engineering　细胞工程
应用细胞生物学和分子生物学的方法，通过类似于工程学的步骤，在细胞整体水平或细胞器水平上，遵循细胞的遗传和生理活动规律，有目的地制造细胞产品的一门生物学技术。

cell-free system　无细胞系统
来源于细胞，不具有完整细胞结构，但包含进行正常生物学反应所需的物质组成的系统。

cell fusion　细胞融合
又称细胞杂交。在实验条件下将不同的体细胞合成一个杂交细胞。利用某些病毒(如 Sendai virus)、乙二醇(PEG)、电脉冲或激光作媒介,可人工诱发细胞融合。

cell hybridization　细胞杂交
在体外条件下,通过人工培养和诱导,将不同种生物或同种生物不同类型的两个或多个细胞合并成一个双核或多核细胞的过程。

cell junction　细胞连接
连接两个细胞之间或连接细胞与其他基质的特化区域。

cell line　细胞系
在培养条件下可进行无限分裂的动物或植物细胞群。

cell locomotion (cell migration)　细胞迁移
一个细胞主动地从一个地方向另一地方运动。

cell memory　细胞记忆
在 DNA 序列没有发生变化的情况下,细胞及其后代保留过去受过影响痕迹的能力,这种影响使基因以不同方式表达。

cell plate　细胞板
在高等植物细胞分裂过程中,新的细胞壁物质在细胞中央集中并逐渐向细胞周边延伸所形成半固体的板状物,它是细胞壁的前体。

cell strain　细胞株
通过选择法或克隆形成法从原代培养物或细胞系中获得的具有特定性质或标记(marker)的培养物称为细胞株。细胞株的特定性质或标记必须在整个培养期间始终保持。

cell surface　细胞表面
细胞质膜外表面结构的总称。

cell theory　细胞学说
由 Schleiden 和 Schwann 于 1838～1839 年提出的基本学说。其主要内容是:一切动、植物均由细胞构成;细胞是多细胞有机体的最小构成成分。Virchow 在 1858 年进一步补充,细胞只能由细胞分裂而来。

cellular oncogene (c-oncogene)　细胞癌基因
简称"c 癌基因",又称原癌基因(proto-oncogene)。编码调节细胞生长或分化有关蛋白质时发生突变或过表达转变为癌基因并引起细胞癌变的正常细胞基因。

centriole　中心粒
为动物细胞中由 9 组微管(每组 3 条)围成的圆筒状细胞器,两颗中心粒在一端相互垂直。在分裂间期中位于核的一侧,细胞分裂时逐渐移向两极,与组建有丝分裂器有关。

centrodesmose　中心体连丝
在有丝分裂时两个分开的中心体间最初出现的连接中心体的细丝,是纺锤体形成的起始结构。

centromere　着丝粒
染色体上同纺锤丝间接相连的区域,着丝粒区 DNA 为高度重复序列。

centromere DNA sequence　着丝粒 DNA 序列
真核细胞染色体着丝粒部位可与动粒结合的 DNA 序列。

centrosome　中心体
动物细胞中位于中央的细胞器,由中心粒和中心粒周围物质构成。

channel protein　通道蛋白
在膜中形成一条窄的穿膜亲水孔的蛋白质,可允许水和小分子被动穿过。

chaperone　分子伴侣
协助其他蛋白质分子肽链正确折叠的蛋白质。

chaperonin　伴侣蛋白
存在于原核生物、线粒体和叶绿体中的分子伴侣蛋白的一个亚组。直接帮助新生肽链和解折叠的蛋白质肽链折叠为具有生物功能构象的蛋白质,包括 GroEL 和热激蛋白 E 60 等。

check point　监控点,检验点
在细胞周期中可使细胞暂时刹车的位点,待条件适宜时才允许细胞进入细胞周期的下一个阶段。

chloroplast　叶绿体
植物细胞中由双层膜围成,含有叶绿素并能进行光合作用的细胞器。

cholesterol　胆固醇
动物细胞膜中的一种主要脂类成分。

chromatid　染色单体
复制染色体中的两条子链。两条染色单体在着丝粒处相连在一起。

chromatin　染色质
真核细胞在间期时,核中由 DNA 和组蛋白质构成脱氧核蛋白复合物,常结合有非组蛋白和少量 RNA。染色质亦称为间期染色体(interphase chromosome)。

chromosome　染色体
真核生物细胞核中含基因的结构,通常多指有丝分裂过程中由染色质凝缩形成的线状或棒状小体,有时也指原核生物和病毒中具有遗传编码的核酸分子。

chromosome scaffold　染色体支架
染色体去掉组蛋白后在电镜下显示的蛋白质性基本构架。

cis　顺面,顺式
部位的近侧,如高尔基体的顺面,基因结构上顺反子(cistrans)中的顺式。

cisterna　潴泡,液泡,囊泡
充有液体内含物由膜围成的扁形囊腔,如内质网池、核间池以及高尔基体的膜囊等。

citric acid cycle　三羧酸循环(CA),Krebs 循环
好气性生物氧化自食物分子中获得的乙酰 CoA 到产生 CO_2 的中心代谢途径。在真核生物中此类反应只局限在线粒体的基质中发生。

clathrin　网格蛋白,成笼蛋白
构成衣被小泡衣被的主要蛋白质,分子呈三分枝状,分子量为 180,000,在衣被小泡形成时装配成笼状结构,起支持网架作用。

coatomer protein Ⅰ(COP Ⅰ) 衣被蛋白Ⅰ
由7个亚基组成的复合物,是组成COP Ⅰ衣被的主要成分。其包被的小泡将蛋白质从高尔基体反面反向运输到内质网,或从反面高尔基扁囊运往顺面的高尔基扁囊。

coatomer protein Ⅱ(COP Ⅱ) 衣被蛋白Ⅱ
由5个亚基组成的复合物,是组成COP Ⅱ衣被的主要成分。其包被小泡将蛋白质从内质网运往高尔基体。

coenzyme 辅酶
全酶中的非蛋白质的小有机分子,可和全酶蛋白结合,使全酶具有活性。

coenzyme A(CoA) 辅酶A
是参加能量传递反应的小有机分子,由三磷腺苷、泛酸和硫基乙胺组成的一种辅酶,通常是作为激活代谢产物(如乙酰)的载体,乙酰辅酶A在三羧酸循环中起着关键性的作用。

collagen 胶原
富有甘氨酸和脯氨酸的纤维蛋白,是动物组织中细胞外基质的主要成分。

collagen fiber 胶原纤维
细胞外基质的骨架成分,由胶原分子有序排列并相互交联构成的纤维,具有很高的抗张力强度。

collage fibril 胶原原纤维
组成胶原纤维的亚单位丝,由三股螺旋的胶原分子通过侧连有序排列成的直径约50～200 nm的纤维结构。

compementary DNA(cDNA) 互补DNA
通过反转录酶催化由转录产生的DNA模板;与mRNA互补的DNA。

complement 补体
一类存在于人和脊椎动物血清中协助免疫反应的一组血清蛋白。可被抗原-抗体复合物或微生物所激活,导致病原微生物裂解或被吞噬。

constitutive heterochromatin 恒定型异染色质
一直处于凝缩状态,为遗传上稳定、复制较晚的染色质,如着丝粒区染色质。

C_3 pathway C_3途径
在卡尔文循环中,将CO_2固定后直接形成三碳分子的途径。

C_4 pathway(Hatch-Slack pathway) C_4途径
又称C_4循环(C_4 cycle)。在某些热带或亚热带起源的植物中,CO_2最初固定于叶肉细胞,在磷酸烯醇式丙酮酸羧化酶的催化下,将CO_2链接到磷酸烯醇式丙酮酸上生成四碳化合物——草酰乙酸,经胞间连丝运向维管束鞘细胞,参与卡尔文循环,合成同化产物的途径。

cristae 嵴
线粒体内膜向基质中突出的内褶物,上面有电子传递链复合物及将氧化磷酸化和电子传递相偶联的酶复合物(F_1+F_0偶联因子)。

cross-linking protein 交联蛋白
具有两个和多个肌动蛋白结合位点,能同时结合多条肌动蛋白丝,将多条肌动蛋白丝交联成为束状或凝胶状网的蛋白。包括成束蛋白和凝溶胶蛋白。

cyanobacteria(cyanophyta) 蓝细菌
过去称为蓝藻(blue-green algae)。系单细胞原核类生物,细胞质中含有光合膜。

cyclic adenine(cAMP) 环腺苷酸
有的激素作用于靶细胞后,在腺苷酸环化酶的催化下,ATP形成的腺苷3,5,-环状磷酸分子。cAMP起信号传递分子的作用,可激活A激酶,调节代谢,故有"第二信使"之称。

cAMP-dependent protein kinase(protein kinase A) 环腺苷酸依赖蛋白激酶A(蛋白激酶A)
一种由环腺苷酸(cAMP)激活、催化,将磷酸基从ATP转移至蛋白质的丝氨酸和苏氨酸残基上的蛋白激酶。

cyclin 细胞周期蛋白
在真核细胞分裂周期中浓度有规律地升高和降低的蛋白质。它可激活CDK依赖蛋白激酶,从而调控细胞周期阶段的变化。

cyclin-dependent kinase(CDK) 细胞周期蛋白依赖性激酶
可与细胞周期蛋白结合成复合物发挥作用的激酶,不同的CDK-周期蛋白能通过专一靶蛋白磷酸化的方式而启动细胞分裂周期的不同阶段。

cytokine 细胞因子
动物受感染后产生的一类信号传递蛋白或肽类物质,在功能上类似于激素,在细胞通信中起局部介体作用,由一个细胞刺激另外的细胞发生反应。细胞因子包括干扰素、白细胞介素和肿瘤坏死因子等。

cytokinin 细胞分裂素
调节植物细胞生长和发育的一个小分子家族。

cytokinesis 胞质分裂
细胞分裂过程中,继细胞核分裂之后发生的细胞分裂,从而形成两个子细胞。

cytoplast 胞质体
细胞经处理(如用细胞松弛素B处理)排除胞核后剩下的细胞质部分,包含有细胞膜。

cytosis 吞排作用
动物细胞内吞作用和外排作用的统称。

cytoskeleton 细胞骨架
细胞中的纤维状结构,如微管、微丝和中间丝等,亦包括核骨架。

cytosol 细胞质溶胶
即细胞质基质,除了膜性细胞器以外的细胞质液相部分,可发生溶胶与凝胶态的变化。又称细胞质溶质。

cytotoxic T lymphocyte 细胞毒性T细胞
能够识别和杀伤外来组织或感染病毒的T细胞。

D

diacylglycerol (DAG) 二酰甘油
由一个甘油分子和两个脂肪酸分子酯化而成的一种甘油酯,起第二信使作用。

dark band (A band) 暗带
又称 A 带。横纹肌肌节中粗肌球蛋白丝所在的区段,具有重折光性。

dedifferentiation 去分化
又称脱分化。分化细胞失去原有的分化结构和功能成为具有未分化细胞特性的过程。

default pathway 缺损(途径)
高尔基体中产生的部分水解酶逃过了 TGN 的分选,分泌到细胞外的途径。

dendritic cell 树突状细胞
一种既具分枝或树突状形态及吞噬功能,又能呈递抗原的细胞。分为髓系和淋巴系两类。

desmosome 桥粒
相邻细胞间的一种斑点状黏着连接结构。其质膜下方有盘状斑,与 10 nm 粗的中间丝相连,使相邻细胞的细胞骨架间接地连成骨架网。

desmotubule 连丝小管
胞间连丝中央由膜围成的一个狭窄的筒状结构。其两端与两个细胞的内质网相连。

diakinesis 终变期
减数分裂前期 1 的最后一个阶段。

differentiation 分化
细胞向特定方向变化的过程。

diplotene 双线期
减数分裂前期 1 的一个阶段,这时联合的染色体分开,只在交叉(chasma)部位相连。

docking protein 停靠蛋白质
又称船坞蛋白质。是存在于内质网膜上的信号识别颗粒受体蛋白。能与结合有信号识别颗粒结合,将正在合成蛋白质的核糖体引导到内质网上。

domain 结构域,功能域
蛋白质中具有一定三级结构的区域,往往具有特定的功能。

dynamin 发动蛋白
一种引起微管在 ATP 介导下滑向正端的微管结合蛋白,也是 G 蛋白的结合蛋白,具有组织特异性,也参与网格小泡的形成。

dynein 动力蛋白
沿微管运动的一种动力蛋白,具有 ATP 酶活性。存在于纤毛和鞭毛外周二联体 A 微管和纺锤体微管等处,可利用水解 ATP 提供的能量,沿微管向其正端移动。

dynein arm 动力蛋白臂
附在纤毛外周二联丝(微管二联体)上的由动力蛋白组成的侧臂。

E

ectodesma [胞]外连丝
贯穿植物表皮细胞外侧壁的一种胞间连丝样纤细通道。从胞腔延伸到胞壁表面,是细胞内、外间物质交流的途径。

effector 效应物
具有调节作用的物质或代谢产物,为诱导物和阻碍物的统称。通过激活或抑制阻碍物分子来影响阻碍物与操纵基因的结合。可改变酶和底物亲和力的代谢产物。

elastic fiber 弹性纤维
由弹性蛋白组成的细胞外基质纤维,具有高度弹性。

elastin 弹性蛋白
动物结缔组织中的一种主要结构蛋白,具有弹性。

electron transport chain 电子传递链
由供体到受体进行电子传递的一组电子载体,链上每一步都伴随有能量的释放。

electrophoresis 电泳
分子、物质颗粒或细胞在电场中的泳动。根据大分子或颗粒的电荷、大小和形状不同,使其在通过电场中的胶体或介质时发生分离的方法。

embryonal carcinoma cell (EC cell) 胚胎癌性细胞
在胚胎发育过程中由畸胎癌(常为睾丸肿瘤)衍生而来的一类多能干细胞。

embryonic germ cell (EG cell) 胚胎生殖细胞
从哺乳动物胚胎生殖嵴原生殖细胞培养成的多能干细胞,性质类似于胚胎干细胞。

embryonic stem cell (ES cell) 胚胎干细胞
是从早期胚胎内细胞团(inner cell mass)或原始胚胎生殖细胞(embryonic germ cell)经体外分化抑制培养分离克隆而成,经人工操作能够发育成一个新个体的全能二倍体细胞。

endoplasmic reticulum (ER) 内质网
真核细胞胞质内广泛分布的膜片层囊,与蛋白质的合成和运输有关。膜囊外表面上结合有核糖体的称为粗面内质网;无核糖体的称为滑面内质网,后者为管形结构。

endosperm 胚乳
被子植物种子中包围在胚外面的组织。为种子萌发提供淀粉或其他营养物。

endosome 内体
动物细胞中的膜围细胞器,含有通过内吞作用新吞入的物质,随之将物质输送到溶酶体进行降解。

enhancer 增强子
调节蛋白结合的 DNA 序列,可调节距离有几千碱基的某个结构基因的转录速率。

enzyme-coupled receptor (enzyme-linked receptor) 酶联受体
又称催化型受体(catalytic receptor)。细胞表面上的主要受体,其细胞质区具有酶活性,或者和细胞质中的酶结合。配体与其结合后,激活酶活性。

enzyme 酶
具有催化作用的蛋白质。酶具有专一性,每

种酶可催化特定的化学反应。

eosinophil 嗜酸性粒细胞
为含嗜酸性颗粒的白细胞。在抗寄生虫感染及I型超敏反应中发挥重要作用。

epidermal growth factor (EGF) 表皮生长因子
由颌下腺等细胞分泌的可刺激上皮细胞和多种细胞增殖的生长因子。

epinemin 丝连蛋白
一种分子量为44500的中间丝结合蛋白。以单聚体存在,在非神经细胞中与波形蛋白结合。

epitope 表位
又称抗原决定簇(antigenic determinant)。抗原分子中决定抗原特异性的特殊化学基团,是与T细胞受体/B细胞受体或抗体特异结合的基本单位。

ER retention protein 内质网驻留蛋白
保留在内质网膜和腔中发挥功能的蛋白质。

ER retention signal 内质网驻留信号
又称KDEL分拣信号(KDEL sorting signal)。驻留在内质网中起作用的蛋白质上的短的氨基酸序列(C端的KDEL序列)。可引导蛋白质由高尔基体返回和驻留在内质网中。

ER retrieval signal 内质网回收信号
某些内质网驻留蛋白肽链的C端所含有的特定氨基酸序列。膜蛋白中为"赖氨酸-赖氨酸-X-X"(KKXX)序列;可溶性蛋白中为"赖氨酸-天冬氨酸-谷氨酸-亮氨酸"(KDEL)序列。当这种蛋白质进入高尔基体以后,可被包装成COPI有被小泡,重新运回内质网。

ER signal sequence 内质网信号序列
引导合成中的蛋白质进入内质网腔的N端信号序列。

eubacteria 真细菌
指大多数的细菌,以示和古细菌区别。

euchromatin 常染色质
在间期核中处于松展状态而在分裂期凝缩的染色体区域。

exocytosis 胞吐(外排)作用
细胞内由膜包围的小泡与质膜融合,然后质膜开口,小泡内物质排出细胞外的作用。

extensin 伸展蛋白
植物细胞壁中富含羟脯氨酸的糖蛋白。通过肽链间的交联形成网络系统,与成纤维素网系相辅,增强细胞壁的韧性和刚性。

extracellular matrix [细]胞外基质
由细胞分泌到细胞间充质中的蛋白质和多糖类大分子物质。构成复杂的网架,连接组织结构,调节组织的发育和细胞生理活动。

expressed sequence tag(EST) 表达序列标记
一段已经测序的cDNA,可用于更快地获得基因组上某个完整的基因。大量收集EST,可展示基因组上编码蛋白质区域。

F

F_0 factor F_0因子
存在于线粒体内膜上$F_1 \sim F_0$偶联因子的膜部,对寡霉素(oligomycin)敏感,符号中之"o"即表示oligomycin之意。

F_1 coupling factor F_1偶联因子
存在于线粒体嵴M面上的颗粒结构,系$F_1 \sim F_0$偶联因子之一,具有ATP酶活性。

facilitated diffusion 协助扩散
为载体分子穿膜被动运输(顺浓度梯度扩散)。当分子浓度达到一定限度时,浓度再增加,扩散速率亦不能再提高。

facultative heterochromatin 兼性异染色质
为生物体在一定时期处于凝缩失活状态的染色质,而在其他时期松展为常染色质。

Feulgen reaction 富尔根反应
专一显示DNA的一种染色法。标本经水解去掉RNA后,DNA的嘌呤-脱氧核糖中的嘌呤被酸水解,暴露出脱氧核糖的醛基,游离的醛基同Schiff试剂反应,呈现紫红色。

fibroblast 成纤维细胞
为结缔组织中的一种普通细胞,分泌细胞外基质、胶原纤维和其他外基质。大分子丰富,在受伤组织和组织培养中其迁移和繁殖迅速。

fibroblast growth factor (FGF) 成纤维细胞生长因子
可促进各类细胞特别是内皮细胞增殖的蛋白质因子家族。

fibronectin 纤连蛋白
细胞外基质中的黏着糖蛋白。可与细胞外基质其他成分、纤维蛋白以及整联蛋白家族细胞表面受体结合,影响细胞活动。

filaggrin 聚丝蛋白
哺乳动物皮肤基底细胞中透明角质颗粒的碱性蛋白质成分。可专门与中间丝结合。

filamentous actin (F-actin) 纤丝状肌动蛋白
简称"F肌动蛋白"。在中性盐溶液中,由球状的G肌动蛋白单体经过组装后形成的纤维状肌动蛋白,是肌肉细肌丝和真核细胞骨架中微丝的主要组分。

filamin 细丝蛋白
一种将肌动蛋白丝横向交联的蛋白质。具有两个肌动蛋白结合位点,把肌动蛋白丝相互交织成网。

flavin adenine dinucleotide (FAD) 黄素腺嘌呤二核苷酸
核黄素的一种辅酶形式。通过接受来自于供体分子的两个电子和来自于溶液的两个质子,参与氧化反应。其氧化还原偶联的还原形式为$FADH_2$。

fimbrin (plastin) 丝束蛋白
为上皮刷状缘微绒毛芯部的肌动蛋白结合蛋白。在肌动蛋白丝之间形成横桥,使肌动蛋白丝连接成紧密的束。

flagellin 鞭毛蛋白
构成细菌鞭毛的主要蛋白质成分,单体的分子量为20,000~40,000。在适当pH值下,鞭动细胞运动。

flavoprotein（FP） 黄素蛋白
由一条多肽结合一个以黄素核苷酸作为辅基的蛋白质。每个辅基能够接受和提供两个质子和两个电子，起递氢体和电子传递体的作用。

flippase 翻转酶
将磷脂分子从内质网膜的胞质面脂单层转移到腔面单层，造成脂分子在脂双层不对称分布的酶。是一个蛋白质家族。又称 flp-frp 重组酶（flp-frp recombinase）。酵母中负责特定 DNA 片段重排的系统。在其存在下，一段 DNA 在 frp 位点被切除，末端再重新连接。

fluid mosaic model 流动镶嵌模型
关于细胞膜结构的一种模型。这种模型认为，磷脂双层分子构成膜的网架，蛋白质分子镶嵌在网格中和脂双层表面。脂双层黏稠度较低，可流动，膜内颗粒也可以移动。

fluorescein 荧光素
在蓝光或紫外线照射下发出绿色荧光的荧光染料。

focal adhesion（focal contact plaque） 黏着斑
通过整联蛋白锚定到细胞外基质上的一种动态锚定型细胞连接。整联蛋白的细胞质端通过衔接蛋白与肌动蛋白丝相连。

focal adhesion kinase（FAK） 黏着斑激酶
在黏着斑复合物形成中发挥关键作用的一种非受体酪氨酸激酶。

freeze fracture technique 冰冻断裂技术
通过速冻和切断断裂面为电镜观察制备标本的方法。在观察之前采用物理法将暴露出的断裂面制成覆膜（replica），制备覆膜之前也要将断裂面进行真空升华蚀刻（etch），故此法又称冰冻蚀刻技术。用此法可制作供观察膜表面或膜内部结构的标本。

fusin 融合病毒蛋白，融合素
淋巴细胞表面趋化因子受体蛋白。是人类免疫缺陷病毒感染 CD4 细胞的最初结合位点，结合后病毒与细胞融合进入细胞。

fusion protein 融合蛋白
由两段或多段基因序列串联形成的融合基因表达所产生的蛋白质。

G

G-banding G 分带
染色体分带的一种方法，经吉姆萨（Giemsa）染色后，染色体上在浅染区间显示出深染带。

gCAP39
一种调节肌动蛋白多聚化的蛋白质因子。

gene chip 基因芯片
固定有寡核苷酸、基因组 DNA 或 mRNA 等的生物芯片。利用这类芯片与标记的生物样品进行杂交，可对样品的基因表达谱生物信息进行快速定性和定量分析。

gene regulation protein 基因调节蛋白
可同基因中的专一序列结合，从而影响基因表达的蛋白质。

germ plasm 生殖质
系 Weismann 于 1883 年首先创用的名词，原指具有一定化学组成并能决定遗传性的遗传物质，可通过生殖细胞一代代传递下去，实际上是指染色体。现在则是指在某些动物种类的卵母细胞中合成的性细胞决定子。受精卵进行卵裂过程中，获得生殖质的细胞将分化为生殖细胞。

ghost cell 血影细胞
溶血后失去细胞质的红细胞外壳（细胞膜）。

globular actin（G actin） 球状肌动蛋白
简称"G 肌动蛋白"。为由一条多肽链构成的球形分子的单体肌动蛋白。

glial cell 胶质细胞
全称神经胶质细胞（neuroglial cell）。广泛分布于中枢和周围神经系统中的支持细胞。包括脊椎动物中枢神经系统中的少突胶质细胞和星形胶质细胞以及周围神经系统中的神经鞘细胞。起支持、营养和稳定内环境的作用，与神经元、血管之间可进行双向通信。

glyoxysome 乙醛酸循环体
为植物细胞中所特有的一种细胞器，和贮藏脂类的代谢有关。含有乙醛酸循环酶系，在乙醛酸循环中产生的琥珀酸可参加线粒体中的三羧酸循环。

glycosylphosphatidylinositol（GPI） 糖基磷脂酰肌醇
能共价锚定在某些膜蛋白的 C 末端上。

glycogen 糖原
动物细胞中由葡萄糖单元构成的贮存多糖，亦称动物淀粉，肝细胞和肌细胞中含量丰富。

Golgi apparatus（Golgi body） 高尔基体
由成叠的扁囊和小泡组成的细胞器，每叠构成一个高尔基网体，一个细胞中可有多个高尔基网体。其功能是加工和包装分泌蛋白、溶酶体蛋白和质膜蛋白等。

G-protein G 蛋白
全称 GTP 结合蛋白（GTP binding protein）、"鸟嘌呤核苷酸结合蛋白"（guanine nucleotide binding protein）。具有 GTP 酶活性，在细胞信号通路中起信号转换器或分子开关作用的蛋白质。有三聚体 G 蛋白、低分子量的单体小 G 蛋白和高分子量的其他 G 蛋白 3 类。

G-protein coupled receptor（GPCR），G-protein-linked receptor G 蛋白偶联受体
是一种七次跨膜细胞表面受体，同细胞内由 3 个亚基组成的 GTP 结合蛋白（G 蛋白）联合，能为胞外的配体所激活。

granum（复数 grana） 基粒
叶绿体中由成叠的盘状扁膜囊组成的粒状光合膜集合体，膜中含有叶绿素。

green fluorescent protein（GFP） 绿色荧光蛋白
自水母分离的荧光蛋白，广泛用于检测在活细胞蛋白中运动和表达载体的构建。

growth factor 生长因子
细胞外的多肽信号分子，能刺激细胞生长或

繁殖,如表皮生长因子(EGF)、血小板生长因子(PDGF)。

GTPase-activating protein (GAP)　GTP 酶激活蛋白
　　为调节蛋白质家族。与 Ras 蛋白及 G 蛋白 α 亚基结合,激活其 GTP 酶活性,促进其从活化型转变为非活化型,终止信号转导。

H

haemocyte (hemocyte)　血细胞
　　血液中的细胞成分,尤指无脊椎动物中的变形血细胞,相当于哺乳动物中的白细胞。

haemoglobin (hemoglobin, Hb)　血红蛋白
　　存在于脊椎动物、某些无脊椎动物血液和豆科植物瘤中的一组红色含铁的携氧蛋白,具有 4 个亚基。

HAT medium　次黄嘌呤-甲氨蝶呤-胸苷培养基(HAT 培养基)
　　用于选择杂交瘤细胞的培养基,内含有次黄嘌呤(hypoxanthine)、甲氨蝶呤(aminopterin)和胸苷(thymidine)。

heat shock protein (Hsp)　热激蛋白
　　广泛存在于原核细胞和真核细胞中的一类在生物体受到高温等逆境刺激后大量表达的保守性蛋白质家族。在肽链折叠或解折叠变化中发挥作用。目前已知有 Hsp100、Hsp90、Hsp70、Hsp60、Hsp40 和小 Hsp 等亚类。

heavy chain of antibody　抗体重链
　　又称"抗体 H 链"。免疫球蛋白中分子量较大的、含 440 个氨基酸的肽链。根据其恒定区抗原性的不同分为 μ、γ、α、δ 和 ε 5 类,相应的免疫球蛋白分别为 IgM、IgG、IgA、IgD 和 IgE。

heavy meromyosin　重酶解肌球蛋白
　　肌球蛋白经胰蛋白酶水解后,带有肌球蛋白头部的片段。

HeLa cell　海拉细胞
　　一个由宫颈癌组织培养、选育成的细胞系。该宫颈癌组织取自美国 Henrietta Lacks 女士活检标本,取 He 和 La 合并而得名。

helper T cell (Th 细胞)　辅助 T 细胞
　　表达 CD4 分子具有辅助功能的 T 细胞。

hemidesmosome　半桥粒
　　是上皮细胞和其下基膜之间特殊锚定的细胞连接。

heterochromatin　异染色质
　　间期细胞核内凝缩并染色深的染色质。分兼性和恒定性两类。

heterophagy　异体吞噬
　　细胞吞噬感染的病毒、细菌或其他一些颗粒等的过程。

high-mannose oligosaccharide　高甘露糖寡糖
　　细胞糖蛋白上 N 接连寡糖链的一种。

histone octamer　组蛋白八聚体
　　由组蛋白 H2A、H2B、H3 和 H4 各两分子构成的真核细胞染色体中核小体结构的核心颗粒。

holoenzyme　全酶
　　由核心酶蛋白和辅酶(基)组成的具有功能的复合酶。

human leukocyte antigen (HLA)　人白细胞抗原
　　人 T 细胞表面上的一种糖蛋白。在 T 细胞抗原识别方面起关键作用,分为 I 类、II 类和 III 类组织相容性抗原。通常非活化 T 细胞只表达 HLA-I 类抗原,活化的 T 细胞同时表达 HLA-I 和 HLA-II 类抗原。

human leukocyte antigen complex (HLA complex)　人白细胞抗原复合物
　　简称"HLA 复合物"。编码能引起强烈而迅速排斥反应的人主要组织相容性抗原的基因群。

humoral immunity　体液免疫
　　广义是指所有体液免疫因子参与的免疫反应;狭义是指由 B 淋巴细胞分泌抗体介导的免疫反应。

hybridization technique　杂交技术
　　一对碱基互补的单链核苷酸序列通过建立氢键形成双链结构,如 DNA-DNA,DNA-RNA,RNA-RNA。

hybridoma　杂交瘤
　　为了制造单克隆抗体,将分泌抗体的淋巴细胞和瘤细胞融合所形成的杂交细胞。

hydroxylase　羟化酶
　　常指 SER P450 酶系,它可将脂溶性有毒物质代谢为水溶性物质排出体外。

I

I band, light band　明带
　　又称"I 带"。横纹肌肌节中的单折光带,只含细肌丝。

Ig-super family　免疫球蛋白超家族
　　是一种不依赖钙的细胞表面黏着因子。

immune system　免疫系统
　　机体执行免疫应答和免疫功能的组织系统。由免疫器官和组织、免疫细胞和免疫分子组成。

immunity　免疫
　　机体识别和排除抗原性异物,即机体区分自己与非己进而排除异己的功能。通常对机体有利,但在某些条件下也可对机体有害。

immunocytochemistry (ICC)　免疫细胞化学法
　　利用抗原与抗体特异结合的原理,以标记抗体作为探针来显示细胞内抗原成分,主要是多肽与蛋白质(包括受体、酶、分泌物前体等各种基因表达产物),对其进行定位、定性及定量的研究。

immunoglobulin (Ig)　免疫球蛋白
　　一种具有抗体活性或化学结构上与抗体相似的球蛋白,是重要的一类免疫效应分子,多数为丙种球蛋白。由两条相同的轻链和两条相同的重链所组成,根据分子重链不同可分为 IgA、IgD、IgE、IgG 和 IgM 5 类。

importin　核输入蛋白
　　将在细胞质中结合的蛋白质经核孔复合物运进细胞核内的蛋白质。

infinite cell line（continuous cell line） 无限细胞系
又称"连续细胞系"。在体外可以持续生存，具有无限繁殖能力的细胞系。

inflammatory cell 炎症细胞
参与炎症反应的各种细胞，包括巨噬细胞、淋巴细胞、中性粒细胞和嗜酸性粒细胞等。

initiation factor（IF） 起始因子
蛋白质合成起始所需蛋白质，如 IF1、IF2、IF3。30S 核糖体亚单位结合到 mRNA 上，需要有 IF2；fmet-tRNA 需和 IF2 结合，才能结合到 30S 核糖体亚单位和 mRNA 上，形成起始复合物。

insulin-like growth factor（IGF） 胰岛素样生长因子
氨基酸序列与胰岛素类似的蛋白质或多肽生长因子。可促进细胞分裂，包括 IGF-Ⅰ 和 IGF-Ⅱ 两种。

integrin 整联蛋白，整联素
是一种穿膜的细胞表面黏着因子，也是细胞外基质的受体蛋白。

interband 间带
光学显微镜下观察多线染色体所见的一系列交替分布带之间的间隔区。此区因染色质包装程度低而呈浅染。

intercellular adhesion molecule 细胞间黏附分子
广泛分布于各种组织细胞间的免疫球蛋白超家族中的黏附分子。为单体糖蛋白，由弯曲的胞外结构、跨膜区及胞质结构域组成，是淋巴细胞功能相关抗原及白细胞等的配体。

interleukin 白细胞介素
由白细胞产生的在炎症反应中发挥作用的各种物质。

intermediate filament（IF） 中间丝
又称"中间纤维"、"10 nm 丝"。存在于真核细胞中介于微丝和微管之间，直径约 10 nm 的纤丝。是最稳定的细胞骨架成分，主要起支撑作用。因组成的蛋白质不同而有不同的命名。

intermembrane space 膜间隙
存在于线粒体、叶绿体的内膜和外膜之间的空间。

internal membrane 内膜
真核生物非质膜的细胞内部膜结构，主要是指内质网膜、高尔基体等膜结构。

internexin（α-internexin） 丝联蛋白
存在于神经元的中间纤丝蛋白，是中枢神经系统神经元中Ⅳ型中间纤维的亚基。

interphase 间期
在真核细胞的细胞周期中，从一次有丝分裂结束至下一次有丝分裂开始之间的时期，包含 G_1 期、S 期和 G_2 期。

intracellular receptor 细胞内受体
位于细胞质或细胞核内能够与特异性配体结合的受体。其配体包括亲脂素和活化的蛋白激酶 C（PCK）等信号分子，实际上都是配体依赖性转录因子。

intracellular transport 胞内运输
细胞内物质在各细胞器间的移动过程。

intranuclear spindle 核内纺锤体
酵母和原生动物营养阶段进行核内有丝分裂时，在核内形成的纺锤体。纺锤体极端无中心粒，而代之以由电子致密物质构成的纺锤体斑。

inverted microscope 倒置显微镜
物镜置于镜台下方的光学显微镜，适于培养细胞的显微观察和显微操作。

in vitro 体外活体实验，离体实验
细胞、组织或器官在有机体外或离体培养；设计在体外进行的生物过程实验。

in vivo 体内
在有机体内活体条件下进行实验。

ion channel 离子通道
为横过脂膜双层穿膜蛋白组成的通道，特殊无机离子根据电化学梯度扩散通过。

iontropic receptor 离子通道型受体
又称配体门控受体（ligand-gated receptor）、配体门控离子通道（ligand-gated ion channel）。贯穿细胞膜或内质网膜的具有离子通道功能的亲水性蛋白质。在与相应的配体结合后可介导速度很快的信号转导过程，使离子通过。

ion transporter 离子转运蛋白
由蛋白质构成的离子通道。

I region associated antigen（Ia antigen） Ⅰ区相关抗原
简称"Ia 抗原"。由主要组织相容性复合体Ⅰ区基因编码的一组细胞表面糖蛋白抗原。

isozyme 同工酶
能催化同一反应的一种酶的多种形式。但各种同工酶的最适 pH 值和最适底物浓度是不同的。

J

Jak-STAT signaling pathway Jak-STAT 信号传送途径
某些细胞外信号（如干扰素）激活基因表达的快速信号传递途径。包括细胞表面受体、细胞质中的 Jun 激酶以及具有转录作用的转导蛋白和激活蛋白。

Janus kinase（JAK） Janus 激酶
是一种酪氨酸蛋白激酶。

Jun kinase（JNK） Jun 激酶
又称 c-Jun N 端激酶（c-Jun N-terminal kinase）、应激活化的蛋白激酶（stress-activated protein kinase，SAPK）。为促分裂原活化的蛋白激酶（MAPK）信号转导级联反应中的一类蛋白激酶家族。其酪氨酸和苏氨酸残基双重磷酸化而被激活，进入细胞核，调节特定基因表达，参与调节细胞凋亡等生理活动。

junction 连接
由相邻动物细胞质膜上分化出的连接结构，如紧密连接、缝隙连接、桥粒等。

juxtacrine signaling 近分泌信号传递
在细胞与细胞之间直接接触时，穿膜配体与

相应受体的结合所产生的信号传递方式。它是反向信号传导及双向信号转导得以进行的必要形式。

K

karyogram（idiogram） 核型模式图
又称染色体组型图。根据一个细胞中全部染色体的形态特征所描绘并排列而成的模式图。

karyopherin 核转运蛋白
又称核周蛋白。是一组与转运分子穿过核孔复合物进入或输出核有关的蛋白质。既可与运载物结合，也可被核孔蛋白所识别和结合，协助运载物穿过核孔。

karyophilic protein 亲核蛋白
在细胞质内合成后，需要或能够进入细胞核内发挥功能的一类蛋白质。其肽链中带有核定位信号。

keratin 角蛋白
主要存在于上皮细胞中的含硫的中间丝蛋白，是构成毛发、角、指甲、表皮的重要组成成分。

kinectin 驱动结合蛋白
为内质网膜和其他膜上的整合蛋白，可与驱动蛋白结合，是驱动蛋白驱动小泡运动的膜结合点。

kinesin 驱动蛋白
是动力蛋白的一种。驱动蛋白具有 ATP 酶活性，可利用水解 ATP 提供的能量沿微管向其正端移动。

kinetochore 动粒
染色体着丝粒处与纺锤丝相连的盘状蛋白质结构。

kinetosome 基体，毛基体
鞭毛和纤毛的基体。

kinoplasm 动质
细胞质中带有核糖体的成分，表现为细胞质中的嗜碱性区。电镜观察表明动质是聚积在内质网上和细胞质基质中的核糖体，因含 RNA 故呈嗜碱性。

L

lamellipodium 扁平足，片足
薄片状伪足。是细胞表面的薄片状突起的外被质膜，内部有肌动蛋白丝网络支撑，与细胞运动和胞吞有关。

large granular lymphocyte（LGL） 大颗粒淋巴细胞
一类细胞质内含大的嗜苯胺颗粒的淋巴细胞，多为自然杀伤细胞。

lectin 凝集素
植物种子和动物中产生的对专一单糖或寡糖具有结合部位的蛋白质。

leptotene（leptonema） 细线期
减数分类前期Ⅰ的第一阶段，此时染色体呈单一细线形，线上有清晰可见的染色粒。

leucine zipper 亮氨酸拉链
存在于真核生物转录因子中的一种结构域。由两个专一蛋白质分子形成的同或异二聚体中的 α 螺旋组成卷曲螺旋结构域。螺旋肽链中每个重复片段的第七个氨基酸残基均为亮氨酸。

ligand 配体
能与受体蛋白分子专一部位结合并引起细胞反应的分子。

ligase 连接酶
能催化 DNA 单链片段的 5' 末端和另一片段 3' 末端相连接的酶，即多核苷酸连接酶（polynucleotide ligase）。

light chain of antibody 抗体轻链
又称"抗体 L 链"。免疫球蛋白中分子量较小的含 214 个氨基酸的肽链。根据其结构和恒定区抗原性的差异，分为"κ 轻链"（kappa light chain）和"λ 轻链"（lambda light chain）两种类型。

light meromyosin 轻酶解肌球蛋白
肌球蛋白分子经胰蛋白酶水解后，只有尾段的片段。

limited cell line（finite cell line） 有限细胞系
在体外的生存期有限即不能长期传代的细胞系。

lipoprotein 脂蛋白
与脂质相结合的水溶性蛋白质。通常根据其密度分为极低密度脂蛋白、低密度脂蛋白、高密度脂蛋白、极高密度脂蛋白和乳糜微粒。每一种脂蛋白均含有相应的载脂蛋白。

liposome 脂质体
①某些细胞质中的天然脂质小体。②由连续的双层或多层复合脂组成的人工小球囊。借助超声处理，使复合脂质在水溶液中膨胀，即可形成脂质体。它可以作为生物膜的实验模型，或在临床上用于捕获外源性物质（如药物、酶或其他制剂），可更有效地运送到靶细胞，经同细胞融合而释放。

local mediator 局部介质
是一种作用于短距离邻近细胞的自分泌产生的信号分子。

lymphoblast 淋巴母细胞
又称"原淋巴细胞"。受抗原刺激后，可发生活化、增殖和分化的淋巴细胞。

lymphocyte 淋巴细胞
受外源分子（抗原）的刺激产生免疫反应的一类无颗粒白细胞，分为 B 淋巴细胞和 T 淋巴细胞两类。

lymphokine 淋巴因子
由白细胞产生的影响其他细胞的物质。如白细胞介素、γ 干扰素、淋巴毒素等。

lymphokine-activated killer cell（LAK cell） 淋巴因子激活的杀伤细胞
简称"LAK 细胞"。在白细胞介素-2 的诱导下，发育成为能杀伤广谱肿瘤细胞的淋巴细胞。

lymphoma 淋巴瘤
来源于中胚层由淋巴细胞癌变产生的恶性肿瘤。

lymphotoxin (LT)　淋巴毒素
　　曾称"肿瘤坏死因子-β"(tumor necrosis factor-β)。由活化的T细胞所产生的细胞毒性细胞因子。其作用与肿瘤坏死因子类似，是一类重要的炎症介质。

lysosome　溶酶体
　　真核细胞细质中由膜围成的泡状细胞器，含有多种酸性水解酶。

M

macrophage　巨噬细胞
　　专门吞噬颗粒物质的白细胞。

major histocompatability complex (MHC)　主要组织相容性复合体
　　脊椎动物中为细胞表面蛋白质大家族编码的基因复合物，此类蛋白质可同外源蛋白质的肽片段结合，并将其呈递给T淋巴细胞，诱发免疫反应。

major histocompatibility complex antigen (MHC antigen)　主要组织相容性复合体抗原
　　简称"MHC抗原"。对外来移植物发生快速排斥反应的一组质膜糖蛋白。人的组织相容性抗原为HLA，小鼠的为H-2，兔的为RLA。

mast cell　肥大细胞
　　一类胞质内富含嗜碱性颗粒的细胞。颗粒中有组胺、硫酸乙酰肝素和各种酶类，在炎症和免疫反应时，颗粒被释放。

matrix　基质
　　细胞内和细胞器内的无结构物质，由水、介质中悬浮的分子和颗粒组成。

meiosis　减数分裂
　　又称"成熟分裂"(maturation division)。性细胞分裂时，染色体只复制一次，细胞连续分裂两次，染色体数目减半的一种特殊分裂方式。

memory cell　记忆细胞
　　在免疫系统中，对第一次遇到的抗原发生记忆的细胞。当再次与该抗原相遇时能加速第二次免疫反应。

mesenchymal stem cell　间充质干细胞
　　源自未成熟的胚胎结缔组织的细胞，为可形成多种细胞类型的多能干细胞。

mesosome　间体
　　原核生物的质膜向内折叠形成的结构，上面附着DNA，其功能同呼吸和细胞分裂有密切的关系。

messenger RNA (mRNA)　信使核糖核酸
　　确定蛋白质氨基酸序列的RNA分子，由与DNA作用互补合成的RNA转录本剪接而成，传递DNA为蛋白质合成的编码信息。

metacentric　中间着丝粒染色体
　　指着丝粒位于正中的染色体类型。

methionine tRNA (tRNA^{met})　甲硫氨基酸转移RNA
　　能接受甲硫氨基酸的tRNA，在酶的作用下，可形成甲硫氨酰tRNA(met-tRNA)。

m^{7}GpppN　m^{7}甲基鸟嘌呤核苷
　　在mRNA加工过程中，mRNA 5'端形成的7-甲基鸟嘌呤核苷帽子结构的核苷。

microbody　微体
　　为真核细胞中由膜包围的细胞器。细胞中主要的微体有乙醛酸循环体(glyoxysome)、氢化酶体(hydrogenosome)和过氧化物酶体(peroxisome)。

microfilament　微丝
　　主要由肌动蛋白组成的细丝状细胞骨架成分。亦称肌动蛋白丝(actin filament)，直径约6 nm。与原生质流动、保持细胞外形、细胞分裂、质膜皱褶、神经分支、形态发生运动等变化有关。

micropinocytosis　微胞饮
　　质膜内陷形成直径约65 nm的胞饮小泡的内吞作用。

microRNA (miRNA)　微小核糖核酸
　　是一类由内源基因编码的长度约为22个核苷酸的非编码单链RNA分子，它们在动、植物中参与转录后基因表达的调控。

microsome　微粒体
　　在超速离心时分离出的小泡状成分，系有内质网等膜的碎片形成的小泡的外表面，常附有核糖体。是研究粗面内质网功能活动的良好材料。

microtubule　微管
　　由微管蛋白(tubulin)原丝组成的不分支的中空管状结构，直径约25 nm。是细胞骨架成分，与细胞运动现象有关。纺锤体、纤毛、中心粒等均系微管组成的细胞器。

microtubule-associated protein (MAP)　微管相关蛋白质
　　以恒定比例与微管结合的蛋白质。决定不同类型微管的独特属性。

mitochondrial crista　线粒体嵴
　　线粒体内膜向内室折叠构成的双层皱褶或小管。膜上嵌插有F0F1-ATP合酶和氧化磷酸化电子传递链的成分。

mitochondrial inner membrane　线粒体内膜
　　双层线粒体膜的内层膜。膜中装配有氧化磷酸化电子传递链，是合成ATP的主要场所。

mitochondrial matrix　线粒体基质
　　线粒体内膜所围空间中含有的所有物质，包括三羧酸循环的酶系等。

mitochondrial outer membrane　线粒体外膜
　　双层线粒体膜的外层膜。膜上分布有蛋白质转运体。

mitochondrion　线粒体
　　真核细胞中由双层膜围成的制造ATP的细胞器，内膜向内折叠形成嵴。线粒体是进行有氧呼吸代谢的场所，经氧化磷酸化作用后，碳水化合物被完全氧化成CO_2和H_2O，释放出的能量贮存到ATP中。

mitogen　促[细胞]分裂原
　　刺激细胞进行有丝分裂的化合物。

mitosis　有丝分裂
　　核分裂的一种方式，经过分裂产生两个染色

体和遗传性相同的子细胞。在此过程中，染色质在形态上凝缩为丝状染色体。

motif 基序，模体（结构域）
蛋白质的多肽链中可被识别的结构成分，独立折叠成球形结构区域，与蛋白质执行功能有关。

motor protein 动力蛋白，发动蛋白
利用水解 ATP 提供的能量可沿微管或多聚分子移动的蛋白质，如肌球蛋白、驱动蛋白。

M-phase promoting factor（MPF） M 期促进因子
含有周期蛋白和蛋白激酶的复合物，触发细胞进入 M 期（曾称"成熟促进因子"）。

multipotent cell 多[潜]能细胞
在适当的环境条件下，能产生各种细胞类型的细胞。

multiptential stem cell（pluripotent stem cell） 多能干细胞
具有分化成多种分化细胞潜能的干细胞系细胞，如胚胎干细胞和成体干细胞。

multiubiquitylation 多次泛素化
网格蛋白依赖受体的内吞是高度调节的，受体加上多个单泛素分子的泛素化叫做多次泛素化。

multivesicular body 多泡体
膜围绕正在迁移的内体形成内陷的芽并断裂内部小泡，叫做多泡体。

muscle cell 肌肉细胞
具有收缩功能的肌肉组织细胞，包括横纹肌、平滑肌和心肌细胞。

muscle fiber 肌纤维
骨骼肌的单个合胞体细胞，内含有肌原纤维。

myelin sheath 髓鞘
神经膜细胞的质膜沿着轴索的轴心螺旋缠绕形成的多层脂质双层结构。主要成分为髓磷脂，具有高度绝缘性。

myofibroblast 肌成纤维细胞
含有肌动蛋白、肌球蛋白和其他肌肉蛋白的成纤维样细胞。

myofibril 肌原纤维
横纹肌肌纤维细胞质中的肌微丝束。

myofilament 肌[微]丝
组成肌原纤维的细丝。分粗细两种，粗肌丝由肌球蛋白构成，细肌丝由肌动蛋白构成。

myosin 肌球蛋白
由 2 条重链和 4 条轻链组成的纤维蛋白。在横纹肌中是构成粗肌丝的主要成分，头部具有 ATP 酶活性。

N

NADH-cytochrome b_5 reductase NADH-细胞色素 b_5 还原酶
在动物组织脂肪酸饱和电子传递途径中，催化将 NADH 上的氢原子转至该酶辅基 FAD 上（形成 FADH2），从而使传递链中下一个成员细胞色素 b_5 铁卟啉蛋白中的铁离子得以还原。

NANH dehydrogenase complex NADH 脱氢酶复合物
又称"NADH-辅酶 Q 还原酶"（NADH-comezyme Q reductade）。由至少 16 条肽链、辅基黄素单核苷酸（FMN）和铁-硫中心组成的一个传递电子的复合物。NADH 脱氢酶催化将 NADH 上的氢原子传递给予其结合牢固的辅基 FMN，铁-硫中心再将氢原子从辅基上脱下转移给呼吸链中的下一个成员辅酶 Q。

natural killer cell（NK cell） 自然杀伤细胞
简称"NK 细胞"。含有穿孔蛋白和粒酶颗粒的非特异性细胞毒淋巴细胞，是固有免疫系统的主要成员，对杀伤肿瘤细胞和病毒感染细胞起重要作用。

necrosis 坏死
由于损伤、缺血或感染引起的细胞死亡现象，伴有炎症。

nerve growth factor（NGF） 神经生长因子
为蛋白质类的局部介质，促进神经元的存活及其轴突的生长。

nestin 神经[上皮]干细胞蛋白
曾称"巢蛋白"。在神经干细胞中发现的一种 VI 型中间丝蛋白。与轴突生长有关，在上皮细胞中亦存在。

neural stem cell 神经干细胞
存在于成体脑组织中的一种干细胞。可分化成神经元、星形胶质细胞、少突胶质细胞，也可转化为血细胞和骨骼肌细胞。

neurofibril 神经丝蛋白
构成神经细胞轴突中间丝的蛋白质。由 3 种特定的蛋白亚基（NF-L、NF-M、NL-H）组装而成。其功能是提供弹性，使神经纤维易于伸展和防止断裂。

nexin 微管连接蛋白
纤毛轴丝的相邻外周微管二联丝间组成横桥的蛋白质。

nicotinamide adenine dinucleotide（NAD） 烟酰胺腺嘌呤二核苷酸
又称"辅酶 I"。一种脱氢酶的辅酶，是氧化作用中的电子载体。氧化底物时烟酰胺腺嘌呤二核苷酸分子中的烟酰胺环接受一个氢离子和两个电子。

nicotinamide adenine dinucleotide phosphate（NADP） 烟酰胺腺嘌呤二核苷酸磷酸
又称"辅酶 II"。烟酰胺腺嘌呤二核苷酸磷酸化成形式，是光合作用等生物过程中的电子载体。

noncyclic photo phosphorylation 非循环光合磷酸化
叶绿体光系统吸收的光能用于产生 ATP 和 NADPH 的过程。

non-endosymbiotic hypothesis 非内共生学说
又称"细胞内分化学说"。认为原始真核细胞通过质膜的内陷、扩张和分化后形成了线粒体、叶绿体和细胞核的雏形。

nonhistone protein（NHP） 非组蛋白
　　一组极不均一的在细胞内与DNA结合的组织特异蛋白质，参与基因表达调控。
Northern blotting　RNA 印迹法
　　将电泳分离的 RNA 从凝胶中转移到纤维素膜或尼龙膜上，用^{32}P标记的RNA或DNA杂交进行检测的方法。主要用于检测目的基因的转录水平。
nuclear envelope　核被膜
　　真核细胞围核的双层膜，膜上有核孔穿通。外膜的外表面上结合有核糖体，内外双层膜间的空隙称为核周潴泡(池)。
nuclear export signal（NES） 核输出信号
　　某些蛋白质中指导其由核输出信号到细胞质的氨基酸序列。
nuclear import signal　核输入信号
　　又称"核定位信号"（nuclear localization signal, NLS）。凡是细胞质中合成的核蛋白质，其肽链均含有由7个氨基酸组成的序列：脯氨酸-赖氨酸-赖氨酸-赖氨酸-精氨酸-赖氨酸-缬氨酸，此序列称为核输入信号。指导蛋白质从细胞质经核孔复合物输入到细胞核内，起分拣信号功能。
nuclear lamina　核纤层
　　核被膜内表面上由A、B、C 3种核纤层蛋白构成的网络结构。网络结构的外表面与内核膜结合，内表面和染色质相连。核纤层蛋白通过磷酸化和去磷酸化使核纤层解体和重装配，从而对细胞分裂过程中核膜的破裂和重建起调控作用。
nuclear matrix　核基质
　　核液中由非组蛋白质构成的网架结构，与染色质的定位、复制和转录活动有关。
nuclear pore complex　核孔复合体
　　核被膜上沟通核质和细胞质的复杂隧道结构，隧道的内、外口和中央有由核糖核蛋白组成的颗粒。核孔对进出核的物质具有控制作用。
nuclear skeleton　核骨架
　　亦称为核基质，现作为核纤层和核基质的统称。
nucleolar-organizing chromosome　核仁组织染色体
　　含有核糖体RNA基因的染色体，能在rRNA基因部位组织成核仁。
nucleolus　核仁
　　核内转录rRNA和装配核糖体的结构，无膜包围，在电镜下可区分成颗粒区和原纤维区两个区域。
nucleolus organizer region（NOR） 核仁组织区
　　位于染色体次缢痕处，含有多拷贝核糖体RNA基因，具有组织形成核仁能力的染色质区。
nucleoporin　核孔蛋白
　　位于核孔两端、可协助其他蛋白质进出细胞核的蛋白质。
nucleosome　核小体
　　染色体的单位结构，由组蛋白和200个碱基对的DNA双螺旋组成的球形小体。小体的核心由4种组蛋白的8个分子组成八聚体。组蛋白核心的外面缠绕了1.8圈的DNA双螺旋，约165个碱基对，其两核小体结合部有H_1组蛋白分子。小体的直径约10 nm。

O

Okazaki fragment　冈崎片段
　　DNA复制时，后随链在复制叉处以5'→3'方向合成的片段，后经连接酶催化共价连接成连续链，使得后随链沿着3'到5'的方向延伸。
open reading-frame（ORF） 可读框
　　以起始密码子开始，在三联体读框的倍数后出现终止密码子之间的一段序列。可读框有可能编码一条多肽链或一种蛋白质。很多情况下，可读框即指某个基因的编码序列。
operator　操纵基因
　　操纵子中可和阻遏物结合的一段特定核苷酸序列，对相邻结构基因的转录活动具有控制作用。
operon　操纵子
　　操纵子由启动基因、操纵基因和结构基因几部分组成。调节基因的表达产物——阻遏物蛋白，有抑制操纵基因活动的作用，从而关闭操纵子，结构基因停止转录。而诱导物可与阻遏蛋白结合，使其失去活性，重新启动转录。
osmosis　渗透
　　半透膜的两侧，因溶质浓度的不同，仅有水分穿膜的运动，溶质没有穿膜。
osmotic pressure　渗透压
　　在半透膜低溶质浓度一边产生对水分穿膜的渗透阻力。
osteoclast　破骨细胞
　　由分化的巨噬细胞形成的多核细胞，具有破骨功能。
osteocyte　骨细胞
　　存在于骨组织中被自身分泌的骨基质包围的成骨细胞。
outer nuclear membrane　外核膜
　　向着胞质侧的一层核膜。常与粗面内质网相连，其外表面上也常附有大量核糖体颗粒。

P

pachytene　粗线期
　　减数分裂前期Ⅰ的一个阶段，此时形成联会复合体，同源染色体间发生交换。
paracentric inversion　臂内倒位
　　又称"无着丝粒倒位"（akinetic inversion）。发生在染色体一条臂上不包含着丝粒的倒位。
passive transport　被动运输
　　离子或小分子在浓度差或电位差驱动下的穿膜运动。
peptidyl site（P site） 肽酰位
　　简称"P位"。核糖体中肽酰tRNA停留的部位。在蛋白质合成过程中，肽酰位上的肽酰tRNA的肽酰基与氨酰位上的氨酰tRNA的氨基反应，形成新的肽键，增加一个氨基酸。如此循环，直至蛋白质合成结束。

peripheral protein 周边蛋白
存在于脂双层表面的膜蛋白。

peroxisomal-targeting signal (PTS) 过氧化物酶体引导信号
过氧化物酶体蛋白分选信号序列。

phage display 噬菌体展示
全称"噬菌体表面展示"(phage surface display)。将外源基因或随机序列的 DNA 分子群与噬菌体外壳蛋白基因相连接,使外源 DNA 所编码的蛋白质以融合蛋白形式表达在噬菌体外壳表面的方法。

phagosome 吞噬体
指细胞中由质膜内陷包围细胞外物质形成的小颗粒。

phalloidin 鬼笔环肽
由鬼笔鹅膏真菌(*Amanita phalloides*)产生的一种环肽。可与 F 肌动蛋白结合,使 F 肌动蛋白保持稳定。其荧光标记物是鉴定 F 肌动蛋白的重要试剂。

phosphatidylchine (PC) 磷脂酰胆碱(卵磷脂)
细胞膜上主要的磷脂成分之一,胆碱通过磷酸二酯键与磷脂酸结合形成。

phosphatidylinositol (PI) 磷脂酰肌醇
存在于生物膜中的一种磷酸甘油酯。含有通过羧基与磷脂酸相连的肌醇-4-磷酸和 4,5-二磷酸衍生物。在信息传递过程中起调节作用。

phospholipit scramblase 磷脂转位蛋白
具有催化在 ER 膜上新合成的脂类从胞质侧翻转到腔面的作用。

phragmoplast 成膜体
植物细胞分裂时,在晚后期和早末期两组染色体之间形成的结构。成膜体由许多微管和膜泡组成,小泡内的物质用于合成细胞板。细胞板一旦形成,即把成膜体分成两部分,细胞分裂成两个子细胞。

phycobilin 藻胆色素
红藻和蓝藻中所含有的光合作用色素。

phylogenetic tree 系统发生树
又称家系树,表示一组有机体进化历史的关系。

pinocytosis 胞饮作用
细胞内吞液体物质的作用。

plasma cell 浆细胞
能够合成和分泌免疫球蛋白的终末分化阶段的 B 细胞。

plasmolysis 质壁分离
植物细胞在高渗液环境中,液泡内水分外渗出质膜,原生质体收缩,部分质膜与细胞壁脱离的现象。

plasmon 细胞质基因组
细胞质中遗传物质的统称。

plastid 质体
植物细胞中由双层单位膜围成能进行半自主复制的一类细胞器,如叶绿体、有色体(chromoplast)、造油体(oleioplast)、白色体(leukoplast)。

platelet (thrombocyte) 血小板
大量存在于血液中无核盘状小细胞(直径约 3 μm),具有凝血和止血作用。

polymerase chain reaction (PCR) 聚合酶链反应
通过 DNA 互补双链解链和退火聚合多次循环,扩增 DNA 特定区段的技术。

polytene chromosome 多线染色体
又称"唾液腺染色体"(salivary gland chromosome)、"巴尔比亚尼染色体"(Balbiani chromosome)。果蝇等双翅目昆虫细胞的有丝分裂间期核中的一种像电缆样的具有染色带的巨大染色体。是由核内 DNA 多次复制而细胞不分裂,复制后的子染色体有序并行排列而成。1881 年,意大利细胞学家巴尔比亚尼(Balbiani)首先在双翅目摇蚊幼虫的唾液腺中发现。

polyribosome (polysome) 多核糖体
在蛋白质合成时,由 mRNA 链串联起来的多个核糖体。

porin 孔蛋白
存在于细菌或线粒体外膜上充满水的通道。

positive feed back loop 正反馈环
即反应最终产物激化该反应进一步产生。

Pribnow box Pribnow 框
即 TATA 框,是启动子中的小段核苷酸序列,RNA 聚合酶与之结合后开始转录。

primary constriction 主缢痕(初级缢痕)
中期染色体上着丝粒区的狭细部位。

primary culture 原代培养
将机体内的某组织取出,分散成单细胞,在人工条件下培养使其生存并不断生长、繁殖的方法。

primary messenger 第一信使
由细胞产生,可被细胞表面或胞内受体接受、穿膜转导,产生特定的胞内信号的细胞外信使。

primary oocyte 初级卵母细胞
卵原细胞经过有丝分裂增殖后即将进行减数分裂的卵母细胞。

primary structure 一级结构
基本组分间以共价键结合的大分子排列序列。蛋白质的一级结构是指其氨基酸组成的线性序列以及多肽链的氨基酸残基数目。多核苷酸的一级结构是指核苷酸的数目和线性序列。

probe 探针
在分子杂交中用来检测互补序列的带有标记的单链 DNA 或 RNA 片段。

procentriole 前中心粒
新装配尚未成熟的中心粒。

profilin 抑制蛋白
一种微丝的结合蛋白,可抑制纤丝状肌动蛋白的聚合。

programmed cell death 程序性细胞凋亡
细胞正常的温和自杀过程,即细胞凋亡。

prophase 前期

有丝分裂（或减数分裂）的第一阶段，是 DNA 复制后开始出现染色体细线的时期。在此期末，核膜破裂，每条染色体纵分为二（只在着丝粒处相连）。

proteasome 蛋白酶体

细胞质溶质中大的蛋白质复合物，由 10～20 个不同亚基组成桶状结构，降解带有泛素标志的细胞质溶质蛋白质。

protein kinase A（PKA） 蛋白激酶 A

简称"A 激酶"（A kinase）。又称"依赖 cAMP 的蛋白激酶"（CAMP-dependent protein kinase）。一种由环腺苷酸（cAMP）激活，催化剂将磷酸基从 ATP 转移至蛋白质的丝氨酸和苏氨酸残基上的蛋白激酶。

protein kinase C（PKC） 蛋白激酶 C

又称"依赖 Ca^{2+}/钙调蛋白的蛋白激酶"（Ca^{2+}/calmodulin-dependent protein kinase）。一类可使丝氨酸/苏氨酸残基磷酸化的蛋白激酶。有多种亚类，不同的亚类有不同的激活方式，其中典型的亚类可被二酰甘油和 Ca^{2+} 浓度的提高所激活。

protein phosphatase 蛋白磷酸酶

通过水解将蛋白质的磷酸基除掉的酶，具有高度专一性。

protein serine/threonine phosphatase 蛋白质丝氨酸/苏氨酸磷酸酶

特异地水解蛋白质底物上的丝氨酸/苏氨酸磷酸酯键，脱去磷酸，从而调节该蛋白质功能的酶。

protein tyrosine kinase（PTK） 蛋白酪氨酸激酶

往往是酶偶联受体的组成部分，在信号转导中起作用。

protist 原生生物

单细胞真核生物，如鞭毛虫、原生动物等。

purine 嘌呤

核酸中的一类含氮碱基，如腺嘌呤、鸟嘌呤。

pyrimidine 嘧啶

核酸中的一类含氮碱基，如胞嘧啶、胸腺嘧啶、尿嘧啶。

Q

Q-banding Q 分带

染色体的一种分带方法。用荧光染料染色，染色体上显出着色的带状区，可与不着色区分开。

quaternary structure 四级结构

蛋白质四级结构是指各亚基之间的关系，包括亚基数目和空间结构。

quick freeze deep etching 快速冷冻深度蚀刻

在冷冻蚀刻基础上建立起来的一种电镜技术。可对细胞质中的细胞骨架纤维及其结合蛋白进行观察。

quinone 醌

小的脂类分子，是呼吸和光合作用电子传递链存在的电子递体。

quinone cycle 醌循环

脂溶性可移动的泛醌在膜内通过氧化还原反应，反复传递电子和从基质泵出氢的过程。氧化型泛醌接受一对电子，并从基质中摄取质子。每对电子通过泛醌-H_2-细胞色素 C 还原酶复合物有 4 个质子被转运到内膜外。

R

Rab effector Rab 效应子

能与 Rab 蛋白特异性结合的蛋白质。与 Rab 蛋白一起参与运输小泡到靶膜的停靠过程。

receptor mediated endocytosis 受体介导胞吞作用

动物细胞选择吞入的机制，大分子与质膜上的专一受体结合，形成衣被小泡进入细胞。

receptor protein 受体蛋白

可同专一信号分子结合启动细胞发生反应的蛋白质，许多受体蛋白位于质膜上，疏水性信号分子的受体则位于细胞质溶质中。

reduced nicotinamide adenine dinucleotide（NADH） 还原型烟酰胺腺嘌呤二核苷酸

又称"还原型辅酶Ⅰ"。烟酰胺腺嘌呤二核苷酸的还原形式，是氧化磷酸化过程中的电子载体。

respiration 呼吸

细胞内主要产能反应，过程涉及电子由有机分子至 O_2 的传递。

restriction endonuclease（restriction enzyme） 限制性内切核酸酶，限制酶

在一定核苷酸序列上进行切割的 DNA 酶，广泛用于重组 DNA 技术。

restriction point 限制点

存在于哺乳动物细胞周期 G_1 期的重要检查点。通过该点后，细胞周期才能进入下一步运转，进行 DNA 合成和细胞分裂。

retromer 反向复合物

由内体出芽形成的衣被小泡，有收回 MGP 受体的作用。

retrovirus 反转录病毒

一类 RNA 病毒，先通过反转录为双链 DNA 分子，然后整合到宿主基因组上的病毒。

reverse transcription PCR（RT-PCR） 反转录聚合酶链反应

简称"反转录 PCR"。先将 RNA 通过反转录酶的作用合成与之互补的 DNA 链，再以该链作模板进行聚合酶链反应扩增特定 RNA 序列的方法。

rhodopsin 视紫红质

视网膜的光敏色素，为膜整合蛋白，分布于视杆细胞和视锥细胞的盘状区。

Rho factor ρ 因子

为大肠杆菌的一种蛋白质因子，具有识别转录终止信号的能力；和 RNA 聚合酶结合后，能使 RNA 的转录在正确位置终止，并使转录完成的 RNA 解脱出来。

ribozyme 核酶

具有自身催化作用的 RNA。核酶通常具有特

殊的分子结构。

ribulose-1,5-bisphosphate(RuBP) carboxylase 核酮糖-1,5-二磷酸羧化酶
存在于叶绿体基质和类囊体游离外表面上的一种酶,在 C_3 循环中与固定 CO_2 有关。分子由 8 个大亚基和 8 个小亚基组成。大亚基由叶绿体基因组编码,在叶绿体中合成;小亚基由核基因组编码,在细胞质中合成。

RNA editing　RNA 编辑
在初级转录物上增加、删除或取代某些核苷酸而改变遗传信息的过程。是一种遗传信息在 RNA 水平发生改变的过程,可使 RNA 序列不同于基因组模板 DNA 序列。

RNA footprinting　RNA 足迹法
分析 RNA 链与其他分子特异结合部位的方法。将标记的 RNA 与待研究的物质(如某蛋白质或药物)进行结合反应后,用 RNA 酶清除未被结合的部分,再用电泳显示 RNA 链上结合位置的多寡和长度。

RNA interference(RNAi)　RNA 干扰
引起基因沉默的一种技术,将根据基因列制备的双链 RNA 注入体内,可引起该基因的功能。

rough endoplasmic reticulum(rER)　粗面内质网
膜的外表面结合着核糖体的内质网。

S

S period　DNA 合成期(S 期)
细胞分裂间期中 DNA 复制阶段。

sarcolemma　肌纤维膜
肌细胞(纤维)的质膜,亦称肌膜

satellite　随体
由次缢痕与主体隔开的染色体远端一小部分。

satellite DNA　卫星 DNA
用等密度氯化铯梯度离心法分离匀质 DNA 时,在主带附近出现的附带 DNA。一般含有高度 DNA 重复序列,其功能尚不清楚。

scaffold protein　支架蛋白质
它带有多个蛋白质结合域,可把信号转导途径中与该途径相关的蛋白质组织成群的蛋白质。

Schwann cell　施万细胞(许旺细胞)
又称"神经膜细胞"(neurolemmal cell)。脊椎动物体中包绕轴突形成髓鞘的胶质细胞,起绝缘、支持、营养等作用。

secondary lysosome　次级溶酶体
已经进行消化活动的溶酶体。内含溶酶体酶和消化底物,以及消化产物。根据所消化物质的来源不同,分为自噬溶酶体和噬溶酶体。

secretory vesicle　分泌泡
为膜围小细胞器,其含有的分子是准备分泌或暂存待释放的物质。由于染色技术使其颜色变深,故也称为分泌颗粒。

sedimentation coefficient(S)　沉降系数
指悬浮在密度较低的溶剂中的一种溶质分子,在每单位离心场作用下的沉降速率。大多数蛋白质的 S 值介于 1×10^{-12} 秒和 2×10^{-11} 秒,故取 1×10^{-13} 秒为一个漂浮单位(1 S),2×10^{-11} 秒即为 200 S。一种溶剂的温度一定时,S 值是由分子的重量、形状和水化程度决定的。公式为:$S = \dfrac{(dx/dt)}{\omega^2 x}$,式中 dx/dt 为沉降速度,ω 为角速度,x 为沉降带到离心机中心的距离。

selectin　选择蛋白
穿膜的细胞表面黏着因子的一种,为胞外结构域,有凝集素的作用。已知选择蛋白有 3 种,分别是内皮细胞表达的 E,血小板和内皮细胞表达的 P,以及白细胞中表达的 L。

self-assembly　自我装配
在不需要模板和亲体结构指导的条件下,形态结构的自动组建。

severin　截断蛋白,切割蛋白
一种微丝的结合蛋白,有切断微丝的作用。

sialic acid(N-acetylneuraminic acid)　唾液酸
又称 N-乙酰神经氨酸。由 N-乙酰氨基甘露糖和丙酮酸缩合而成的九碳糖。是动物细胞质膜在糖蛋白和糖脂中的重要糖单位。

signal molecule　信号分子
参与细胞信号转导的化学分子,如激素、神经递质、生长因子等。分为亲水性和亲脂性两类。

signal peptide　信号肽
分泌蛋白合成时在 mRNA 前端一组密码子指导下首先合成的一段氨基酸序列,有引导肽链穿经内质网膜的作用。

signal recognition particle(SRP)　信号识别颗粒
由 RNA 和蛋白质构成的颗粒状复合物,对信号肽具有识别能力。

signal sequence　信号序列
蛋白质肽链中指导该蛋白进入细胞的特定部位(如内质网、核或线粒体等)的氨基酸序列。

signal transduction　信号转导
细胞对细胞外信号发生反应的过程,即将胞外信号转变为细胞内信号的传递。

signaling cascade　信号传递级联
为关联的蛋白质反应序列,一般包括磷酸化和去磷酸化反应;在细胞内起传递信号的作用。

simple diffusion　简单扩散
即自由扩散(free diffusion)。

sister chromatid　姊妹染色单体
染色体经复制后存在于同一染色体中的两个相同的成分。

skin stem cell　皮肤干细胞
存在于表皮基底层和毛囊基部的干细胞。

small nuclear RNA(snRNA)　小核 RNA
在 GU-AG 和 AU-AC 内含子剪接和其他 RNA 加工过程中起作用的一种真核小 RNA。

small nucleolar RNA(snoRNA)　小核仁 RNA
在 rRNA 化学修饰中起作用的真核小 RNA。

smooth endoplasmic reticulum（sER） 滑面内质网
不与核糖体结合的内质网部分,主要进行脂类合成和解毒。

smooth muscle cell 平滑肌细胞
一类长梭状单核肌肉细胞,无肌节结构。

sodium channel 钠通道
又称钠钾泵或钠钾 ATP 酶。协助钠离子穿膜主动运输的结构,为一种需要 ATP 的 ATP 酶。

solenoid 螺线管
真核细胞染色质包装的二级结构。有组蛋白 H1 存在时,直径 10 nm 的核小体念珠结构螺旋盘绕,形成外径 30 nm、内径 10 nm、螺距 11 nm 的染色质结构,每螺旋圈有 6 个核小体。组蛋白 H1 对这一结构的稳定起重要作用。

somatic cell 体细胞
动、植物机体中除生殖细胞之外的各种细胞。

somatic cell hybrid 体细胞杂种
两种体细胞融合形成的异核体。

somatic cell nuclear transfer 体细胞核移植
将体细胞的细胞核移植入去核的卵母细胞中以获得新的胚胎的技术。

somatic recombination 体细胞重组
在体细胞有丝分裂时,由染色体交换而发生的遗传重组。

sorting signal 分拣信号
在细胞内被转运的蛋白质上面的特异序列。

Southern blotting DNA 印迹法
在多种限制性酶切片段的背景中检测特定 DNA 片段的技术。

spectrin 血影蛋白（红细胞膜内蛋白）
红细胞膜内表面上的一种边周蛋白。

sperm（spermatozoon） 精子
动物和低等植物成熟的雄配子。大小、形态因种类而异。

spermatid 精[子]细胞
次级精母细胞经减数分裂产生的成熟单倍体细胞,将变态为成熟精子。

spermatocyte 精母细胞
在精原细胞有丝分裂增殖过程中产生的某些最终能分化为成熟精子的细胞。分为初级精母细胞和次级精母细胞。

spermatogonium 精原细胞
雄性哺乳动物曲细精管上皮中能经过多次有丝分裂产生精母细胞的干细胞。

spherosome 圆球体
植物细胞中来源于内质网的小泡,是贮藏脂质的细胞器。

spningomyelin 鞘磷脂
为膜脂的成分之一。

spliceosome 剪接体
在 GU-AG 和 AU-AC 内含子剪接过程中起作用的蛋白质 RNA 复合物。

spongioblast（glioblast） 成胶质细胞
从室管膜细胞向外迁移而发展的一种过渡性胶质细胞。将进一步分化为大的神经胶质细胞,包括星形胶质细胞和少突胶质细胞。

stathmin 抑微管装配蛋白
又称"微管去稳定蛋白"、"癌蛋白 18"（oncoprotein 18, Op18）。增殖细胞的细胞质中广泛存在的磷蛋白。可与微管蛋白二聚体结合,抑制微管蛋白向微管上添加,从而促进微管解聚。

stem cell 干细胞
即没有分化的细胞,一定条件下可进一步分化为多种细胞。

stress activated channel 压力门通道
作为特殊离子通道的膜蛋白,需在机械力作用下打开通道。

stress fiber 应力纤维
由肌动蛋白丝和肌球蛋白 II 丝组成的可收缩丝束。具有肌节结构,一端与穿膜整联蛋白连接,与细胞运动有关。

stroma 基质
叶绿体中大的内部空间,内浸着类囊体,基质中含有在光合作用下使 CO_2 参入糖中的酶。

stroma-thylakoid 基质类囊体
在叶绿体基质中,连接于基粒之间的不发生堆叠的较大的类囊体。

suppressor T cell 抑制性 T 细胞
具有免疫抑制作用的 T 细胞。

survival factor 存活因子
细胞存活所必需的细胞外信号分子,可避免细胞发生程序性死亡。

survivin 存活蛋白
为肿瘤细胞中抑制细胞凋亡的蛋白质,具有抑制蛋白酶的活性。

suspension culture 悬浮培养
细胞在培养液中呈悬浮状态生长与增殖的培养技术。

symbiotic hypothesis（endosymbiotic hypothesis） 共生假说（内共生假说）
关于真核细胞起源的一种学说。这一学说主张真核细胞是由原始原核生物在偶然机会中建立了特定共生关系产生的。线粒体是由一种较大的异养原核生物吞入的游离生活的好气细菌变来的;叶绿体是由吞入的一种蓝细菌进化来的。

symport 同向运输
膜载体蛋白将两种溶质分子以同方向穿膜运输。

synaptonemal complex（SC） 联会复合物
减数分裂 I 的偶线期中配对的同源染色体间形成的蛋白质复合结构。

syncolin 微管成束蛋白
在鸡红细胞中发现的与微管相关的蛋白质,可将微管结合成束。

synemin 联丝蛋白
与结蛋白和波形蛋白结合在一起的一种中间丝结合蛋白。

synergid cell（synergid） 助细胞
为被子植物胚囊珠孔端的两个细胞。与卵细

胞组合成卵器,对受精有重要作用。

T

T cell receptor (TCR)　T 细胞受体
　　T 细胞表面可被抗原识别的受体,能特异地识别组织相容性复合物分子。

target cell　靶细胞
　　受到信号分子的作用发生反应的细胞。

targeting transport　靶向运输
　　蛋白质在细胞基质中合成后,按其氨基酸序列分拣信号的有无以及分拣信号的性质被选择性地送到细胞不同部位的过程。

TATA box　TATA 框
　　又称"戈德堡-霍格内斯框"(Goldberg-Hognes box)。真核生物启动子中可以与 RNA 聚合酶紧密结合的序列。存在于转录起始点前的约 25 个核苷酸处,其共有序列为 5'-TATAAAA-3',决定转录起始点的准确位置。

T lymphocyte　T 淋巴细胞
　　可产生细胞毒(lymphokine,淋巴因子)、专一杀死抗原细胞的淋巴细胞。

T system　T 系统
　　横纹肌肌纤维质膜内陷形成的管状结构系统,即横小管系统(transverse tubular system),功能是在肌纤维内部传递冲动信号。

telephase　末期
　　有丝分裂过程最后的核重建阶段。

telomere　端粒
　　染色体的末端结构,含有特定的 DNA 序列,以特有的方式进行复制。

tertiary structure　三级结构
　　生物大分子的三维折叠结构。

thick myofilament (thick filament)　粗肌丝
　　又称"肌球蛋白丝"(myosin filament)。横纹肌中的肌球蛋白Ⅱ丝,直径 12~14 nm,仅存于肌肉细胞中。

thin myofila ment (thin filament)　细肌丝
　　又称"肌动蛋白丝"(actin filament)。横纹肌中与 Z 盘相连、直径 7~9 nm 的纤丝。由 F 肌动蛋白、原肌球蛋白和肌钙蛋白组成。

thioester bond　硫酯键
　　通过酸(酯)基团和硫基(-SH)缩合反应形成高能键,例如乙酰(CoA)和许多酶底物复合物。

transmission electron microscope (TEM)　透射电子显微镜
　　在一个高真空系统中,由电子枪发射电子束,穿过被研究的样品,经电子透镜聚焦放大,在荧光屏上显示高度放大的物象,还可作摄片记录的一类最常见的电子显微镜。

thrombin　凝血酶
　　是一种蛋白水解酶,使血小板聚集,有凝血作用。

thylakoid　类囊体
　　叶绿体或原核生物细胞质中具有光合作用的光反应系统的封闭膜囊结构。

thymocyte　胸腺细胞
　　骨髓中的前 T 细胞经血流进入胸腺即成为胸腺细胞。

thymosin　胸腺素
　　一种调节肌动蛋白聚合的蛋白因子。

tight junction　紧密连接
　　相邻上皮细胞质膜间的一种连接结构,此外两个细胞的质膜整合蛋白彼此结合,形成密封条索,封闭细胞间隙。又称封闭连接(occluding junction)。

tissue-specific promoter　组织特异性启动子
　　只在特定组织中具有活性的启动子。

Toll-like recepter (TLR)　Toll 样受体
　　是一类细胞表面和细胞内受体。可识别各种微生物产物,与配体结合后可起始信号传递途径,因不同细胞起不同反应。

tonofilament　张力丝
　　与桥粒相连的直径为 10 nm 的中间丝,被中间丝结合蛋白交联为中间丝束。

totipotent cell　全能细胞
　　具有发育为完整个体或分化出各种细胞潜能的细胞。

totipotent stem cell (TSC)　全能干细胞
　　可分化成各种类型组织细胞的干细胞。哺乳动物中只有受精卵才是全能干细胞。

transcellular transport　跨细胞运输
　　溶质从上皮细胞或内皮细胞一侧穿过质膜被吸收进入细胞内,随后穿过细胞质从另一侧被送到细胞外间隙的移动过程。实际上是穿越细胞的运输方式。

topoisomerase　拓扑酶
　　与 DNA 复制有关的一类酶,通过对 DNA 双螺旋切口和封口反应催化 DNA 发生超螺旋象转变。拓扑酶Ⅰ催化 DNA 双螺旋解负超螺旋;拓扑酶Ⅱ(促超螺旋酶)的作用则相反,催化展开的 DNA 形成负超螺旋,能量由 ATP 水解提供。

transcriptase　转录酶
　　以 DNA 为模板的 RNA 聚合酶。

trans Golgi network (TGN)　反面高尔基网
　　高尔基体中离内质网最远的部分,所产生的蛋白质和脂类将抵达溶酶体、分泌泡或细胞表面。

transfer RNA (tRNA)　转移 RNA
　　在核糖体上合成蛋白质的过程中,可携带一种氨基酸运转到生长的多肽链上的 RNA。每一种 tRNA 具有特定的反密码子,可识别 mRNA 上的特定密码子。

transforming growth factor (TGF)　转化生长因子
　　最初从转化细胞培养液分离得到的可作用于细胞生长、转化等的生长因子。包括转化生长因子 α 和转化生长因子 β 两类。

translation　翻译
　　在核糖上按照 mRNA 的密码子序列合成具有一定氨基酸序列多肽链的过程。

translation initiation factor 翻译起始因子
和核糖体及 mRNA 结合,起始蛋白合成的蛋白因子,也称起始因子。

translocase 移位酶
蛋白质合成过程中催化肽基 tRNA 由 A 位转移到 P 位的酶。

transmembrane protein 穿膜蛋白
又称"跨膜蛋白"。一类双亲性的蛋白质。具有亲水区段和疏水区段,起疏水区段穿膜与脂分子双层内部的疏水尾部互相作用,而其亲水区段则暴露于膜两侧。

transport protein 运输蛋白
又称"转运蛋白"、"膜运输蛋白"(membrane transport protein)。介导一种或多种专一性离子或小分子转膜移动的任何膜整合蛋白的统称,不涉及运输机制。

transport vesicle 转移小泡
细胞内携带蛋白质膜泡,从一个区间到另一区间。例如,从 ER 到高尔基体的膜泡。

transposon (transposable element) 转座子(转座元件)
可在 DNA 中的一个位置转移到另一个位置的遗传元件。

transverse tubule (T-tubule) 横小管
简称"T 小管"。横纹肌纤维在暗带与明带交界处,肌膜内陷,环绕肌原纤维形成的小管。与导电兴奋和引起肌肉收缩有关。

treadmilling 踏车现象
微丝或微管在一定条件下,其正端有亚基不断添加的同时,负端有亚基不断脱落,使一纤维在一端延长而另一端缩短的交替现象。

tricarboxylic acid cycle 三羧酸循环
又称柠檬酸循环(citric acid cycle)、克雷布斯循环(Krebs cycle)。体内物质糖、脂肪或氨基酸有氧氧化的主要过程。通过生成的乙酰辅酶 A 与草酰醋酸缩合成三羧酸(柠檬酸)开始,通过一系列氧化步骤产生 CO_2、NADH 及 $FADH_2$,最后仍生成草酰醋酸,进行再循环,从而为细胞提供降解乙酰基及产生能量的基础,由克雷布斯(Krebs)于 20 世纪 30 年代最先提出。

triplet code 三联体密码
由 3 个相邻的核苷酸组成的 mRNA 基本编码单位。有 64 种三联体密码,其中有 61 种氨基酸密码(包括起始密码)及 3 个终止密码,由它们决定多肽链的氨基酸种类和排列顺序的特异性以及翻译的起始和终止。

tropomodulin 原肌球蛋白
是一种微丝结合蛋白质成分,通过同肌钙蛋白相互作用,调节横纹肌收缩。

troponin 肌钙蛋白
结合在横纹肌细肌丝上的一种调节蛋白,可被一定浓度的 Ca^{2+} 激活,在横纹肌收缩中起着开关的作用。

tryntopham repressor 色氨酸阻遏子
一种细菌蛋白。在色氨酸存在时,结合到 DNA 特定区域,阻止色氨酸合成酶的产生。

tubulin 微管蛋白
微管的主要蛋白质成分,分为 α、β 和 γ 微管蛋白。

γ-tubulin ring γ-微管蛋白环
中心体基质中由 γ-微管蛋白构成的环状结构,是细胞中微管生长的起始点。

tumor necrosis factor (TNF) 肿瘤坏死因子
主要由活化的单核/巨噬细胞产生,抗感染,引起发热,诱导肝细胞急性期蛋白合成,促进髓样白血病细胞向噬细胞分化,促进细胞增殖和分化,是最重要的炎症介质,并参与某些自身免疫病的病理损伤。

tumor-suppressor gene 肿瘤抑制基因(抑癌基因)
在正常细胞中限制细胞增殖和浸润行为的基因。当二倍体细胞中基因的双副本均丢失或失活时,细胞便失去控制,变成了癌细胞。

twinfilin 双解丝蛋白
酵母肌动蛋白聚合因子,含两个肌动蛋白解聚因子同源功能,位于皮层肌动蛋白细胞骨架中,可与 G 肌动蛋白以 1:1 比例形成紧密的复合物,在肌动蛋白细胞骨架动态变化中起调节作用。

tyrosine kinase 酪氨酸激酶
一种使特殊蛋白质酪氨酸残基磷酸化的酶。

tyrosine kinase-linked receptor 酪氨酸激酶偶联受体
缺少细胞内催化活性的酶联受体,其配体多为细胞因子,此受体的细胞内区无蛋白激酶活性,而是通过偶联方式激活 Janus 蛋白激酶活性,随之通过信号级联反应调节相关基因的表达。

U

ubiquinone 泛醌
又称辅酶 Q(coenzyme Q)。呼吸和光合作用的电子传递链中的一种小的带有长的异戊二烯侧链的脂溶性醌类,可移动电子载体分子,是唯一的一种非蛋白质结合的辅基。

ubiquitin 泛素
在真核细胞中存在的一个小的高度保守蛋白质。通过与蛋白质的赖氨酸残基连接,形成聚泛素,导致被结合的蛋白质被细胞溶胶中的蛋白酶体所识别并降解。但一些特定抑制剂可抑制其降解。

ubiquitinoylation 泛素化
在蛋白质分子一个位点上结合单个或多个泛素残基的现象。

uniport 单向转运
小分子顺浓度梯度穿膜的蛋白质介导的协助扩散。同一膜上,一种物质穿膜的转运与另一种物质跨越此膜转运无关的现象。负责单向转运的是一类穿膜转运蛋白。

unit membrane 单位膜
在电镜下看到的暗-明-暗两层式膜。Robertson 提出的膜结构模型,主张膜是由一磷脂双层和内外表面各附有一层展开的 β 构型蛋白所组成,现只取其电镜下的形态含义。

univalent, monovalent 单价体
　　减数分裂时因没有同源染色体而不能联会的单条染色体。

U-small nuclear ribonucleoprotein (U-snRNP) U-小核核糖核蛋白
　　一系列含尿嘧啶(U)较多的小核 RNA 与蛋白质结合的复合物,参与 RNA 剪接。

V

vascular endothelial growth factor (VEGF) 血管内皮生长因子
　　血小板衍生生长因子家族中的一员,具有刺激血管内皮细胞发生有丝分裂和血管生成的功能。由多数肿瘤细胞、伤口中角质细胞和巨噬细胞产生,其受体仅表达于血管内皮细胞表面,能增加血管通透性,促进血管内皮细胞增殖及血管形成。

vector 介体,载体
　　通常以噬菌体、质粒用于携带 DNA 片段进行基因克隆。

vesicle 小泡
　　由膜围成的球形小泡,内充有含蛋白质的液体。

villin 绒毛蛋白
　　一种钙调的微丝结合蛋白,低钙时形成微管束,高钙时将微丝切成片段,常见于微绒毛中。

vimentin 波形蛋白
　　存在于间充质来源细胞(如成纤维细胞)和体外培养的细胞中的一种中间纤维蛋白。其一端与核膜相连,另一端与细胞表面的桥粒或半桥粒相连,将细胞核和细胞器维持在特定的空间。

viroid 类似毒
　　仅由 240~350 个核苷酸构成的极小植物病毒。无蛋白质外壳,为裸露的感染颗粒。

virus 病毒
　　由核酸包以蛋白质外壳构成的感染性粒子,可在感染的宿主细胞中,利用宿主细胞的系列机构进行复制。

vital stain (vital dye) 活性染料
　　可使活细胞呈染色反应的物质,如中性红、尼罗兰。某些染色剂对细胞器染色具有选择性,如詹纳斯绿可专一性地使线粒体着色。

voltage-gated channel 电压门通道
　　有选择性地允许离子穿膜的膜蛋白,通道的开放受膜电位变化的影响。此类通道主要存在于电兴奋细胞,如神经细胞、肌细胞。

W

W chromosome W 染色体
　　性染色体之一。在 ZW 性别决定的物种中,只在异配性别即雌性细胞中出现的性染色体。

Western blotting 蛋白质印迹法
　　又称免疫印迹法(immunoblotting)。总蛋白质进行 SDS-聚丙烯酰胺凝胶电泳,然后将蛋白质转至尼龙膜,用特定蛋白质抗体进行免疫反应,显色后可显示该特定蛋白的存在与表达量。用来检测在不均一的蛋白质样品中是否存在目标蛋白的一种方法。

X

X body X 小体
　　又称性染色质体(sex chromatin body)、X 染色质(X chromatin)、巴氏小体(Barr body)。哺乳动物体在细胞有丝分裂间期核内失活的 X 染色体经异固缩形成的浓缩异染色质化的小体。其数目与 X 染色体数目相关。1949 年由巴尔(Barr)首先在雌猫神经细胞核中发现。

X chromosome X 染色体
　　性染色体之一,在雌性哺乳动物和雄性两栖动物中成对存在。

X inactivation X 失活
　　雌性成体细胞中两条 X 染色体中的一条在遗传性状的表达上丧失功能的现象。

Y

Y chromosome Y 染色体
　　性染色体之一,XX 为女性,XY 为男性。

yeast artificial chromosome (YAC) 酵母人工染色体
　　由四膜虫的端粒为酵母的着丝粒和自主复制序列剪接而成的人造载体。可容纳达 1,000 kb 的 DNA 大片段,用来在酵母细胞中扩增基因组 DNA 并对其进行测序。

yeast-one-hybridsystem 酵母单杂交系统
　　用酵母细胞检测蛋白质与 DNA 之间相互作用的一种系统。

Z

Z chromosome Z 染色体
　　性染色体之一。在 ZW 性别决定的物种中,在雌性、雄性细胞中都出现的性染色体。

Z disc Z 盘
　　又称 Z 线(Z line)。肌原纤维相邻两肌节交接处的盘状结构。为 α 辅肌动蛋白集中处,与细肌丝相连。

zinc finger 锌指
　　主要存在于转录因子 DNA 结合结构域中的一些蛋白质中的结构域。由多肽链绕一个锌原子折曲成发夹结构。

zygote 合子
　　雌雄配子结合,产生受精卵。

zygotene 偶线期
　　减数分裂的前期 I 中同源染色体进行联会的阶段。随着染色体的配对,染色体间形成复合物。

(沈大棱)

图书在版编目(CIP)数据

细胞生物学=Cell Biology:英文/李瑶,吴超群,沈大棱主编. —2 版. —上海:
复旦大学出版社,2013.3(2020.7 重印)
ISBN 978-7-309-09460-2

Ⅰ.细… Ⅱ.①李…②吴…③沈… Ⅲ.细胞生物学-高等学校-教材-英文 Ⅳ.Q2

中国版本图书馆 CIP 数据核字(2013)第 005672 号

细胞生物学 Cell Biology(第二版)
李 瑶 吴超群 沈大棱 主编
责任编辑/宫建平

复旦大学出版社有限公司出版发行
上海市国权路 579 号 邮编:200433
网址:fupnet@fudanpress.com http://www.fudanpress.com
门市零售:86-21-65102580 团体订购:86-21-65104505
外埠邮购:86-21-65642846 出版部电话:86-21-65642845
宁波市大港印务有限公司

开本 787×1092 1/16 印张 21.25 字数 555 千
2020 年 7 月第 2 版第 5 次印刷

ISBN 978-7-309-09460-2/Q·84
定价:88.00 元

如有印装质量问题,请向复旦大学出版社有限公司出版部调换。
版权所有 侵权必究

复旦大学出版社向使用本社 Cell Biology(第二版)作为教材进行教学的教师免费赠送教学光盘。欢迎完整填写下面表格来索取光盘。

教师姓名:_____

任课课程名称:_____

任课课程学生人数:_____

联系电话:(O)_____ (H)_____ 手机:_____

e-mail 地址:_____

所在学校名称:_____ 邮政编码:_____

所在学校地址:_____

学校电话总机(带区号):_____ 学校网址:_____

系名称:_____ 系联系电话:_____

每位教师限赠送光盘一个。

邮寄光盘地址:_____

邮政编码:_____

请将本页完整填写后,剪下邮寄到上海市国权路 579 号

复旦大学出版社宫建平收

邮政编码:200433　　联系电话:(021)65654724　13651938710

联系邮件:jpgong59@vip.sina.com